The Enzymes

VOLUME XXVIII

STRUCTURE, FUNCTION AND REGULATION OF TOR COMPLEXES FROM YEASTS TO MAMMALS

PART B

THE ENZYMES

Edited by

Fuyuhiko Tamanoi

Department of Microbiology,
Immunology, and Molecular Genetics
Molecular Biology Institute
University of California, Los Angeles
Los Angeles, CA 90095, USA

Michael N. Hall

Biozentrum, University of Basel
Basel, Switzerland

Volume XXVIII

STRUCTURE, FUNCTION AND REGULATION OF TOR COMPLEXES FROM YEASTS TO MAMMALS

PART B

ELSEVIER

AMSTERDAM • BOSTON • HEIDELBERG • LONDON
NEW YORK • OXFORD • PARIS • SAN DIEGO
SAN FRANCISCO • SINGAPORE • SYDNEY • TOKYO
Academic Press is an imprint of Elsevier

Academic Press is an imprint of Elsevier
32 Jamestown Road, London NW1 7BY, UK
Radarweg 29, PO Box 211, 1000 AE Amsterdam, The Netherlands
Linacre House, Jordan Hill, Oxford OX2 8DP, UK
30 Corporate Drive, Suite 400, Burlington, MA 01803, USA
525 B Street, Suite 1900, San Diego, CA 92101-4495, USA

First edition 2010

ISBN: 978-0-12-381005-2
ISSN: 1874-6047

For information on all Academic Press publications
visit our website at elsevierdirect.com

Printed and bound in USA

10 11 12 13 10 9 8 7 6 5 4 3 2 1

Contents

4. mTOR Signaling by Amino Acid Nutrients: Involvement of MAP4K3

Lijun Yan and Richard F. Lamb

5. mTORC2: The Other Facet of mTOR

Caroline Treins and Julian Downward

6. TORC2 and Chemotaxis in *Dictyostelium discoideum*

Yoichiro Kamimura, Huaqing Cai, and Peter N. Devreotes

7. The TOR-Mediated Regulation of Autophagy in the Yeast *Saccharomyces cerevisiae*

Yoshiaki Kamada and Yoshinori Ohsumi

8. Conservation of the Tsc/Rheb/TORC1/S6K/S6 Signaling in Fission Yeast

Akio Nakashima and Fuyuhiko Tamanoi

9. The Systemic Control of Growth, Physiology, and Behavior by TOR Signaling in *Drosophila*

Nathalie Arquier, Renald Delanoue, and Pierre Léopold

10. Cell-Intrinsic Functions and Regulation of TOR Signaling in *Drosophila*

Thomas P. Neufeld

11. TOR Signaling and Cell Death

Tao Wang and Bruce A. Edgar

12. Elucidating TOR Signaling in *Chlamydomonas reinhardtii*

María Esther Pérez-Pérez and José L. Crespo

13. mTORC1 and mTORC2 in Energy Homeostasis

Marion Cornu and Michael N. Hall

14. TOR Signaling and Aging

Malene Hansen and Pankaj Kapahi

15. mTOR Signaling and Human Cancer

Naphat Chantaravisoot and Fuyuhiko Tamanoi

16. Systems Biology and TOR: Past, Present, and Future

Soyeon I. Lippman and James R. Broach

Preface

This volume (Part B) is a continuation of Volume 27, "Structure, Function and Regulation of TOR Complexes from Yeasts to Mammals, Part A." The two volumes capture recent developments in the study of TOR signaling. They should be of interest to a wide range of researchers, including biochemists, developmental biologists, molecular biologists, and cancer researchers.

Volume 27 of "The Enzymes" focused on basic mechanisms of the TOR complexes and TOR signaling. In Volume 28, we extend the discussion to include topics such as chemotaxis, autophagy, and cell death. We also discuss the involvement of TOR signaling in energy homeostasis, aging, and cancer. Finally, a systems biology approach to TOR signaling is discussed. However, we realize that there are many more topics to be covered. These will be the focus of future volumes on TOR signaling.

We could not have put together this volume in a timely fashion without the efforts of a dedicated group of chapter authors. We are grateful to the authors. We thank Lisa Tickner and the production team at Elsevier for advice and encouragement. We also thank Gloria Lee of UCLA for assistance in communication, preparation, and editing of the chapters.

Fuyuhiko Tamanoi
Michael N. Hall
June 2010

1

mTORC1-Mediated Control of Protein Translation

JAMIE M. DEMPSEY[a,b] • SARAH J. MAHONEY[a,b] •
JOHN BLENIS[a]

[a]*Department of Cell Biology*
Harvard Medical School
Boston, Massachusetts, USA

[b]*Program in Biological and Biomedical Sciences*
Harvard Medical School
Boston, Massachusetts, USA

I. Abstract

Eukaryotic cells expend vast amounts of resources during protein translation, and have therefore evolved mechanisms to police and regulate this process. At the center of this regulation lies the mTORC1 pathway. After sensing inputs from nutrients, growth factors, energy and cellular stresses, mTORC1 and its downstream effectors control overall protein synthesis from the generation of ribosomes to the direct regulation of translation initiation and elongation factors. This chapter reviews both recent and well-established examples of mTORC1 signaling and regulation of protein synthesis.

II. Introduction

Upon detecting changes in the extracellular environment, cells respond appropriately by altering gene expression. After transcription, an mRNA molecule is translated into a protein by ribosomes, and this protein

THE ENZYMES, Vol. XXVIII

ISSN NO: 1874-6047

DOI: 10.1016/S1874-6047(10)28001-0

synthesis can be broken down into three steps: translation initiation, elongation, and termination. In eukaryotes, translation initiation generally occurs by cap-dependent or internal ribosomal entry site (IRES)-dependent mechanisms. The majority of eukaryotic translation is initiated at the mRNA cap complex, which has led to a focus on the regulation and function of the proteins associated with the mRNA cap. A select group of eukaryotic messages can also be initiated when the 40S ribosomal subunit directly associates with an IRES sequence that is usually found within the 5′-untranslated region (UTR) (for reviews, see Ref. [1]). While IRES-mediated translation initiation is also regulated, this chapter will focus on the regulation of protein translation by the mTORC1 pathway via cap-dependent mechanisms.

Safeguarding cells from expending energy and resources on translation under conditions where there is no cellular need is achieved by precisely regulating protein synthesis and ribosomal biogenesis. The rate and timing of protein synthesis is carefully controlled in mammalian cells by signaling pathways that integrate and interpret extracellular cues. These cues affect multiple aspects of translation and the translational machinery, such as protein stabilization and localization, protein–protein interactions, and catalytic activity. Depending on the exact combination of signals, translational control occurs on a global, cap-dependent, or message-specific scale. Although many signaling pathways regulate protein synthesis, the mTORC1 pathway appears to be a master regulator, as it coordinates growth factor, amino acid, and energy inputs (see Chapters 3, 4, and 6 of Volume 27).

A. Translation: The Regulated Steps of Translation Initiation and Elongation

The platform and catalytic activity for translation is provided by ribosomes, which are composed of a large (60S) and a small (40S) subunit in eukaryotes. The building of each of the subunits, which consists of both rRNA and ribosomal proteins (RPs), is a complex process involving all three RNA polymerases and RP translation. The assembly of the 40S and 60S ribosomal subunits with the initiator tRNA, as well as certain eukaryotic translation initiation factors (eIFs) at the initiation codon of an mRNA transcript is referred to as translation initiation. Assembly of the translation initiation complex at the 5′-mRNA cap requires the recruitment of multiple proteins and preassembled protein complexes in a series of ordered events (Figure 1.1). The ternary complex consists of an initiator Met-tRNA$_i^{Met}$ bound to eIF2·GTP, which mediates the association of Met-tRNA$_i^{Met}$ with the 40S ribosomal subunit. The 43S preinitiation complex (PIC) is formed by binding of this charged 40S ribosomal subunit complex with the eIF3

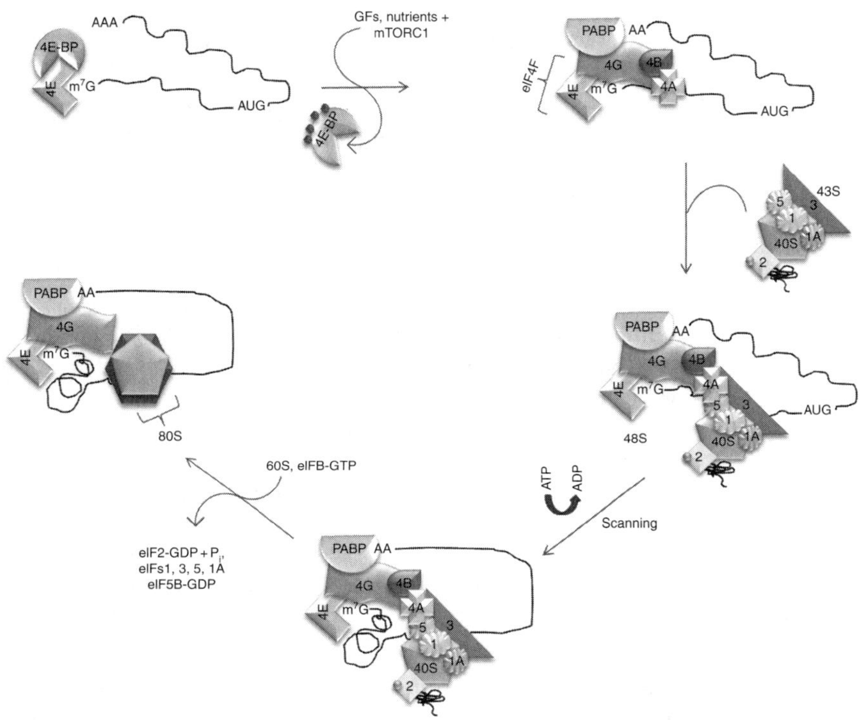

FIG. 1.1. Initiation of eukaryotic translation occurs after the activation of mTORC1 by growth factors and nutrients, resulting in the phosphorylation of 4E-BP1 and subsequent disassociation from eIF4E. The eIF4F protein complex binds to the mRNA creating docking sites for the 43S preinitiation complex (PIC), which is composed of the ternary complex, along with eIF1A, eIF1, eIF3, eIF5 bound to the 40S ribosomal subunit. After binding of the 43S PIC to the cap, the protein complex is termed the 48S PIC. Secondary structures in the 5′-untranslated region (UTR) are unwound in an ATP-dependent process by eIF4A as the 48S PIC scans to locate the AUG initiation codon. Once located, eIF2 hydrolyzes GTP, certain initiation factors are released and the 60S binds to the 40S subunit forming the translation-ready 80S ribosome.

complex (composed of five core subunits and seven to eight associated subunits), eIF1A, and eIF5 initiation factors. In this structure, the eIF3 complex also functions as a scaffold for signaling enzymes (reviewed in Refs. [2, 3]).

At the 5′ end of mRNAs with an m^{7}G (7-methylguanosine) cap, the cap binding protein (CBP) eIF4E is bound during steady-state translation, and is part of the eIF4F complex. The eIF4F complex also contains eIF4A and eIF4G. eIF4A is the RNA helicase responsible for unwinding secondary structures in the 5′-UTR, which is especially important for transcripts with highly structured 5′-UTRs. eIF4G serves as a scaffold for eIF4A, polyA binding proteins (PABPs), and the eIF3 complex. PABPs bind to the polyA tail, helping to form an efficient translating machine by circularizing the mRNA–protein (mRNP) complex. The 48S PIC is formed after the 43S PIC binds to the eIF4F complex (Figure 1.1). Translation can be inhibited by the eIF4E-binding proteins (4E-BPs), which antagonize eIF4G's ability to bind to eIF4E and thus preventing assembly of the 48S PIC (reviewed in Refs. [2–6]).

Once the 48S PIC is formed at the mRNA cap complex, the 5′-UTR is scanned to locate the initiation codon (AUG). After the release of several initiation factors from the 40S ribosomal subunit, the 60S ribosomal subunit is able to bind, forming the functional 80S ribosome. Aided by eIF5, eIF2 hydrolyzes GTP and is subsequently released from the $tRNA_i^{Met}$. The eIF2 protein is recycled when eIF2B catalyzes the exchange of GDP for GTP on eIF2 (reviewed in Ref. [3]).

Translation elongation follows the initiation phase, and is supported by the eukaryotic elongation factors (eEF). After the 80S ribosome is assembled at the initiation codon, the initiation tRNA is in place so that the aminoacyl-tRNA that corresponds to the next codon can bind. The 60S ribosomal peptidyl transferase activity catalyzes peptide bond formation between the growing polypeptide and the aminoacyl-tRNA. Translocation of the ribosome relative to the tRNAs and mRNA complex is aided by the binding of eEF2·GTP to the ribosome. The activity of eEF2 can be suppressed allowing cells to control the elongation step of translation [7]. Phosphorylation by eEF2 kinase (eEF2K) results in this suppression, and multiple signaling pathways regulate the activity of eEF2K, including the mTORC1 pathway [8].

It is imperative that cells police protein translation, and each of the multitude of proteins involved in ribosome biogenesis, translation initiation, and elongation could hypothetically serve as points of regulation for cellular signaling pathways. Indeed, reversible protein phosphorylation regulates the function of most of the aforementioned translation proteins. For protein synthesis, the regulation of translation initiation serves as the rate-limiting step, but the mechanism detailing the order of factor addition has yet to be fully elucidated. Recent landmark experiments have identified protein kinases at the cap complex, which temporally control the initiation of translation (Figure 1.2).

The mechanisms for certain aspects of eukaryotic translation initiation, such as scanning and the order of initiation factor assembly at the cap, are not fully understood. It is known that many factors such as eIF4G, mTORC1, and eIF4B associate with the cap complex in a growth factor-sensitive manner [9] (Figure 1.2), which happens concurrently with the dissociation of translational repressors like 4E-BP1. How signaling events modulate initiation factor association and dissociation will likely be clarified in future studies helping to build on the model of translation initiation.

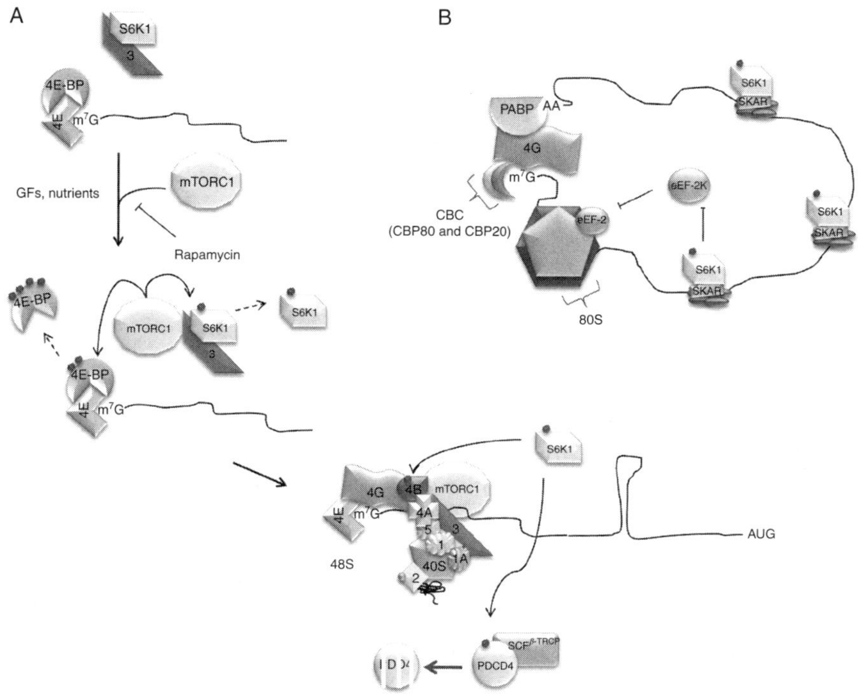

FIG. 1.2. mTORC1 and S6K1 associate with mRNA during translation. (A) The mTORC1 pathway is activated by growth factors and nutrients, resulting in the phosphorylation and assembly of initiation factors onto the mRNA cap complex. There is a pool of S6K1 associating with eIF3 that is not bound to the cap complex in unstimulated cells. S6K1 and eIF3 disassociate after growth factor stimulation. Concomitantly, 4E-BP1 and S6K1 are phosphorylated by an activated mTORC1. Many substrates involved in translation are then phosphorylated by S6K1, correlating with an increase in cap-dependent translation. These substrates include eIF4B and PDCD4. (B) During the pioneer round of translation, the scaffolding protein SKAR associates with the EJCs, and recruits an activated S6K1, correlating with an enhancement of the translation efficiency of these new mRNAs.

Although countless signaling events feed into regulation, this chapter will focus on mTORC1's contribution to regulation of translation initiation, elongation, and ribosome biogenesis.

III. mTORC1 Targets and Control of Translation

As discussed in great detail in Chapter 1 by Hall et al. in Volume 27, mTOR exists in two distinct protein complexes: mTORC1 and mTORC2 [10–15]. Briefly, mTORC1 contains mTOR, raptor (*r*egulatory-*a*ssociated *p*rotein of m*TOR*), and mLST8 (also known as GβL) [16]. This complex directly phosphorylates 4E-BP1 and S6 kinase (S6K) to regulate ribosomal biogenesis, protein synthesis, and cell growth [17] in response to a variety of upstream growth factor, nutrient, and stress signals [18]. Raptor directly interacts with these substrates through the *TO*R *s*ignaling (TOS) motif, an essential phenylalanine followed by four alternating acidic and hydrophobic residues (FDIDL in S6K1 and FEMDI in 4E-BP1) [18–21].

A. mTORC1 Association with and Assembly of the Preinitiation Complex

The majority of the eIFs as well as many RPs [19–22] contain identified phosphorylation sites. Although the exact functions for many of these phosphorylation events are not fully understood, phosphorylation of many translation initiation factors positively or negatively regulates protein synthesis through interesting mechanisms. It is anticipated that protein kinases must be in close proximity to their appropriate targets in order for specific phosphorylation events to occur within the translation initiation complex. Recent work has demonstrated that the eIF3 complex acts as a conduit between mTORC1 activity, 4E-BP1 and S6K1 phosphorylation, and assembly of the translation initiation complex [9, 23] (Figure 1.2).

When the cell does not require enhanced protein synthesis (low energy, amino acids, and/or growth factors and hormones), an inactive S6K1 is bound to the eIF3 complex. Cap-binding assays and sucrose gradient analysis suggest that this complex is not bound to the cap complex, but is part of a free eIF3·S6K1 complex. After stimulation of the cells with insulin, mTORC1 interacts with eIF3, while S6K1 dissociates from the complex. Sucrose density gradients show mTORC1 associated with a larger mRNP complex [9], and this evidence as well as mTORC1 association with the mRNA cap pinpoints the mTORC1·eIF3 complex at the mRNA cap in stimulated cells (Figure 1.2). This interaction correlates with release of S6K1 from the eIF3 complex, as well as phosphorylation of S6K1 at its hydrophobic motif (Thr389). mTORC1 activity has also been reported to induce the stable recruitment of an eIF3 complex subunit, eIF3j [24]. In addition, eIF3j is reported to mediate the recruitment of the 40S subunit suggesting that mTORC1 binding may initiate the assembly of the 43S PIC[25]. mTORC1 also phosphorylates 4E-BP1 and its subsequent release from the cap complex also correlates with localization of mTORC1 at the cap. This event then initiates the assembly of the 48S PIC (see below).

B. mTORC1 Phosphorylation of 4E-BP1

The suppression of cap-dependent translation occurs when the hypophosphorylated form of the 16 kDa 4E-BP1 binds eIF4E and competitively inhibits the binding of eIF4G to eIF4E. After activation, mTORC1 phosphorylates 4E-BP1 releasing it from the cap complex so that eIF4E is free to interact with eIF4G, initiating translation (reviewed in Ref. [26]) (Figure 1.2). The phosphorylation of 4E-BP1 (Thr37, Thr46, Ser65, and Thr70) is proline-directed, and occurs in an ordered manner. The Thr37/Thr46 phosphorylation sites prime 4E-BP1 for subsequent phosphorylation at Ser65/Thr70. *In vitro* studies have shown mTORC1 *directly* phosphorylating the Thr37 and Thr46 sites, however, the phosphorylation of these sites is not always blocked by rapamycin, suggesting that rapamycin-insensitive kinases may contribute to these phosphorylation events *in vivo*. The phosphorylation of Ser65/Thr70 is exclusively mTORC1-dependent in cells even though mTORC1 does not directly phosphorylate Ser65/Thr70 *in vitro*. An mTORC1-regulated kinase *or* phosphatase activity may control the phosphorylation status of these sites ([27] and reviewed in Ref. [28]). Inactivation of 4E-BP1 via mutations of the phosphosites to alanines or treatment with rapamycin reduces, but does not completely abolish, cap-dependent translation, indicating a low level of redundancy in translation signaling pathways (reviewed in Ref. [26]). The role of 4E-BP1 phosphorylation in promoting translation initiation may be message-specific, as overexpression of eIF4E results in increased translation of particular mRNAs with highly structured 5′-UTRs [29].

C. S6K and Targets Involved in the Regulation of Translation

The other well-known mTORC1 targets are the S6Ks, which belong to a family of basophilic serine/threonine kinases known as AGC kinases (reviewed by Jacinto in Chapter 7, Volume 27), which phosphorylate at basophilic motifs, particularly RXRXXS*/T* for S6K1. The most well-known S6K target is ribosomal protein S6 (rpS6), although the function of rpS6 phosphorylation remains unclear. Occasionally S6K will phosphorylate noncanonical basophilic motifs, such as that found in the substrate SKAR [30]. Other targets of S6K that are involved in the regulation of protein translation include eIF4B, PDCD4, and eEF2K.

1. rpS6

Upon activation by mTORC1, S6K1 is then able to phosphorylate many proteins at the cap complex to affect different kinds of activities (Figure 1.2). rpS6, which is a component of the 40S ribosomal subunit, was the first identified target of S6K1. Since rpS6 is a highly abundant protein that can be purified with ribosomes and phosphorylated in a regulated manner at multiple sites [31], it has served as a model for kinase activity and general phosphorylation for many years. Despite the relatively long and illustrious history of rpS6 phosphorylation as a model for regulated kinase activity, the actual function of rpS6 phosphorylation remains contested, with rpS6 phosphorylation correlating with increased translation under certain stimulation conditions and contexts [32], but not others [33]. Phosphorylated rpS6 has been shown to bind with more affinity to mRNA [34] and the 5′ cap [35] *in vitro*. None of the aforementioned studies provide definitive evidence that rates of translation are changed upon introduction of phosphomimetic (aspartate) or nonphosphorylatable (alanine) mutants to cells. Interestingly, alanine mutations in the yeast homologs of rpS6 abolished phosphorylation but gave no discernable effects on growth rate [36], weakening the model that S6Ks regulate translation through phosphorylation of rpS6.

Some of the outstanding questions of the *in vivo* function of rpS6 phosphorylation were cleared up when Ruvinsky and colleagues generated a knock-in/knockout mouse that effectively abolished the ability of rpS6 to be phosphorylated by substituting the five known C-terminal phosphorylated serines with alanines. Upon investigation, they surprisingly found that some tissues of these mice exhibited a mild *increase* in global translation rates with no significant change in 5′-TOP translation, and that MEFs and pancreatic beta-cells were smaller, suggesting a role for rpS6 phosphorylation in cell size regulation and glucose homeostasis [37]. Clearly, more research is necessary to elucidate the biological significance of phosphorylation of rpS6, which may turn out to be cell- and tissue type-specific. Regardless of the function, rpS6 phosphorylation has been and will continue to serve as a valuable, biochemical readout for activation of the mTORC1 pathway.

2. *eIF4B*

In 1978, John Hershey's group showed that eIF4B, among other translation initiation factors, became phosphorylated *in vitro* in rabbit reticulocytes through activity of a cAMP-independent protein kinase [38]. eIF4B is a critical regulatory subunit of eIF4A helicase, and in combination, these proteins work to unwind complex secondary structures found in the 5′-UTR of some mRNAs, allowing for more efficient scanning by the 40S ribosome for the start codon. Later studies indicated that not only was eIF4B phosphorylated, but its phosphorylation status was regulated by serum levels in HeLa cells [39], and upon further investigation into the specific regulators of eIF4B phosphorylation, it was found that phorbol esters, insulin, and

EGF could all stimulate the phosphorylation of eIF4B [36, 37, 40]. *In vitro* kinase studies showed that several purified kinases, including S6K, could phosphorylate eIF4B, although specific sites were not identified [41].

The early studies of eIF4B phosphorylation showed that multiple kinases could potentially be responsible for phosphorylating eIF4B, but none of them showed a direct effect of phosphorylation on either general or specific translation efficiency. Recently, eIF4B Ser422 was shown to be phosphorylated by S6K1 in cells [42], and this phosphorylation event was shown to promote association with the PIC [9] (Figure 1.2). Holz and colleagues showed that the overexpressed wild-type eIF4B associated with eIF3 in an insulin- and rapamycin-dependent manner. They also found that a phosphomimetic mutant of eIF4B (Ser422Asp) associated constitutively with eIF3 even in the presence of rapamycin, and that the alanine mutant at this site did not associate with eIF3, even in stimulated cells [9]. This study also showed a correlation between these mutations and the efficiency of cap-dependent translation in an *in cellulo* luciferase reporter assay.

3. *PDCD4*

Programmed cell death 4 (PDCD4) is another target of S6K, and like eIF4B, PDCD4 also binds the helicase eIF4A [43], though instead of activating translation, it inhibits it [44]. Akt was first shown to be able to phosphorylate PDCD4 *in vitro* at Ser67 and Ser457 [45]. Dorrello et al. found later that not only was Ser67 phosphorylated by S6K1 in cells, but that this phosphorylation promoted recruitment of a $SCF^{\beta\text{-TRCP}}$ ubiquitin ligase complex to PDCD4, leading to ubiquitination and subsequent degradation of PDCD4 [46] (Figure 1.2). Since PDCD4 binds to and inhibits eIF4A, this degradation of PDCD4 releases repression of translation at eIF4A [46].

4. *eIF4G*

The "scaffold" of the cap complex, eIF4G, provides a docking site for many translation initiation factors such as eIF4E, eIF4A, eIF3, and PABP. There are at least three serine residues in the C-terminal third of eIF4GI that are serum- and mitogen-dependent and rapamycin-sensitive. Raught and colleagues suggest that these are not direct mTORC1 target sites, but rather sites whose accessibility may be regulated by mTORC1 signaling [47]. Functions for these phosphorylation events have not yet been determined, although one might speculate that phosphorylation may affect the ability of proteins to dock onto eIF4G.

5. SKAR, CBP80, and the Pioneer Round of Translation

The "pioneer" round of translation is described as the first passage of the ribosome along the length of a newly transcribed mRNA, and many of the players involved in steady-state translation have also been proposed to play a part in this process [48]. During transcription, the pre-mRNA is outfitted with a 5′-m7GpppN cap, which is then bound by the cap-binding complex (CBC) and not eIF4E. The CBC is composed of the cap-binding proteins (CBP) 80 and 20, and is required for the proper splicing of the newly transcribed mRNA [49]. During the splicing process, proteins are deposited on the newly processed mRNA near the CBC to form the TREX complex [50], and approximately 20 nucleotides upstream of each exon–exon junction to form exon junction complexes (EJCs) [51]. These complexes are engaged and removed during the pioneer round [52, 53], and it has been suggested that the EJC proteins are involved in regulating the efficiency of this first and subsequent rounds of translation through a mechanism that remains unclear [54].

SKAR, a downstream target of S6K1 involved in cell growth control [30], was recently shown to be a novel EJC-interacting protein, linking mTORC1/S6K1 signaling to the pioneer round of translation. Like the EJC, SKAR contributes to an increase in the translational yield of spliced messages, and serves as a scaffolding protein, recruiting activated S6K1 to the newly generated mRNAs, where it phosphorylates several proteins on the CBP80-bound mRNA species [55]. S6K1 can also phosphorylate CBP80 *in vitro*, and may be involved in cdc42-regulated effects on mRNA splicing [56]. The biological significance of these phosphorylation events and how they affect the pioneer and subsequent rounds of translation will undoubtedly be the subject of future studies.

D. mTORC1 Signaling in Elongation: Phosphorylation of eEF2 and eEF2 Kinase

While many factors are regulated during translation initiation, one translation elongation factor has been identified as also regulated by mTORC1 signaling. eEF2 is a GTP-binding protein that catalyzes movement of the ribosome relative to the mRNA transcript, and phosphorylation of eEF2 at Thr56 in the GTP-binding domain renders it unable to bind ribosomes for efficient translation [57]. eEF2 kinase (eEF2K) is responsible for this inhibitory phosphorylation, and can be inhibited through a number of phosphorylation events by various kinases, including S6K resulting in the progression of translation (Figure 1.2). Redpath et al. showed that insulin signaling and serum stimulation caused downregulation of eEF2 phosphorylation and eEF2K activity in Chinese hamster ovary (CHO) cells, while rapamycin treatment blocked this effect, implicating mTOR signaling. They also showed that eEF2K could act as a substrate of S6K1 *in vitro* [58] and that phosphorylation of this site resulted in an inhibition of eEF2K phosphotransferase activity. Context-specific phosphorylation has also been shown, as cdc2/cyclin B can phosphorylate eEF2K specifically during early mitosis in an amino acid- and rapamycin-sensitive manner, which may control translation rates in mitotic cells through an mTORC1-dependent mechanism [59]. Finally, the rapamycin-sensitive phosphorylation of eEF2K at Ser78, near the calmodulin-binding domain, decreases its affinity for calmodulin [60].

E. mTORC1 Signaling and Ribosomal Biogenesis

Maintaining dynamic protein levels is dependent upon ribosome biogenesis. The mTORC1 pathway tightly regulates this energy-consuming process after sensing cues from the cellular environment. At least 60% of total cellular transcription and a large fraction of translation can be attributed to ribosome biogenesis in rapidly growing yeast cells [61]. A growing body of evidence indicates that mTORC1 signaling controls ribosome biogenesis on numerous levels, including the regulation of transcription by all three RNA polymerases (Pols), as well as the regulation of translation of RPs (Figure 1.3).

1. *Regulation of RNA Polymerase I Transcription*

The transcription of rDNA by Pol I into 45S pre-rRNA is thought to limit ribosome biogenesis [62]. The 400+ rDNA genes in humans sit in tandem repeats on five pairs of chromosomes within nucleoli, small rRNA production units with transcription equipment (reviewed in Ref. [63]). Growth factor stimulation and phosphorylation events control both 45S precursor rRNA transcription [64, 65] and processing [66]. In addition, serum induces the phosphorylation of proteins directly bound to Pol I [67]. Multiple protein kinases and phosphoproteins have been shown to bind the rDNA promoter region and regulate rRNA production [68–70]. mTOR translocates into and out of the nucleus [71], and studies have shown yeast Tor1 to be dynamically associating with the rDNA promoter [72], placing it near the Pol I transcriptional machinery.

Several transcription factors are required to recruit Pol I to the rDNA promoter. Upstream binding factor (UBF) can be found directly bound to the rDNA promoter, recruiting Pol I by interacting with a Pol I subunit. The multisubunit SL1 (promoter selectivity factor, known as TIF-IB in mouse)

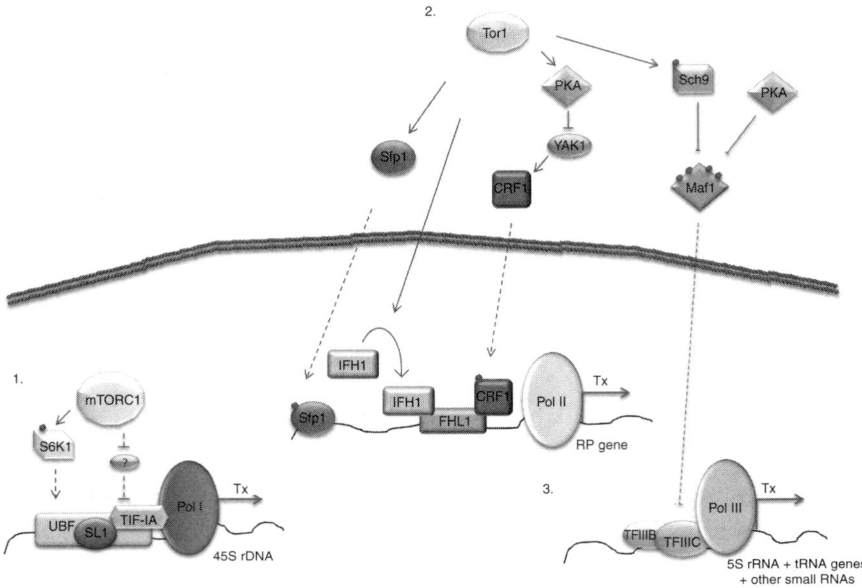

FIG. 1.3. The Tor pathway's role in regulation of ribosomal biogenesis. An overview of the Tor pathway's inputs into the regulation of ribosomal biogenesis is compiled from work in yeast model systems (Tor1), as well as mammalian cell lines (mTORC1). 1. rRNA production is regulated through UBF and TIF-1A. Pol I-driven transcription and production of rRNA is either increased or decreased after the phosphorylation of UBF and TIF-1A following activation of mTORC1 by nutrients, growth factors, and the energy status of the cell. The phosphorylation of TIF-1A is reduced following activation of mTORC1 suggesting negative regulation of a kinase or activation of a phosphatase. Lack of certainty for the responsible kinase is represented by a dashed line. 2. The regulation of ribosomal protein (RP) mRNA production in yeast occurs after Tor1 induces activator IFH1 to bind to the Pol II machinery at RP promoters. Tor1 also causes the inhibitor CRF1 to be localized to the cytoplasm. PKA is phosphorylated by Tor1, which leads to the inhibition of YAK1, resulting in CRF1 in the cytoplasm. Sfp1 is found in the nucleus bound to RP promoters, leading to an enhancement of the transcription of RP genes after phosphorylation by TORC1. Activation is indicated by solid arrows, while dashed arrows indicate protein movement. 3. The regulation of Pol III transcription in yeast by Tor1 via Maf1. In times of nutrient deprivation, an unphosphorylated Maf1 binds to and inhibits the Pol III complex, as well as prevents an interaction between TFIIIB and TFIIIC. An activated Tor1 leads to the phosphorylation of Sch9 and PKA, which then phosphorylate Maf1, resulting in disassociation from the Pol III complex and the cytoplasmic localization of the protein.

complex, which consists of TATA-binding protein (TBP) and several TBP-associated factors (TAFs) also binds the rDNA promoter. Another protein required for Pol I recruitment is transcription intermediary factor IA (TIF-IA, also known as RRN3) and is thought to function as a bridge between Pol I and SL1 (reviewed in Ref. [73]).

Phosphorylation of UBF induced by growth factor stimulation promotes binding to TBP and transactivation of Pol I [74, 75], and mTORC1 signaling has been implicated in UBF phosphorylation and regulation [76]. Hannan et al. reported increased rRNA production after phosphorylation of the UBF C-terminal domain by S6K [77] (see Figure 1.3). Serum and growth factors also induce phosphorylation of the TIF-IA rRNA transcription factor [78], while rapamycin treatment of growing cells decreases phosphorylation at Ser44, an activating modification. CDK2-cyclin E may be responsible for the mTORC1-induced TIF-IA Ser44 phosphorylation and a subsequent increase in rRNA production [79]. Though the mechanism is unknown, rapamycin was also shown to enhance phosphorylation of Ser199 to impair TIF-IA binding to Pol I. Taken together, mTORC1 positively regulates an enhancing phosphorylation on TIF-IA and negatively regulates a suppressing phosphorylation.

2. *Transcription of Ribosomal Protein mRNAs and TORC1 Signaling*

There are approximately 79 distinct RPs in each ribosome. The coordinated synthesis of near equimolar amounts of RPs is needed for efficient ribosome production, and appears to be tightly regulated at the transcriptional level (reviewed in Ref. [80]). Though the RP genes are scattered throughout the genome, they contain similar promoter regions, which permit their synchronized expression [80]. To control ribosome biogenesis, cell signaling pathways stimulated by growth factors regulate the rate of RP mRNA transcription [81].

There have been several TORC1-dependent phosphorylation events that regulate transcription of RP genes identified in studies carried out in yeast (Figure 1.3). Sfp1 is a zinc finger transcription factor shown to be phosphorylated downstream of TORC1, and this transcription factor regulates RP gene expression, cell proliferation, and cell size [82–85]. After phosphorylation, Sfp1 translocates to the nucleus and binds to the RP promoter more efficiently [86]. Though *in vitro* and *in vivo* phosphorylation of Sfp1 was shown to be decreased by $>50\%$ after mutation of Sfp1's seven putative TORC1 phosphorylation sites, little effect on cell size or growth rate [86] was observed suggesting TORC1 likely regulates RP gene expression and cell growth via both Sfp1 phosphorylation and supplementary mechanisms.

Indeed, FHL1, a Forkhead-like transcription factor that is found at RP gene promoters in yeast has been shown to be regulated by TOS. The RP gene transcriptional coactivator IFH1 (*i*nteracting with *f*ork*h*ead 1) binds to FHL1 at the *RPL30* promoter after induction by TOR [87–89] (Figure 1.3). In addition, TOR downregulates the corepressor CRF1 binding to FHL1 at the *RPL30* promoter [89]. TOS does not appear to control FHL localization to the *RPL30* promoter, so the mechanisms by which IFH1 and CRF1 activate and repress RP gene transcription are unknown [89]. It also remains to be determined whether these activities are conserved in metazoans, although similar mechanisms are possible, since the TOR/mTOR pathways are highly conserved in eukaryotes [16].

3. *TOR Signaling and the Regulation of RNA Polymerase III Transcription*

RNA polymerase III transcribes both tRNA and the 5S rRNA component of the large ribosomal subunit, and several signaling-regulated factors are involved in this transcription. Of these, the best studied Pol III regulators are the TFIIIB subunits and Maf1.

Pol III transcription of tRNA is negatively regulated by MafI, which was first identified in a genetic screen as a modifier of tRNA activity [89, 90]. MafI was shown to interact with RNA Pol III, and a loss-of-function allele exhibits increased tRNA and 5S rRNA production [91]. Along with other signaling pathways, the mTOR pathway regulates Maf1 activity (Figure 1.3). Phosphorylation of Maf1 basophilic sites by PKA and the yeast S6K homolog, Sch9, blocks Maf1 nuclear localization [92]. Yeast TORC1 also interacts with and phosphorylates Maf1 [93]. Upon rapamycin treatment, Maf1 is dephosphorylated in a PP2A-dependent manner. Dephosphorylated Maf1 binds Pol III and Pol III-associated genomic loci, such as the 5S rDNA promoter, and represses Pol III transcription [93–96]. The Maf1 studies show that the TORC1 signaling pathway can regulate Maf1 activity on a variety of levels including subcellular localization and direct association with chromatin. Although most of the preliminary Maf1 studies have been carried out in yeast, Maf1 and its ability to downregulate Pol III activity is conserved in higher eukaryotes [97, 98].

4. *Regulation of Ribosomal Protein Translation*

In order to complete the set of building blocks for ribosomal biogenesis, mTORC1 signaling events also help control RP translation. RPs are preferentially translated in serum-stimulated cells over non-RPs [99]. mTOR signaling and nutrient supplementation specifically upregulate translation of mammalian mRNA transcripts containing a 5′-terminal oligopyrimidine (5′-TOP) motif, found at the transcriptional start site of most RP gene transcripts and many translation elongation factors ([100] and reviewed in Ref. [101]). The 5′-TOP sequence is characterized by a cytosine residue followed by 4–14 pyrimidines [101]. Regulation of 5′-TOP mRNA translation has been attributed to mTOR signaling [102], although the role of S6K is under question. No S6K substrate has been found to regulate 5′-TOP-specific translation and *s6k1*$^{-/-}$/*s6k2*$^{-/-}$ mice continue to exhibit rapamycin-sensitive 5′-TOP translation [103]. However, other AGC kinases such as RSK could be compensating for the lack of S6K as S6 and eIF4B phosphorylation also persists in the *s6k1*$^{-/-}$/*s6k2*$^{-/-}$ mice [103].

How the many steps of ribosome biogenesis are coordinated to control the synthesis of equimolar amounts of rRNA and RPs remains unresolved. Further, the potential regulation of rRNA or RP processing events and ribosome assembly has been mostly overlooked. Since not all rDNA transcripts are actively transcribed at the same time (reviewed in Ref. [104]), it will be interesting to see whether signaling events can modulate epigenetic states of rDNA and other genes involved in ribosome biogenesis. Future research will hopefully answer these questions and elucidate more detailed mechanisms of the complex regulation of ribosome biogenesis.

IV. Conclusion

Multiple independent mechanisms downstream of mTORC1 signaling have evolved to regulate protein translation. Specifically, overall rates of cap-dependent translation are affected by mTORC1's phosphorylation of 4E-BPs. Through S6K phosphorylation and enhancement of RNA helicase activity, the translation efficiency of a subset of mRNAs with highly structured 5′-UTRs is increased. In addition, ribosomal biogenesis is directly affected at multiple steps by mTORC1 signaling. Understanding the molecular events downstream of mTORC1 signaling that exert control over protein translation will aid in our understanding of players that could be scientifically, as well as clinically, important.

References

1. Baird, S.D., et al. (2006). Searching for IRES. *RNA* 12(10):1755–1785.
2. Sonenberg, N., and Hinnebusch, A.G. (2009). Regulation of translation initiation in eukaryotes: mechanisms and biological targets. *Cell* 136(4):731–745.
3. Pestova, T.V., and Hellen, C.U. (2001). Functions of eukaryotic factors in initiation of translation. *Cold Spring Harb Symp Quant Biol* 66:389–396.
4. Pestova, T.V., et al. (2001). Molecular mechanisms of translation initiation in eukaryotes. *Proc Natl Acad Sci USA* 98(13):7029–7036.
5. Sonenberg, N., and Dever, T.E. (2003). Eukaryotic translation initiation factors and regulators. *Curr Opin Struct Biol* 13(1):56–63.
6. Hernandez, G., and Vazquez-Pianzola, P. (2005). Functional diversity of the eukaryotic translation initiation factors belonging to eIF4 families. *Mech Dev* 122(7–8):865–876.

7. Ryazanov, A.G., Shestakova, E.A., and Natapov, P.G. (1988). Phosphorylation of elongation factor 2 by EF-2 kinase affects rate of translation. *Nature* 334(6178):170–173.
8. Wang, X., et al. (2001). Regulation of elongation factor 2 kinase by p90(RSK1) and p70 S6 kinase. *EMBO J* 20(16):4370–4379.
9. Holz, M.K., et al. (2005). mTOR and S6K1 mediate assembly of the translation preinitiation complex through dynamic protein interchange and ordered phosphorylation events. *Cell* 123(4):569–580.
10. Kim, D.H., et al. (2003). GbetaL, a positive regulator of the rapamycin-sensitive pathway required for the nutrient-sensitive interaction between raptor and mTOR. *Mol Cell* 11(4):895–904.
11. Loewith, R., et al. (2002). Two TOR complexes, only one of which is rapamycin sensitive, have distinct roles in cell growth control. *Mol Cell* 10(3):457–468.
12. Kim, D.H., et al. (2002). mTOR interacts with raptor to form a nutrient-sensitive complex that signals to the cell growth machinery. *Cell* 110(2):163–175.
13. Hara, K., et al. (2002). Raptor, a binding partner of target of rapamycin (TOR), mediates TOR action. *Cell* 110(2):177–189.
14. Sarbassov, D.D., et al. (2004). Rictor, a novel binding partner of mTOR, defines a rapamycin-insensitive and raptor-independent pathway that regulates the cytoskeleton. *Curr Biol* 14(14):1296–1302.
15. Jacinto, E., et al. (2004). Mammalian TOR complex 2 controls the actin cytoskeleton and is rapamycin insensitive. *Nat Cell Biol* 6(11):1122–1128.
16. Jacinto, E., and Hall, M.N. (2003). Tor signalling in bugs, brain and brawn. *Nat Rev Mol Cell Biol* 4(2):117–126.
17. Hay, N., and Sonenberg, N. (2004). Upstream and downstream of mTOR. *Genes Dev* 18(16):1926–1945.
18. Zeng, Z., et al. (2007). Rapamycin derivatives reduce mTORC2 signaling and inhibit AKT activation in AML. *Blood* 109(8):3509–3512.
19. Loeb, J.E., and Blat, C. (1970). Phosphorylation of some rat liver ribosomal proteins and its activation by cyclic AMP. *FEBS Lett* 10(2):105–108.
20. Kabat, D. (1970). Phosphorylation of ribosomal proteins in rabbit reticulocytes. Characterization and regulatory aspects. *Biochemistry* 9(21):4160–4175.
21. Trauch, J.A., Mumby, M., and Traut, R.R. (1973). Phosphorylation of ribosomal proteins by substrate-specific protein kinases from rabbit reticulocytes. *Proc Natl Acad Sci USA* 70(2):373–376.
22. Kaerlein, M., and Horak, I. (1976). Phosphorylation of ribosomal proteins in HeLa cells infected with vaccinia virus. *Nature* 259(5539):150–151.
23. Harris, T.E., et al. (2006). mTOR-dependent stimulation of the association of eIF4G and eIF3 by insulin. *EMBO J* 25(8):1659–1668.
24. Miyamoto, S., Patel, P., and Hershey, J.W. (2005). Changes in ribosomal binding activity of eIF3 correlate with increased translation rates during activation of T lymphocytes. *J Biol Chem* 280(31):28251–28264.
25. Fraser, C.S., et al. (2004). The j-subunit of human translation initiation factor eIF3 is required for the stable binding of eIF3 and its subcomplexes to 40 S ribosomal subunits in vitro. *J Biol Chem* 279(10):8946–8956.
26. Clemens, M.J. (2001). Translational regulation in cell stress and apoptosis. Roles of the eIF4E binding proteins. *J Cell Mol Med* 5(3):221–239.
27. Gingras, A.C., et al. (2001). Hierarchical phosphorylation of the translation inhibitor 4E-BP1. *Genes Dev* 15(21):2852–2864.

28. Fingar, D.C., and Blenis, J. (2004). Target of rapamycin (TOR): an integrator of nutrient and growth factor signals and coordinator of cell growth and cell cycle progression. *Oncogene* 23(18):3151–3171.
29. Koromilas, A.E., Lazaris-Karatzas, A., and Sonenberg, N. (1992). mRNAs containing extensive secondary structure in their 5′ non-coding region translate efficiently in cells overexpressing initiation factor eIF-4E. *EMBO J* 11(11):4153–4158.
30. Richardson, C.J., et al. (2004). SKAR is a specific target of S6 kinase 1 in cell growth control. *Curr Biol* 14(17):1540–1549.
31. Krieg, J., Hofsteenge, J., and Thomas, G. (1988). Identification of the 40 S ribosomal protein S6 phosphorylation sites induced by cycloheximide. *J Biol Chem* 263(23):11473–11477.
32. Thomas, G., et al. (1982). The effect of serum, EGF, PGF2 alpha and insulin on S6 phosphorylation and the initiation of protein and DNA synthesis. *Cell* 30(1):235–242.
33. Montine, K.S., and Henshaw, E.C. (1990). TPA stimulates S6 phosphorylation but not protein synthesis in Ehrlich cells. *Biochem Biophys Res Commun* 166(3):1340–1345.
34. Burkhard, S.J., and Traugh, J.A. (1983). Changes in ribosome function by cAMP-dependent and cAMP-independent phosphorylation of ribosomal protein S6. *J Biol Chem* 258(22):14003–14008.
35. Roux, P.P., et al. (2007). RAS/ERK signaling promotes site-specific ribosomal protein S6 phosphorylation via RSK and stimulates cap-dependent translation. *J Biol Chem* 282(19):14056–14064.
36. Morley, S.J., and Traugh, J.A. (1990). Differential stimulation of phosphorylation of initiation factors eIF-4F, eIF-4B, eIF-3, and ribosomal protein S6 by insulin and phorbol esters. *J Biol Chem* 265(18):10611–10616.
37. Wolthuis, R.M., et al. (1993). Epidermal growth factor stimulates phosphorylation of eukaryotic initiation factor 4B, independently of protein kinase C. *Biochim Biophys Acta* 1177(2):160–166.
38. Benne, R., et al. (1978). Phosphorylation of eukaryotic protein synthesis initiation factors. *Proc Natl Acad Sci USA* 75(1):108–112.
39. Duncan, R., and Hershey, J.W. (1985). Regulation of initiation factors during translational repression caused by serum depletion. Covalent modification. *J Biol Chem* 260(9):5493–5497.
40. Morley, S.J., and Traugh, J.A. (1989). Phorbol esters stimulate phosphorylation of eukaryotic initiation factors 3, 4B, and 4F. *J Biol Chem* 264(5):2401–2404.
41. Tuazon, P.T., Merrick, W.C., and Traugh, J.A. (1989). Comparative analysis of phosphorylation of translational initiation and elongation factors by seven protein kinases. *J Biol Chem* 264(5):2773–2777.
42. Shahbazian, D., et al. (2006). The mTOR/PI3K and MAPK pathways converge on eIF4B to control its phosphorylation and activity. *EMBO J* 25(12):2781–2791.
43. Goke, A., et al. (2002). DUG is a novel homologue of translation initiation factor 4G that binds eIF4A. *Biochem Biophys Res Commun* 297(1):78–82.
44. Yang, H.S., et al. (2003). The transformation suppressor Pdcd4 is a novel eukaryotic translation initiation factor 4A binding protein that inhibits translation. *Mol Cell Biol* 23 (1):26–37.
45. Palamarchuk, A., et al. (2005). Akt phosphorylates and regulates Pdcd4 tumor suppressor protein. *Cancer Res* 65(24):11282–11286.
46. Dorrello, N.V., et al. (2006). S6K1- and betaTRCP-mediated degradation of PDCD4 promotes protein translation and cell growth. *Science* 314(5798):467–471.
47. Raught, B., et al. (2000). Serum-stimulated, rapamycin-sensitive phosphorylation sites in the eukaryotic translation initiation factor 4GI. *EMBO J* 19(3):434–444.

48. Maquat, L.E. (2004). Nonsense-mediated mRNA decay: splicing, translation and mRNP dynamics. *Nat Rev Mol Cell Biol* 5(2):89–99.
49. Izaurralde, E., et al. (1994). A nuclear cap binding protein complex involved in pre-mRNA splicing. *Cell* 78(4):657–668.
50. Reed, R., and Cheng, H. (2005). TREX, SR proteins and export of mRNA. *Curr Opin Cell Biol* 17(3):269–273.
51. Le Hir, H., Moore, M.J., and Maquat, L.E. (2000). Pre-mRNA splicing alters mRNP composition: evidence for stable association of proteins at exon–exon junctions. *Genes Dev* 14(9):1098–1108.
52. Ishigaki, Y., et al. (2001). Evidence for a pioneer round of mRNA translation: mRNAs subject to nonsense-mediated decay in mammalian cells are bound by CBP80 and CBP20. *Cell* 106(5):607–617.
53. Wang, W., et al. (2001). The role of Upf proteins in modulating the translation read-through of nonsense-containing transcripts. *EMBO J* 20(4):880–890.
54. Nott, A., Le Hir, H., and Moore, M.J. (2004). Splicing enhances translation in mammalian cells: an additional function of the exon junction complex. *Genes Dev* 18(2):210–222.
55. Ma, X.M., et al. (2008). SKAR links pre-mRNA splicing to mTOR/S6K1-mediated enhanced translation efficiency of spliced mRNAs. *Cell* 133(2):303–313.
56. Wilson, K.F., Wu, W.J., and Cerione, R.A. (2000). Cdc42 stimulates RNA splicing via the S6 kinase and a novel S6 kinase target, the nuclear cap-binding complex. *J Biol Chem* 275 (48):37307–37310.
57. Proud, C.G. (1994). Peptide-chain elongation in eukaryotes. *Mol Biol Rep* 19(3):161–170.
58. Redpath, N.T., Foulstone, E.J., and Proud, C.G. (1996). Regulation of translation elongation factor-2 by insulin via a rapamycin-sensitive signalling pathway. *EMBO J* 15(9):2291–2297.
59. Smith, E.M., and Proud, C.G. (2008). cdc2-cyclin B regulates eEF2 kinase activity in a cell cycle- and amino acid-dependent manner. *EMBO J* 27(7):1005–1016.
60. Browne, G.J., and Proud, C.G. (2004). A novel mTOR-regulated phosphorylation site in elongation factor 2 kinase modulates the activity of the kinase and its binding to calmodulin. *Mol Cell Biol* 24(7):2986–2997.
61. Warner, J.R. (1999). The economics of ribosome biosynthesis in yeast. *Trends Biochem Sci* 24(11):437–440.
62. Mayer, C., and Grummt, I. (2006). Ribosome biogenesis and cell growth: mTOR coordinates transcription by all three classes of nuclear RNA polymerases. *Oncogene* 25 (48):6384–6391.
63. Raska, I., et al. (2004). The nucleolus and transcription of ribosomal genes. *Biol Cell* 96 (8):579–594.
64. Grummt, F., Paul, D., and Grummt, I. (1977). Regulation of ATP pools, rRNA and DNA synthesis in 3T3 cells in response to serum or hypoxanthine. *Eur J Biochem* 76(1):7–12.
65. de Capoa, A., et al. (1985). The transcriptional activity of individual ribosomal DNA gene clusters is modulated by serum concentration. *J Cell Sci* 74:21–35.
66. Isfort, R.J., and Cody, D.B. (1992). Serum and growth factors stimulate ribosomal RNA processing in Syrian hamster embryo cells: divergence of this signalling pathway from immediate-early gene expression. *Cell Signal* 4(6):665–674.
67. Freiburg, M. (1987). Phosphorylation of RNA polymerase I-associated polypeptides of *Tetrahymena pyriformis*. *J Biochem* 101(2):465–475.
68. Kuehn, G.D., et al. (1979). Polyamine-mediated phosphorylation of a nucleolar protein from *Physarum polycephalum* that stimulates rRNA synthesis. *Proc Natl Acad Sci USA* 76(6):2541–2545.

69. Morrow, B.E., Ju, Q., and Warner, J.R. (1990). Purification and characterization of the yeast rDNA binding protein REB1. *J Biol Chem* 265(34):20778–20783.
70. Zhang, J., Niu, H.F., and Jacob, S.T. (1991). E1BF is an essential RNA polymerase I transcription factor with an intrinsic protein kinase activity that can modulate rRNA gene transcription. *Proc Natl Acad Sci USA* 88(19):8293–8296.
71. Kim, J.E., and Chen, J. (2000). Cytoplasmic-nuclear shuttling of FKBP12-rapamycin-associated protein is involved in rapamycin-sensitive signaling and translation initiation. *Proc Natl Acad Sci USA* 97(26):14340–14345.
72. Li, H., et al. (2006). Nutrient regulates Tor1 nuclear localization and association with rDNA promoter. *Nature* 442(7106):1058–1061.
73. White, R.J. (2005). RNA polymerases I and III, growth control and cancer. *Nat Rev Mol Cell Biol* 6(1):69–78.
74. Kihm, A.J., et al. (1998). Phosphorylation of the rRNA transcription factor upstream binding factor promotes its association with TATA binding protein. *Proc Natl Acad Sci USA* 95(25):14816–14820.
75. Voit, R., et al. (1992). The nucleolar transcription factor mUBF is phosphorylated by casein kinase II in the C-terminal hyperacidic tail which is essential for transactivation. *EMBO J* 11(6):2211–2218.
76. Wu, A., et al. (2005). Regulation of upstream binding factor 1 activity by insulin-like growth factor I receptor signaling. *J Biol Chem* 280(4):2863–2872.
77. Hannan, K.M., et al. (2003). mTOR-dependent regulation of ribosomal gene transcription requires S6K1 and is mediated by phosphorylation of the carboxy-terminal activation domain of the nucleolar transcription factor UBF. *Mol Cell Biol* 23(23):8862–8877.
78. Russell, J., and Zomerdijk, J.C. (2005). RNA-polymerase-I-directed rDNA transcription, life and works. *Trends Biochem Sci* 30(2):87–96.
79. Mayer, C., et al. (2004). mTOR-dependent activation of the transcription factor TIF-IA links rRNA synthesis to nutrient availability. *Genes Dev* 18(4):423–434.
80. Perry, R.P. (2007). Balanced production of ribosomal proteins. *Gene* 401(1–2):1–3.
81. Geyer, P.K., et al. (1982). Regulation of ribosomal protein mRNA content and translation in growth-stimulated mouse fibroblasts. *Mol Cell Biol* 2(6):685–693.
82. Blumberg, H., and Silver, P. (1991). A split zinc-finger protein is required for normal yeast growth. *Gene* 107(1):101–110.
83. Jorgensen, P., et al. (2002). Systematic identification of pathways that couple cell growth and division in yeast. *Science* 297(5580):395–400.
84. Marion, R.M., et al. (2004). Sfp1 is a stress- and nutrient-sensitive regulator of ribosomal protein gene expression. *Proc Natl Acad Sci USA* 101(40):14315–14322.
85. Jorgensen, P., et al. (2004). A dynamic transcriptional network communicates growth potential to ribosome synthesis and critical cell size. *Genes Dev* 18(20):2491–2505.
86. Lempiainen, H., et al. (2009). Sfp1 interaction with TORC1 and Mrs6 reveals feedback regulation on TOR signaling. *Mol Cell* 33(6):704–716.
87. Wade, J.T., Hall, D.B., and Struhl, K. (2004). The transcription factor Ifh1 is a key regulator of yeast ribosomal protein genes. *Nature* 432(7020):1054–1058.
88. Schawalder, S.B., et al. (2004). Growth-regulated recruitment of the essential yeast ribosomal protein gene activator Ifh1. *Nature* 432(7020):1058–1061.
89. Martin, D.E., Soulard, A., and Hall, M.N. (2004). TOR regulates ribosomal protein gene expression via PKA and the Forkhead transcription factor FHL1. *Cell* 119(7):969–979.
90. Boguta, M., Czerska, K., and Zoladek, T. (1997). Mutation in a new gene MAF1 affects tRNA suppressor efficiency in *Saccharomyces cerevisiae*. *Gene* 185(2):291–296.
91. Pluta, K., et al. (2001). Maf1p, a negative effector of RNA polymerase III in *Saccharomyces cerevisiae*. *Mol Cell Biol* 21(15):5031–5040.

92. Moir, R.D., et al. (2006). Protein kinase A regulates RNA polymerase III transcription through the nuclear localization of Maf1. *Proc Natl Acad Sci USA* 103(41):15044–15049.
93. Wei, Y., Tsang, C.K., and Zheng, X.F. (2009). Mechanisms of regulation of RNA polymerase III-dependent transcription by TORC1. *EMBO J* 28(15):2220–2230.
94. Desai, N., et al. (2005). Two steps in Maf1-dependent repression of transcription by RNA polymerase III. *J Biol Chem* 280(8):6455–6462.
95. Oficjalska-Pham, D., et al. (2006). General repression of RNA polymerase III transcription is triggered by protein phosphatase type 2A-mediated dephosphorylation of Maf1. *Mol Cell* 22(5):623–632.
96. Roberts, D.N., et al. (2006). Dephosphorylation and genome-wide association of Maf1 with Pol III-transcribed genes during repression. *Mol Cell* 22(5):633–644.
97. Reina, J.H., Azzouz, T.N., and Hernandez, N. (2006). Maf1, a new player in the regulation of human RNA polymerase III transcription. *PLoS One* 1:e134.
98. Johnson, S.S., et al. (2007). Mammalian Maf1 is a negative regulator of transcription by all three nuclear RNA polymerases. *Mol Cell* 26(3):367–379.
99. Tushinski, R.J., and Warner, J.R. (1982). Ribosomal proteins are synthesized preferentially in cells commencing growth. *J Cell Physiol* 112(1):128–135.
100. Jefferies, H.B., et al. (1994). Rapamycin selectively represses translation of the "polypyrimidine tract" mRNA family *Proc Natl Acad Sci USA* 91(10):4441–4445.
101. Meyuhas, O. (2000). Synthesis of the translational apparatus is regulated at the translational level. *Eur J Biochem* 267(21):6321–6330.
102. Jefferies, H.B., et al. (1997). Rapamycin suppresses 5'TOP mRNA translation through inhibition of p70s6k. *EMBO J* 16(12):3693–3704.
103. Pende, M., et al. (2004). $S6K1^{(-/-)}/S6K2^{(-/-)}$ mice exhibit perinatal lethality and rapamycin-sensitive 5'-terminal oligopyrimidine mRNA translation and reveal a mitogen-activated protein kinase-dependent S6 kinase pathway. *Mol Cell Biol* 24(8):3112–3124.
104. Stefanovsky, V., et al. (2006). Growth factor signaling regulates elongation of RNA polymerase I transcription in mammals via UBF phosphorylation and r-chromatin remodeling. *Mol Cell* 21(5):629–639.

2

The TSC1–TSC2 Complex: A Key Signal-Integrating Node Upstream of TOR

CHRISTIAN C. DIBBLE • BRENDAN D. MANNING

Department of Genetics and Complex Diseases
Harvard School of Public Health
Boston, Massachusetts, USA

I. Abstract

The TSC1 (hamartin) and TSC2 (tuberin) proteins function as a heterodimer that integrates diverse extracellular and intracellular signals to regulate the two TOR complexes (TORC1 and TORC2; mTORC1 and mTORC2 in mammals) and the processes of cell growth and proliferation. The TSC1–TSC2 complex negatively regulates TORC1 through its GTPase-activating protein (GAP) activity toward the small G-protein Rheb (Ras homolog enriched in brain), an essential activator of TORC1. In contrast, TORC2 in mammalian cells is positively regulated by the TSC1–TSC2 complex through both Rheb and TORC1-dependent and independent mechanisms. The regulatory relationship between the TSC1–TSC2 complex and TORC1 appears to be conserved in most eukaryotes, including species of yeast. Multisite phosphorylation of both TSC1 and TSC2 promotes or inhibits the ability of the complex to inhibit Rheb downstream of a wide variety of kinases and signaling pathways. Several of these pathways include important oncogene products and tumor suppressors that can promote inhibition of the TSC1–TSC2 complex and likely contribute to the aberrantly elevated

THE ENZYMES, Vol. XXVIII

ISSN NO: 1874-6047

DOI: 10.1016/S1874-6047(10)28002-2

levels of mTORC1 signaling detected in the majority of human cancers. Inactivating mutations in the *TSC1* and *TSC2* genes give rise to a multifaceted tumor syndrome known as tuberous sclerosis complex. However, such mutations are not frequent in sporadic cancers. As the TSC1–TSC2 complex lies at a pivotal position within a signaling network that controls physiological growth and proliferation and that is frequently misregulated in human diseases, increasing our understanding of how the complex is regulated and the mechanisms by which it regulates TOR has widespread implications.

II. Introduction

Mapping of the *TSC1* and *TSC2* genes was driven by efforts to understand the molecular basis of tuberous sclerosis complex (TSC), a disease that results from inactivating mutations in these genes. TSC is a multifaceted genetic disorder that is characterized by a high incidence of epilepsy, autism spectrum disorders, and tumors affecting multiple organ systems, including the brain, skin, heart, kidney, and lung [1]. The cloning of *TSC2* in 1993 and *TSC1* in 1997 touched off a burst of research into the molecular function of the two proteins. Although the disease that results specifically from mutations in *TSC1* and *TSC2* is relatively rare, estimated at 1 in 6000 to 10,000 births [2, 3], understanding the signaling network involving their gene products has had implications far beyond TSC, including important insights into complex diseases such as cancer, diabetes, and autism. Through the past decade of research, TSC1 and TSC2 have emerged as a critical signal-integrating node that senses and responds to extracellular and intracellular cues to regulate TOR. This chapter discusses molecular characteristics and functions of the TSC1 and TSC2 proteins in the context of the signaling network that regulates and is regulated by these proteins.

A. TSC1 and TSC2 Genes, Transcripts, and Proteins

Despite their shared nomenclature, distinct genes encode TSC1 and TSC2 (also known as hamartin and tuberin, respectively), and their sequences share no similarity to each other. The human *TSC1* gene, comprised of 23 exons (21 of which are coding exons), is located at 9q34. The full-length *TSC1* transcript is approximately 8.6 kb and encodes a protein of 1164 amino acids with a predicted molecular weight of 130 kDa [4, 5]. The *TSC2* gene, comprised of 42 exons (41 of which are coding exons), is located at 16p13.3 [6, 7]. The full-length *TSC2* transcript is 5.7 kb and encodes a protein of 1807 amino acids with a predicted molecular weight of 201 kDa.

Very few studies have been carried out on transcriptional regulation of these genes, but transcript variants and protein isoforms have been detected in cells and tissues. While three mRNA splice variants of the human *TSC1* transcript exist, these affect only the 5′-untranslated region and not the protein product [8]. Four protein isoforms of human TSC2 have been identified, with isoform 1 being the full-length product. Isoform 2 lacks exon 25 (amino acids 946–988), while isoform 3 lacks exon 25 plus the first codon of exon 26. Isoform 4 is a common variant of 1784 amino acids lacking residues 1272–1294. Although the isoforms are differentially expressed in specific tissues, it is not known whether they exhibit functional differences [9, 10]. In mammals, TSC1 and TSC2 are ubiquitously expressed in all cell types and at all stages of the cell cycle speaking to their fundamental importance in controlling basic cellular processes.

B. TSC1 and TSC2 Proteins Form a Heterodimer and Act as a Functional Unit

In all species examined, the TSC1 and TSC2 proteins directly interact in what is predominantly thought to be a stable 1:1 heterodimer [11]. There is some tenuous evidence for their interaction at ratios of 1:2 or 2:1 based on sucrose gradient fractionations that suggest they participate in a complex that is up to 450 kDa [12] and separate experiments that show epitope-tagged fragments of TSC2 transgenically expressed in rat can coimmunoprecipitate endogenous TSC2 from tissue lysates [13]. As an indication of the strength of the TSC1–TSC2 interaction, the association between the proteins is resistant *in vitro* to detergents, high salt concentrations, and extreme pH conditions [12, 14–17]. This interaction is important for the stability of TSC2, which is thought to undergo ubiquitination and proteosomal degradation when dissociated from TSC1, as in *TSC1* null cells [18–20]. The stability of TSC1, on the other hand, is not greatly affected by TSC2. Although other proteins have been proposed to interact with the complex, none have been shown to do so in a stoichiometric manner [21].

C. TSC1 and TSC2 Evolutionary Conservation

TSC1 and TSC2 are broadly conserved across multiple eukaryotic kingdoms, with the possible exception of plants for which it is not clear that orthologs exist. The vast majority of metazoans whose genomes have been sequenced encode TSC1 and TSC2 orthologs, including vertebrates, insects, worms, and even the amoeba-like *Trichoplax adhaerens*, one of the most evolutionarily basal metazoans known. There are also extensive examples of orthologs in the fungi kingdom, including both multicellular molds and unicellular yeast, plus the phylogenetically distinct slime molds, such as *Dictyostelium discoideum*. Noteworthy exceptions to the broad conservation of TSC1 and TSC2 in metazoans and fungi are the lack of clear orthologs in the nematode *Caenorhabditis elegans* and the budding yeast *Saccharomyces cerevisiae*, two important model organisms. Interestingly, TSC1 and TSC2 orthologs do exist in the budding yeast *Candida albicans*, the fission yeast *Schizosaccharomyces pombe*, and the nematode *Brugia malayi*, suggesting that TSC1 and TSC2 existed in the last common ancestor of metazoans, fungi, and slime molds, potentially following the split from plants.

D. TSC1 and TSC2 Protein Domains

TSC1 and TSC2 share little sequence homology with other proteins, but each contains at least one domain with known functions in other proteins. A C-terminal region spanning amino acids 719–998 of human TSC1 is

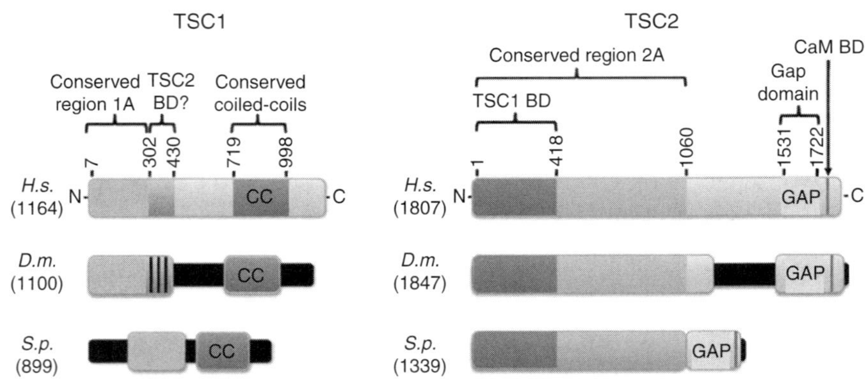

Fig. 2.1. TSC1 and TSC2 protein domains and evolutionary conservation. Amino acid sequence conservation was determined using BLAST and TCOFFEE alignment programs. Species and amino acid length are shown to the left of each ortholog; *Homo sapiens* (*H.s.*), *Drosophila melanogaster* (*D.m.*), and *Schizosaccharomyces pombe* (*S.p*). Poorly conserved regions (black), conserved regions (colored) are shown. All numbering is for full-length human proteins. In TSC1, two domains have previously been reported: a putative TSC2-binding domain (a.a. 302–430, blue) that is only weakly conserved in *D. melanogaster* (indicated by striping in region of *D.m.* ortholog) and is not clearly conserved in *S. pombe*; and a predicted coiled-coil region (a.a. 719–998, CC, dark green) that is conserved down to yeast. A third well-conserved N-terminal region has not been assigned a function (a.a. 7–302, conserved region 1A, light green). In TSC2, three domains have previously been reported, the putative TSC1-binding domain (a.a. 1–418, red), the catalytic GAP domain (a.a. 1531–1722, yellow), and a putative calmodulin (CaM) binding domain (a.a. 1740–1757, purple line). These regions are all conserved down to yeast. The TSC1 binding domain lies within a larger conserved region (a.a. 1–1060, conserved region 2A, red and orange). (For interpretation of the references to color in this figure legend, the reader is referred to the Web version of this chapter.)

strongly predicted to form a large contiguous region of coiled-coil secondary structure (Figure 2.1) [5], a feature well known to mediate protein–protein interactions in other proteins (coiled-coils reviewed in Ref. [22]). Although the exact molecular function of the conserved coiled-coil domain remains to be elucidated, it has been proposed to be required for TSC2 binding based on coimmunoprecipitation experiments in *Drosophila* [17] and yeast two-hybrid assays [15]. However, this data is in conflict with a later yeast two-hybrid study that found the coiled-coil region did not bind TSC2 [23]. This second study instead identified an N-terminal "TSC2-binding domain" that roughly mapped to amino acids 302–430 as sufficient for binding TSC2 [23]. However, the contribution of this domain to TSC2 binding is uncertain, as it lacks clear conservation in the *S. pombe* ortholog of TSC1 and is only weakly conserved in the *Drosophila melanogaster* ortholog, despite the known association between TSC1 and TSC2 in these organisms [17, 24]. Interestingly, studies in mammalian cells and tissues support a role for both the N-terminal and C-terminal halves of TSC1 in binding TSC2 [25]. Although the N-terminus of TSC1 was initially reported to contain a potential transmembrane domain [5], it is predominantly cytosolic and not thought to be an integral membrane protein [12]. In addition to the coiled-coil domain, TSC1 orthologs are readily identifiable through a conserved approximately 300 amino acid region of homology in their N-termini (Figure 2.1, conserved region 1A). This region does not share homology with other known proteins, and its function is currently unknown.

The C-terminus of TSC2 contains the heterodimer's only characterized functional domain, a GAP domain extending from amino acids 1531–1722 (Figure 2.1), which is discussed in detail in Section III. Two very small regions of predicted coiled-coil also exist on TSC2 (amino acids 346–371 and 1008–1021), but these exhibit a relatively low probability of forming coiled-coils (as determined using PCOILS program [26]). The more N-terminal of these putative coiled-coils (346–371) was found to interact with TSC1 by yeast two-hybrid [15], and also falls within a larger region found in a second study to associate with TSC1 (amino acids 1–418) [23]. However, studies in transgenic rats expressing deletion mutants of TSC2 suggest that both N-terminal and C-terminal regions of TSC2 cooperate in TSC1 binding [13]. Therefore, the interactions between TSC1 and TSC2 may be more extensive than the currently defined binding regions suggest. The proposed N-terminal TSC1-binding domain is well conserved in general as is the entire N-terminal half of TSC2 (amino acids 1–1060; Figure 2.1, conserved region 2A) though the region between amino acids 419–1060 has not been assigned a function. A region C-terminal to the GAP domain in TSC2 has been proposed to contain several putative features including a calmodulin binding domain (amino acids 1740–1757) [27], an overlapping region that binds estrogen receptor-α (amino acids 1734–1807) [28], and a nuclear localization signal (amino acids 1743–1755) [28, 29]. However, a clear role for this region in the regulation and function of the TSC1–TSC2 complex awaits further investigation.

III. Downstream Functions: Regulation of the TOR Complexes by the TSC1–TSC2 Complex

A. The TSC1–TSC2 Complex Negatively Regulates Growth and Proliferation

Since the genetic loss of TSC1 and TSC2 that underlies TSC results in tumor formation in multiple organs, these genes have long been classified as tumor suppressors. An important milestone in understanding the function of the TSC1–TSC2 complex came when its tumor suppressor activity was linked at the molecular level to the GAP domain of TSC2 [30, 31] and to control over a basic physiological process—cell growth. Germline ablation of either *Tsc1* or *Tsc2* in rodents leads to embryonic lethality [32–35]. However, using *Drosophila* genetics, it was found that elimination of *Tsc1* or *Tsc2* in adult tissues, such as the eye or wing, results in overgrowth phenotypes due not only to an increase in cell number but, most dramatically, to a cell autonomous increase in cell size [16, 17, 36, 37]. Reciprocally, a decrease in cell size and number resulted from co-overexpression of TSC1 and TSC2 [16, 17]. Importantly, the genetic interactions described in some of these studies suggested that TSC1 and TSC2 might act within the Insulin/IFG-1 signaling pathway, which was previously known to play a critical role in organ and organismal size control [16, 17].

B. The TSC1–TSC2 Complex Inhibits mTORC1 by Acting as a GAP for Rheb

It is now well recognized that the TSC1–TSC2 complex negatively regulates cell growth and proliferation, in large part, by inhibiting mTOR complex 1 (mTORC1), whose components and role as a master regulator of cell growth are evolutionarily conserved throughout eukaryotes. A combination of *Drosophila* genetics and biochemical studies in mammalian cells demonstrated that the TSC1–TSC2 complex inhibits mTORC1 [16, 38–42] by acting as the specific GAP for Ras homolog enriched in brain (Rheb) [43–51]. Rheb, a member of the Ras small G-protein superfamily, is essential for mTORC1 activation and can directly stimulate mTORC1 kinase activity when in its GTP-bound active state [52, 53]. A second Rheb homolog termed Rheb2 or Rheb-like 1 (RhebL1) also exists in mammals and similarly activates mTORC1 downstream of the TSC1–TSC2 complex [53–55]. Relative to other Ras family members, Rheb has low intrinsic GTPase activity [56, 57] and is therefore more dependent on its GAP for GTP hydrolysis. Hence, TSC2 inhibits mTORC1 activation by stimulating the intrinsic GTPase activity of Rheb, thereby converting it to its GDP-bound state, which cannot activate mTORC1 (Figure 2.2). In addition to stabilizing TSC2, TSC1 binding also greatly enhances the GAP activity of TSC2 toward Rheb [49], further supporting genetic the evidence that the TSC1–TSC2 complex acts as a functional unit to inhibit mTORC1 and cell growth.

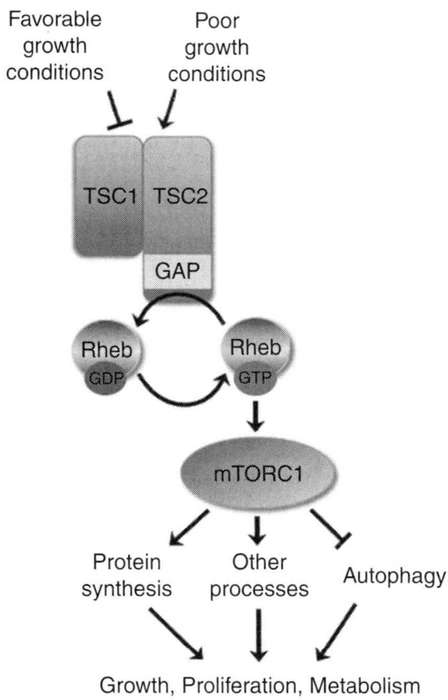

FIG. 2.2. The TSC1–TSC2 complex is a pivotal inhibitor of mTORC1. The TSC1–TSC2 heterodimer inhibits mTORC1 by acting as the specific GAP for Rheb. TSC2-GAP activity greatly accelerates conversion of Rheb to its inactive GDP-bound state. Since Rheb-GTP is essential for mTORC1 activation, the TSC1–TSC2 complex inhibits mTORC1. In general, favorable growth conditions inhibit the TSC1–TSC2 complex and therefore activate mTORC1 while poor growth conditions activate the TSC1–TSC2 complex and inhibit mTORC1. See Figure 2.5 for specific conditions that regulate the complex.

The GAP activity of TSC2 has been shown to be highly specific for Rheb, both *in vitro* and *in vivo* [47–49, 51]. Although the TSC2-GAP domain was first shown to have *in vitro* GAP activity toward two other distinct small G-proteins, Rap1 and Rab5 [58, 59], the measured GAP activity was quite low, making it unlikely that these are physiological substrates. The specificity of the TSC2-GAP domain for Rheb is derived from at least two properties of these proteins. First, key residues within the catalytic pocket of Rheb differ from the corresponding residues in Ras and the mechanism by which TSC2 and the related Rap1GAP assist in the hydrolysis of GTP is quite distinct from that of GAPs for Ras and another small G-protein Rho. Ras and Rho GAPs utilize an "arginine finger" to assist the catalytic glutamine of Ras and Rho in hydrolyzing GTP [60, 61]. In contrast, Rap1GAP and TSC2-GAP provide an "asparagine thumb" that catalytically substitutes for the nonfunctional "catalytic" glutamine of Rap1 and Rheb [57, 62–65]. Second, residues within a hydrophobic patch (e.g., F1666 of human TSC2) and a separate looped region (e.g., L1594) that flank the asparigine thumb within the TSC2-GAP domain participate in unique binding interactions with Rheb [57]. Therefore, it is likely that Rheb and Rheb2/RhebL2 are the only direct targets of the TSC2-GAP domain.

Like TSC1, TSC2, and the components of mTORC1, Rheb appears to be ubiquitously expressed in mammalian tissues [43, 66]. Rheb orthologs also exist in all species in which the TSC1–TSC2 complex is conserved, suggesting a conserved role for TSC1–TSC2 in regulating Rheb. In *S. pombe*, a conserved relationship between the TSC1, TSC2, and Rheb orthologs has been established, as has their regulation of TOR functions in this yeast ([67, 68]; reviewed in Ref. [69]). Interestingly, a Rheb ortholog exists in *S. cerevisiae*, which lacks orthologs of TSC1 and TSC2, and certain functions of Rheb, such as the control of arginine uptake, are conserved between *S. pombe* and *S. cerevisiae* [67, 70, 71]. However, to date, there is no evidence that Rheb regulates TOR functions in *S. cerevisiae*, where Rheb is not essential and its disruption does not lead to a clear growth phenotype [70, 72]. The functional differences between Rheb from *S. pombe* and *S. cerevisiae* were nicely highlighted by experiments in which expression of human Rheb but not *S. cerevisiae* Rheb was able to rescue a growth-arrest phenotype in Rheb-deficient *S. pombe* [73]. Therefore, in eukaryotes that lack TSC1 and TSC2 orthologs, such as *S. cerevisiae,* the relationship between Rheb and TORC1 may be quite different, perhaps suggesting that tight regulation of Rheb by the TSC1–TSC2 complex coevolved with a critical role for Rheb in TORC1 activation. The regulation of TORC1 by Rheb and the downstream functions of TORC1 in various organisms are discussed in detail in other chapters.

C. The TSC1–TSC2 Complex is Required for Full mTORC2 Activity

While the functions of mTOR are best studied within mTORC1, mTOR also functions within a second well-conserved multiprotein complex termed mTOR complex 2 (mTORC2). To date, little is known regarding the upstream regulation of mTORC2. However, mTORC2 has been shown to phosphorylate conserved regulatory sites on multiple AGC family kinases, including Akt, SGK1, and specific isoforms of PKC, and is required for the full activation of these kinases ([74–77]; reviewed in Ref. [78]). Interestingly, the kinase activity of mTORC2 is stimulated by growth factors [74, 79], but mTORC2 is required for both basal (e.g., Akt-T450) and growth factor-stimulated (e.g., Akt-S473) phosphorylation events on these kinases. How mTORC2 achieves these dual functions is unknown. Unlike mTORC1, mTORC2 does not depend on Rheb for its activation, and it is not negatively regulated by TSC2-GAP activity [53, 79, 80]. Surprisingly, the TSC1–TSC2 complex was found to promote mTORC2 activity and thus plays opposing roles in the regulation of the two mTOR complexes (Figure 2.3). In striking contrast to the constitutive activation of mTORC1 in *TSC1*$^{-/-}$ and *TSC2*$^{-/-}$ cells, loss of the TSC1–TSC2 complex results in severe attenuation of mTORC2 kinase activity [79]. This is reflected in the reduced phosphorylation of kinases downstream of mTORC2, including both growth factor-dependent and -independent regulatory sites [81].

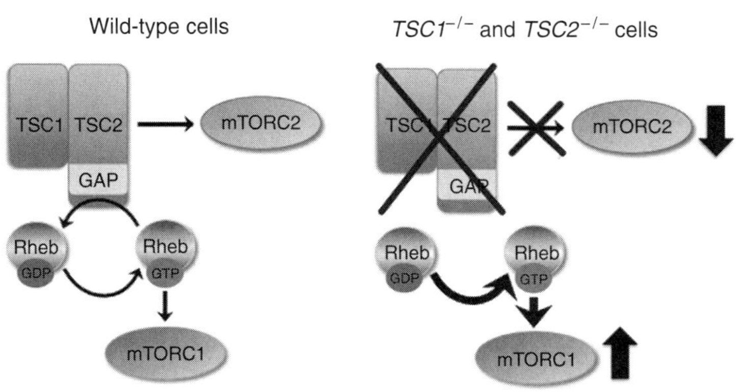

Fig. 2.3. The TSC1–TSC2 complex has opposing effects on mTORC1 and mTORC2. In wild-type cells, the TSC1–TSC2 complex inhibits mTORC1 but promotes mTORC2 activity. In *TSC1*$^{-/-}$ and *TSC2*$^{-}$/$^{-}$ cells, loss of TSC2-GAP activity leaves Rheb in its active GTP-bound state leading to constitutive activation of mTORC1. In contrast, loss of the TSC1–TSC2 complex results in severe attenuation of mTORC2 activity and this is at least in part independent of mTORC1-dependent feedback inhibition.

The TSC1–TSC2 complex can promote mTORC2 activity and signaling through at least two distinct mechanisms: (1) inhibition of Rheb and mTORC1 leading to relief of mTORC1-dependent feedback mechanisms that inhibit mTORC2 either directly or upstream of its activation; (2) a more direct mechanism that is independent of Rheb and mTORC1 and likely involves direct binding to mTORC2. mTORC1-dependent feedback mechanisms that inhibit the cellular response to insulin and PDGF have been found to be potently activated in cells lacking the TSC genes [11, 82, 83] and may contribute to the loss of growth-factor stimulated mTORC2 activity in these cells [79]. Furthermore, an inhibitory phosphorylation site (T1135) on the mTORC2 core component Rictor has been found to be phosphorylated by S6K1 downstream of mTORC1 [84]. However, the positive regulatory effects of the TSC1–TSC2 complex on mTORC2 can be separated from its inhibitory effects on Rheb and mTORC1 [79], and both growth factor-dependent and -independent substrates of mTORC2 are affected by loss of the TSC genes [81]. Furthermore, while S6K1-mediated phosphorylation of T1135 on Rictor affects the ability of mTORC2 to phosphorylate Akt-S473 in cells, it does not grossly affect mTORC2 kinase activity [84], which is blunted upon loss of the TSC1–TSC2 complex [79]. Therefore, the TSC1–TSC2 complex positively regulates mTORC2, at least in part, through a Rheb and mTORC1-independent mechanism. While the molecular mechanism of this regulation remains unknown, the TSC1–TSC2 complex has been found to directly associate with mTORC2 [79] and can stimulate mTORC2 kinase activity *in vitro* [81]. The association between the TSC1–TSC2 complex and mTORC2 involves relatively weak, salt-sensitive interactions that depend on TSC2 and Rictor [79, 81]. It is clear that more work is needed to understand the mechanisms controlling mTORC2 activity and the role of the TSC1–TSC2 complex in this regulation.

IV. Upstream Regulation: The TSC1–TSC2 Complex Integrates Diverse Signals to Regulate mTORC1

As a critical upstream regulator of mTORC1, the TSC1–TSC2 complex has emerged as a signal-integrating hub that senses cell growth conditions. In general, conditions favorable to cell growth and proliferation inhibit the TSC1–TSC2 complex to activate Rheb and mTORC1, while poor growth conditions activate the TSC1–TSC2 complex to inhibit Rheb and mTORC1. Multisite phosphorylation of specific serine and threonine residues on TSC1 and TSC2 is the primary mechanism by which the complex senses these growth conditions. We and others have mapped over 50 distinct phosphorylation sites combined on TSC1 and TSC2 and many of these are phosphorylated in response to specific cellular cues (specific sites are discussed in detail below). Therefore, the TSC1–TSC2 complex can be thought of as a "molecular switchboard" controlling mTORC1 activation [85]. It is worth noting that while studies with phosphorylation-site mutants of TSC1 and TSC2 demonstrate that these sites affect the ability of the TSC1–TSC2 complex to regulate Rheb and mTORC1 in response to specific stimuli, the molecular mechanism by which individual sites regulate the TSC1–TSC2 complex is largely unknown. For instance, none of these sites have been shown to directly affect the *in vitro* GAP activity of the TSC1–TSC2 complex toward Rheb. Proposed *in vivo* mechanisms by which specific phosphorylation sites might regulate the TSC1–TSC2 complex will be discussed below (specific sites and kinases are summarized in Table 2.1 and Figure 2.4 and upstream stimuli in Figure 2.5).

TABLE 2.1

PHOSPHORYLATION SITES ON TSC1 AND TSC2[a]

Site[b]	Kinase[c]	Function[d]	Conservation[e]	Reference(s)
TSC1				
T417	CDK1	Inhibit	Vertebrate[f]	[103]
S487	IKKb	Inhibit	Mammal	[101]
S511	IKKb	Inhibit	Vertebrate	[101]
S584	CDK1	Inhibit	Vertebrate	[103]
S1047	CDK1	Inhibit	Mammal	[103]
TSC2				
S540	ERK	Inhibit	Vertebrate	[100]
S664	ERK	Inhibit	Vertebrate	[100]
S939	Akt/RSK1	Inhibit	Metazoan	[39, 41, 98]
S981	Akt/RSK1	Inhibit	Vertebrate	[91, 98]
S1130	Akt	Inhibit	Vertebrate	[39]
S1132	Akt	Inhibit	Vertebrate	[39]
S1254	MAPKAPK2	Unknown	Vertebrate	[120]
T1271	AMPK	Activate	Vertebrate	[134]
S1371	GSK3	Activate	Human	[112]
S1375	GSK3	Activate	Mammal	[112]
S1379	GSK3	Activate	Mammal	[112]
S1383	GSK3	Activate	Metazoan	[112]
S1387	AMPK	Activate	Metazoan	[134]
T1462	Akt/RSK1	Inhibit	Metazoan	[39, 41, 98]
S1798	RSK1	Inhibit	Vertebrate	[98]

[a]Only sites with some *in vivo* evidence of phosphorylation by known kinases are included.
[b]T, threonine; S, serine; amino acid numbering based on full-length human proteins.
[c]Based on conclusions in reference(s).
[d]Inhibits or activates the TSC1–TSC2 complex, generally with phosphoylation of mTORC1 substrates as a readout. Based on conclusions in reference(s).
[e]Based on individual BLAST alignments of orthologs with human protein.
[f]In vertebrates other than humans, such as mouse and fish, T417 is conserved but the +1 proline (which is required for targeting by CDK1) is not conserved.

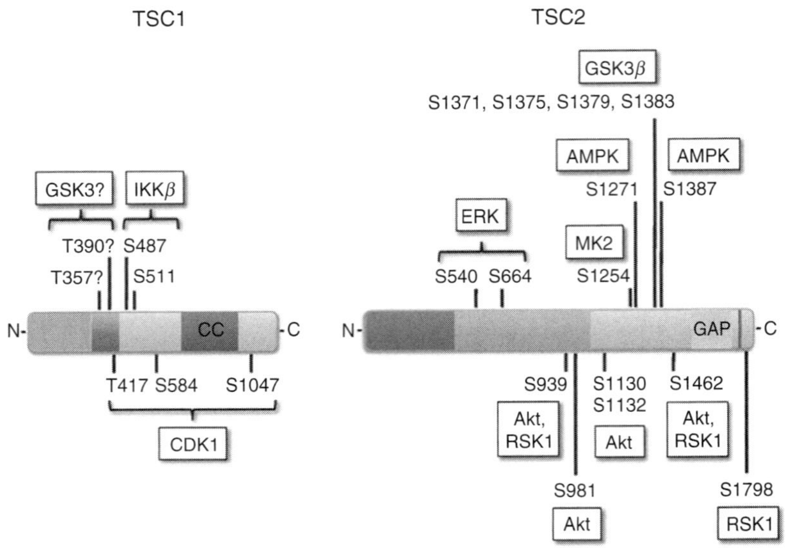

Fig. 2.4. Regulatory phosphorylation sites on TSC1 and TSC2. Phosphorylation of both TSC1 and TSC2 on multiple residues and by a variety of kinases is a key mechanism by which the complex is regulated. For each phosphorylated site, the residue (S, serine, or T, threonine), amino acid numbering for the human orthologs, and kinase are designated. A question mark (?) indicates that the event requires further confirmation. Regions are colored and marked as in Figure 2.1. See Table 2.1 for conservation of each site and relevant references. (For interpretation of the references to color in this figure legend, the reader is referred to the Web version of this chapter.)

A. Growth Factor Signaling: The PI3K–Akt Pathway

The Ser/Thr kinase Akt is activated downstream of class I phosphoinositide-3 kinase (PI3K) and phosphorylates a number of downstream targets to promote cell survival, growth, and proliferation (reviewed in Ref. [86]). Akt was the first kinase found to directly phosphorylate the TSC1–TSC2 complex, and it has been shown to phosphorylate TSC2 in both *Drosophila* and mammals [38, 39, 41]. In mammals, Akt is thought to phosphorylate TSC2 on five residues, S939, S981, S1130, S1132, and T1462, although the strength of the evidence for each site varies. Of these sites, S939 and T1462 have been definitively shown to be *in vivo* Akt substrates [39, 41], and these are conserved and phosphorylated in flies (corresponding to *D. melanogaster* S924 and T1518, respectively [38]). Consistent with Akt activating mTORC1 through the phosphorylation and inhibition of TSC2, overexpression of TSC2 mutants containing nonphosphorylatable alanine substitutions at different combinations of these sites can dominantly block the

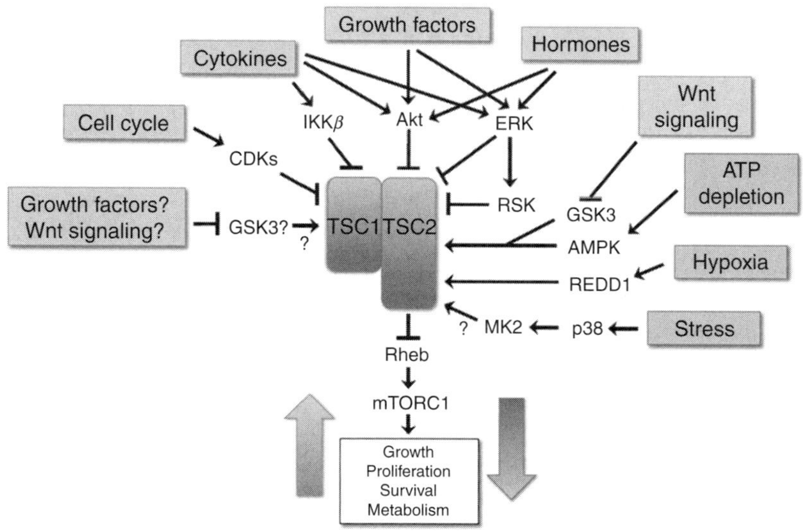

FIG. 2.5. Physiological signaling inputs integrated by the TSC1–TSC2 complex. The TSC1–TSC2 complex integrates inputs from numerous stimuli, signaling pathways, and kinases to control mTORC1 and downstream processes. Stimuli that inhibit the TSC1–TSC2 complex and activate mTORC1 (green) or that activate TSC1–TSC2 complex and inhibit mTORC1 (red) are shown with the downstream kinases that are thought to directly phosphorylate TSC1 or TSC2. REDD1 is not a kinase but is thought to activate the TSC1–TSC2 complex by blocking the ability of Akt-targeted sites to inhibit the complex. (For interpretation of the references to color in this figure legend, the reader is referred to the Web version of this chapter.)

stimulation of mTORC1 signaling by growth factors [39, 41, 87]. Importantly, a study in which Akt phosphorylation-site mutants of TSC2 were stably expressed at near endogenous levels in *Tsc2*$^{-/-}$ MEFs demonstrated that both the evolutionarily conserved (S939 and T1462) and vertebrate-specific sites (S981, S1130, and S1132) on TSC2 were essential for insulin to activate mTORC1 [88]. Therefore, it appears that in mammalian cells, the phosphorylation of these five sites acts cooperatively to inhibit the TSC1–TSC2 complex for maximal mTORC1 activation downstream of growth factors that activate Akt. However, the importance of TSC2 phosphorylation by Akt might vary by tissue and species. For instance, in *Drosophila* where Akt has also been found to phosphorylate and inhibit TSC2 [38], expression of a TSC2 mutant lacking the Akt phosphorylation sites in an otherwise TSC2-null background demonstrated that this regulation is dispensable for normal fly development [89, 90]. Similar knock-in experiments in mammals are underway and will help determine the role of Akt-mediated phosphorylation of TSC2 in physiology and disease.

Although the molecular mechanism by which Akt phosphorylation inhibits the TSC1–TSC2 complex is not fully understood, several mechanisms have been proposed. Dissociation of the TSC1–TSC2 complex in response to phosphorylation by Akt was observed in one *Drosophila* study [38], but this inhibitory mechanism has not been supported by other studies in *Drosophila* [89]. Furthermore, Akt-mediated phosphorylation of TSC2 has not been found to disrupt the TSC1–TSC2 complex in the majority of studies in mammalian cells [39, 41]. Degradation of TSC2 in response to its phosphorylation by Akt has also been proposed as a mechanism of inhibition [91, 92], but the time scale of this effect (hours) does not match that of mTORC1 activation in response to Akt (minutes). However, it is possible that this mechanism could contribute to longer term inhibition of the TSC1–TSC2 complex by sustained Akt signaling. Phosphorylation of S939, S981, and T1462 has been suggested to create binding sites on TSC2 for 14-3-3 proteins [87, 93–95]. Although this binding has been invariably proposed to inhibit the TSC1–TSC2 complex, the mechanism is not clear. While one study concluded that 14-3-3 binding sequesters TSC2 from an uncharacterized membrane fraction containing TSC1 [87], independent studies have found that the interaction of TSC1 with TSC2 was unaffected by 14-3-3 binding [94, 95]. Therefore, further studies are needed to determine how phosphorylation of these sites on TSC2 by Akt and, perhaps, subsequent binding to 14-3-3 proteins acutely block the ability of the TSC1–TSC2 to act as a GAP for Rheb allowing mTORC1 activation. It is likely that, as proposed in specific studies [87], this regulation involves rapid changes in the subcellular localization of the TSC1–TSC2 complex and/or Rheb. However, the bona fide localization of the TSC1–TSC2 complex in cells is currently unknown.

B. Growth Factor Signaling: The ERK–RSK Pathway

Two other growth factor-induced protein kinases, ERK and its downstream target RSK (reviewed in Ref. [96]), also phosphorylate and inhibit TSC2 to activate mTORC1 signaling. ERK and RSK can promote activation of mTORC1 in response to stimuli that do not activate Akt, such as the phorbol ester PMA [97]. RSK phosphorylates S1798 in the extreme C-terminus of TSC2 and also appears to phosphorylate two sites overlapping with Akt, S939 and T1462 [98]. Alanine mutations of the three RSK-targeted sites on TSC2 can dominantly block PMA-induced activation of mTORC1 signaling. Using mass spec analyses, other sites on TSC2 have been shown to be phosphorylated in response to PMA, including S664, which lies within a consensus motif preferentially phosphorylated by ERK [99]. In a separate study, ERK was shown to inactivate the TSC1–TSC2 complex by phosphorylating S664 and a second site S540 [100]. Interestingly, ERK-dependent phosphorylation not only induced dissociation of the TSC1–TSC2 complex in cells but also caused dissociation of co-immunoprecipitated TSC1 and TSC2 *in vitro*, suggesting that phosphorylation directly interferes with the TSC1–TSC2 interaction. However, PMA-induced ERK activation was not found to disrupt the TSC1–TSC2 complex in other studies [97–99], and this mechanism warrants further study. Therefore, upon activation of the ERK pathway, both ERK and RSK phosphorylate distinct sites on TSC2 that lead to the stimulation of mTORC1 signaling.

C. Cytokine Signaling: IKKβ

The proinflammatory cytokine tumor necrosis factor α (TNFα) stimulates mTORC1 signaling in certain cell types and has been shown to do so, at least in part, through activation of inhibitory κB kinase β (IKKβ) downstream of the TNF receptor [101]. IKKβ phosphorylates TSC1 on S487 and S511, and these sites are required for TNFα-induced activation of mTORC1. Although the sequence surrounding S511 conforms to the consensus motif targeted by IKKβ and is conserved throughout vertebrates, S487 does not lie within a typical IKKβ recognition motif and is conserved only in mammalian orthologs of TSC1. The phosphorylation of TSC1 on these sites was suggested to interfere with TSC2 binding, as stimulation with TNFα or expression of phospho-mimetic (Ser to Asp) mutants affecting these two sites on TSC1 reduced the amount of TSC2 coimmunoprecipitating with TSC1 [101]. However, the effects on the TSC1–TSC2 interaction were partial, suggesting that other inhibitory mechanisms might also contribute. It is worth noting that cytokines such as TNFα can also activate the Akt and ERK pathways in many cell types and, therefore, the phosphorylation of TSC2 by these kinases will likely cooperate with IKKβ-induced phosphorylation of TSC1 to inhibit the TSC1–TSC2 complex and activate mTORC1.

D. The Cell Cycle: Cyclin-Dependent Kinases

Our current understanding of how the TSC1–TSC2 complex is regulated throughout the cell cycle is poor. However, there is some evidence for direct regulation by cyclin-dependent kinases (CDKs) at both the G1/S and G2/M transitions. Progression through the restriction point in G1 and entry into S phase is controlled in part by the CDK4/6–cyclin D complex. In one study, co-overexpressed cyclin D and TSC2 were found to coimmunoprecipitate [102]. Overexpression of CDK4/6–cyclin D increased the phosphorylation of the mTORC1-substrates S6K and 4E-BP1, and this correlated with electrophoretic mobility shifts of TSC1 and TSC2, suggestive of phosphorylation. However, whether this apparent phosphorylation was the direct effect of CDK4/6–cyclin D activity and, if so, what the relevant sites are remains unknown. It will be important to determine whether endogenous CDK–cyclin complexes phosphorylate TSC1 or TSC2 at the G1/S transition and whether this affects the role of mTORC1 in promoting cell cycle entry.

The TSC1–TSC2 complex has also been reported to be regulated by the CDK1–cyclin B complex which controls progression through mitosis [103, 104]. Cells arrested at G2/M or progressing unperturbed through the G2/M transition exhibited increased TSC1 phosphorylation, as detected by a decrease in electrophoretic mobility or by recognition with a phosphothreonine antibody [103]. CDK1 was found to phosphorylate TSC1 *in vitro* and to induce its phosphorylation when overexpressed in cells. A candidate approach was used in which three consensus CDK phosphorylation sites (T417, S584, S1047) were mutated to alanine, and this mutant was not phosphorylated by CDK1. Furthermore, when overexpressed in cells, this triple alanine mutant of TSC1 decreased S6K1 activity and 4E-BP1 phosphorylation, suggesting that phosphorylation of these sites might inhibit the TSC1–TSC2 complex.

In a related study, it was shown that polo-like kinase 1 (Plk1), another important mitotic kinase, binds TSC1 in a manner that is blocked by mutation of T310, a possible CDK1 phosphorylation site [25]. Plk1 binding was proposed to promote mTORC1-dependent translation of proteins required for entry into mitosis, such as cyclin B1. However, whether T310 is actually phosphorylated by CDKs, or at all, is unknown and a role for Plk1 in regulating the TSC1–TSC2 complex, or vice versa, requires further investigation. In general, a model that CDK1 activates mTORC1 at the G2/M transition is complicated by the fact that cap-dependent translation, a processes promoted by mTORC1 signaling, is known to be globally suppressed as cells enter mitosis [105, 106].

E. ATP Depletion: AMPK

Anabolic processes that underlie cell growth, such as protein synthesis, are energy costly, and therefore, mechanisms have evolved to inhibit master regulators of growth, such as mTORC1, under conditions of energy stress. mTORC1 activity is attenuated when the intracellular levels of ATP are decreased [107] and this inhibition has been found to be mediated primarily by AMP-dependent protein kinase (AMPK), a major sensor of cellular energy status (reviewed in Ref. [108]). In response to a rise in the ratio of AMP to ATP, AMPK is activated and phosphorylates TSC2 on at least two sites (T1271 and S1387). Through an unknown molecular mechanism involving this phosphorylation of TSC2, AMPK activates the TSC1–TSC2 complex to inhibit Rheb and mTORC1 [112, 109]. Emphasizing the importance of mTORC1 inhibition in the downstream functions of AMPK, energy depletion also results in phosphorylation of the core mTORC1 component Raptor by AMPK, leading to a decrease in mTORC1 activity [110].

F. Wnt Signaling: GSK3

The Wnt signaling pathway directs embryonic development and adult tissue homeostasis in metazoans by regulating cell proliferation, differentiation, and polarity (reviewed in Ref. [111]). Consistent with the Wnt pathway's role in regulating proliferation, exposure of cells to the Wnt ligand was found to stimulate mTORC1 signaling in diverse cell types [112]. In the absence of Wnt signaling, the protein kinase glycogen synthase kinase-3β (GSK3β), which is a negative regulator of growth and proliferation, was found to phosphorylate TSC2 on adjacent motifs at S1371, S1375, S1379, and S1383. Interestingly, this required priming phosphorylation by AMPK at S1387, and so GSK3-mediated phosphorylation of these sites cooperates with AMPK to activate the TSC1–TSC2 complex under conditions of energy depletion. Upon stimulation of the Wnt pathway, GSK3β is inhibited allowing inactivation of the TSC1–TSC2 complex and subsequent activation of mTORC1. Although Akt is also known to inhibit GSK3, the pool of GSK3β that phosphorylates TSC2 was suggested to lie exclusively downstream of the Wnt pathway. However, this requires further study, as do the physiological circumstances that might require mTORC1 to be controlled by cross-talk between Wnt signaling and the energy sensing AMPK pathway.

TSC1 has also been reported to be an *in vitro* substrate of GSK3 and overexpression of GSK3 can induce TSC1 phosphorylation in cells [113]. The phosphorylation of TSC1 by GSK3 *in vitro* was moderately reduced by mutating two candidate sites (T357 and T390 of human TSC1) to alanine. As with TSC2, Wnt signaling was shown to inhibit an association of TSC1 with GSK3. However, it remains unclear whether endogenous GSK3 phosphorylates endogenous TSC1 on these or other sites and what effects this phosphorylation has on the TSC1–TSC2 complex and mTORC1. In a separate study, the TSC1–TSC2 complex was suggested to negatively regulate Wnt signaling [114]. While more work is needed, these studies suggest the potential for interesting cross-regulation between the Wnt pathway and the TSC1–TSC2 complex.

G. Hypoxia: REDD1

mTORC1 signaling is also sensitive to the levels of intracellular oxygen, and hypoxia potently inhibits mTORC1 [115]. Activation of the TSC1–TSC2 complex through at least two separate mechanisms has been proposed to mediate mTORC1 inhibition in response to hypoxia. First, through a decrease in mitochondrial oxidative phosphorylation, hypoxia causes energy stress and activates AMPK, which, as detailed above, can phosphorylate and activate the TSC1–TSC2 complex [116]. Second, a protein with no apparent functional domains called REDD1 (also known as DDIT4 and RTP801) can mediate inhibition of mTORC1 in response to hypoxic conditions in both flies and mammals [117, 118]. In response to hypoxia, REDD1 expression is induced by the transcription factor HIF1 and has been shown to be essential for the inhibitory effects of hypoxia on mTORC1 signaling. It has been proposed that REDD1 activates the TSC1–TSC2 complex by binding to 14-3-3 proteins and dissociating them from the two major Akt-phosphorylation sites on TSC2, S939 and T1462 [93]. Therefore, in this model, REDD1 activates the TSC1–TSC2 complex by reversing its inhibition by Akt. Interestingly, REDD1 has also been reported to be induced by energy stress and to cooperate with AMPK in activating the TSC1–TSC2 complex [119]. Collectively, these studies demonstrate that the TSC1–TSC2 complex plays a key role in oxygen sensing upstream of mTORC1.

H. Other Phosphorylation Sites

Many other phosphorylation sites on TSC1 and TSC2 have been identified, suggesting that other stimuli and pathways will be found that regulate mTORC1 by activating or inhibiting the TSC1–TSC2 complex. However, the regulatory mechanisms and physiological significance of these sites awaits further study. For instance, S1254 on TSC2 has been suggested to be phosphorylated by MAP Kinase Activated Protein Kinase 2 or MAP-KAPK2 (MK2) in response to p38 overexpression or anisomycin treatment [120]. Furthermore, targeted mass spectrometry and phospho-proteomic studies have revealed many sites on both TSC1 [121–126] and TSC2 [99, 122, 125–131] that have not been characterized.

V. Aberrant Inhibition of the TSC1–TSC2 Complex Leading to Activation of mTORC1 in the Majority of Human Tumors

The complexity and breadth of the signaling network upstream of the TSC1–TSC2 complex is a testament to the importance of proper regulation of mTORC1 in the tight control over cellular growth and proliferation.

Aberrantly elevated mTORC1 signaling is detected in the majority of genetic tumor syndromes and sporadic cancers [132]. Several of the most commonly activated oncogene products and inactivated tumor suppressors lie upstream of the TSC1–TSC2 complex (Figure 2.6). Amplifications and activating mutations in several receptor tyrosine kinases, PI3K, Ras, Raf, and Akt or functional inactivation of the tumor suppressors PTEN, NF1, NF2, and LKB1 leads to abnormal inactivation of the TSC1–TSC2 complex

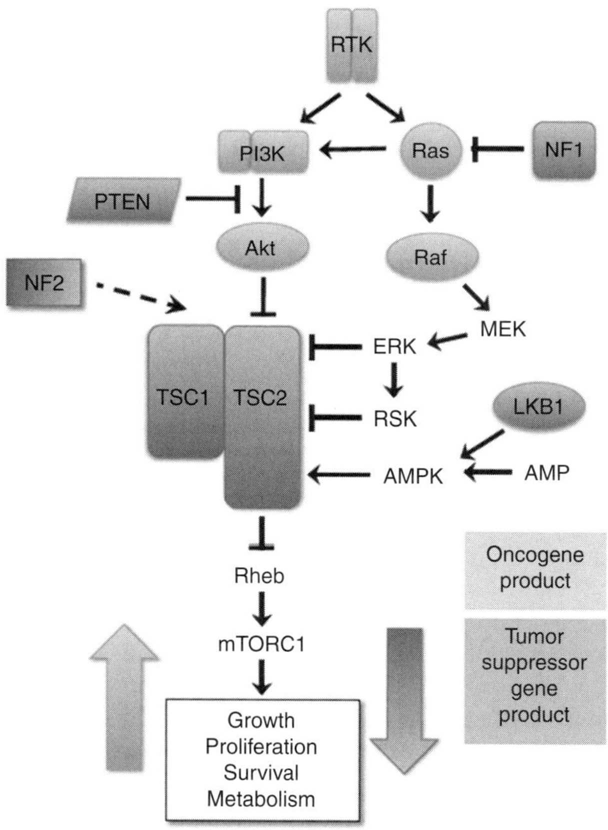

FIG. 2.6. Multiple oncogene products and tumor suppressors lie upstream of the TSC1–TSC2 complex and mTORC1. Mutations leading to the misregulated activation of oncogene products (green) and functional inactivation of tumor suppressor gene products (red) lead to inactivation of the TSC1–TSC2 complex and activation of mTORC1. Uncolored kinases are not considered to be oncogenes or tumor suppressors. RTK, receptor tyrosine kinase; PI3K, phosphoinositide-3 kinase; PTEN, phosphatase and tensin homolog; Ras; Raf; NF1, neurofibromin 1; NF2, neurofibromin 2; AMP, adenosine monophosphate. (For interpretation of the references to color in this figure legend, the reader is referred to the Web version of this chapter.)

and activation of mTORC1. However, mTORC1 activation is only one consequence of genetic lesions affecting these upstream pathways, partially explaining the phenotypic differences of tumor syndromes and cancers in which one or another part of the signaling network is perturbed. Distinct oncogenic mutations are also likely to lead to differences in the levels and constitutive nature of mTORC1 signaling. This is particularly true with loss of the TSC tumor suppressors themselves, which leads to the highest level of mTORC1 activation. However, unlike upstream tumor suppressors, TSC gene loss has not been found to be frequent in sporadic cancers, and the TSC disease is not a cancer predisposition syndrome like diseases caused by mutations in PTEN (Cowden syndrome) or LKB1 (Peutz Jeghers syndrome). A likely explanation for this is the existence and strength of mTORC1-dependent feedback mechanisms triggered by loss of the TSC genes, as well as loss of mTORC2 signaling [79, 81, 133].

VI. Important Outstanding Questions Concerning the TSC1–TSC2 Complex

The enormous increase in our understanding of the TSC1–TSC2 complex and upstream regulation of mTORC1 over the past decade raises several remaining fundamental questions. First, a comprehensive understanding of how phosphorylation regulates the complex needs to be developed. This should include determining the complete repertoire of kinases and pathways upstream of the complex, identifying all phosphorylation sites, and characterizing the hierarchy of the phosphorylation sites and how they influence each other. This would be greatly aided by structural information regarding the positioning of these regulatory sites within the quaternary structure of the complex. Second, the molecular mechanism(s) by which specific phosphorylation events regulate the ability of TSC2 to act as a GAP toward Rheb within cells is currently unknown. The answer to this problem will likely require more detailed knowledge regarding the subcellular localization of the TSC1–TSC2 complex and Rheb. Third, the molecular nature of the positive regulation of mTORC2 by the TSC1–TSC2 complex, and how specific phosphorylation sites on the complex might affect this regulation remains to be elucidated. Finally, in addition to mTORC2, are there other major cellular targets of the TSC1–TSC2 complex that are independent of the Rheb-mTORC1 circuit? Given the widespread importance of the signaling network centered around the TSC1–TSC2 complex, the answers to these questions will broadly impact our understanding of basic cell physiology and human disease.

References

1. Crino, P.B., Nathanson, K.L., and Henske, E.P. (2006). The tuberous sclerosis complex. *N Engl J Med* 355:1345–1356.
2. O'Callaghan, F.J., Shiell, A.W., Osborne, J.P., and Martyn, C.N. (1998). Prevalence of tuberous sclerosis estimated by capture–recapture analysis. *Lancet* 351:1490.
3. Osborne, J.P., Fryer, A., and Webb, D. (1991). Epidemiology of tuberous sclerosis. *Ann N Y Acad Sci* 615:125–127.
4. Fryer, A.E., Chalmers, A., Connor, J.M., Fraser, I., Povey, S., Yates, A.D., Yates, J.R., and Osborne, J.P. (1987). Evidence that the gene for tuberous sclerosis is on chromosome 9. *Lancet* 1:659–661.
5. van Slegtenhorst, M., de Hoogt, R., Hermans, C., Nellist, M., Janssen, B., Verhoef, S., Lindhout, D., van den Ouweland, A., Halley, D., Young, J., et al. (1997). Identification of the tuberous sclerosis gene TSC1 on chromosome 9q34. *Science (New York, NY)* 277:805–808.
6. Kandt, R.S., Haines, J.L., Smith, M., Northrup, H., Gardner, R.J., Short, M.P., Dumars, K., Roach, E.S., Steingold, S., Wall, S., et al. (1992). Linkage of an important gene locus for tuberous sclerosis to a chromosome 16 marker for polycystic kidney disease. *Nat Genet* 2:37–41.
7. Consortium, E.C.T.S. (1993). Identification and characterization of the tuberous sclerosis gene on chromosome 16. *Cell* 75:1305–1315.
8. Ali, M., Girimaji, S.C., and Kumar, A. (2003). Identification of a core promoter and a novel isoform of the human TSC1 gene transcript and structural comparison with mouse homolog. *Gene* 320:145–154.
9. Xu, L., Sterner, C., Maheshwar, M.M., Wilson, P.J., Nellist, M., Short, P.M., Haines, J.L., Sampson, J.R., and Ramesh, V. (1995). Alternative splicing of the tuberous sclerosis 2 (TSC2) gene in human and mouse tissues. *Genomics* 27:475–480.
10. Xiao, G.H., Jin, F., and Yeung, R.S. (1995). Identification of tuberous sclerosis 2 messenger RNA splice variants that are conserved and differentially expressed in rat and human tissues. *Cell Growth Differ* 6:1185–1191.
11. Zhang, H., Cicchetti, G., Onda, H., Koon, H.B., Asrican, K., Bajraszewski, N., Vazquez, F., Carpenter, C.L., and Kwiatkowski, D.J. (2003). Loss of Tsc1/Tsc2 activates mTOR and disrupts PI3K–Akt signaling through downregulation of PDGFR. *J Clin Invest* 112:1223–1233.
12. Nellist, M., van Slegtenhorst, M.A., Goedbloed, M., van den Ouweland, A.M., Halley, D.J., and van der Sluijs, P. (1999). Characterization of the cytosolic tuberin–hamartin complex. Tuberin is a cytosolic chaperone for hamartin. *J Biol Chem* 274:35647–35652.
13. Momose, S., Kobayashi, T., Tada, N., Itoyama, S., and Hino, O. (2007). N-terminal hamartin-binding and C-terminal GAP domain of tuberin can separate in vivo. *Biochem Biophys Res Commun* 356:693–698.
14. Plank, T.L., Yeung, R.S., and Henske, E.P. (1998). Hamartin, the product of the tuberous sclerosis 1 (TSC1) gene, interacts with tuberin and appears to be localized to cytoplasmic vesicles. *Cancer Res* 58:4766–4770.
15. van Slegtenhorst, M., Nellist, M., Nagelkerken, B., Cheadle, J., Snell, R., van den Ouweland, A., Reuser, A., Sampson, J., Halley, D., and van der Sluijs, P. (1998). Interaction between hamartin and tuberin, the TSC1 and TSC2 gene products. *Hum Mol Genet* 7:1053–1057.
16. Gao, X., and Pan, D. (2001). TSC1 and TSC2 tumor suppressors antagonize insulin signaling in cell growth. *Genes Dev* 15:1383–1392.

17. Potter, C.J., Huang, H., and Xu, T. (2001). *Drosophila* Tsc1 functions with Tsc2 to antagonize insulin signaling in regulating cell growth, cell proliferation, and organ size. *Cell* 105:357–368.
18. Han, S., Witt, R.M., Santos, T.M., Polizzano, C., Sabatini, B.L., and Ramesh, V. (2008). Pam (protein associated with Myc) functions as an E3 ubiquitin ligase and regulates TSC/mTOR signaling. *Cell Signal* 20:1084–1091.
19. Hu, J., Zacharek, S., He, Y.J., Lee, H., Shumway, S., Duronio, R.J., and Xiong, Y. (2008). WD40 protein FBW5 promotes ubiquitination of tumor suppressor TSC2 by DDB1-CUL4-ROC1 ligase. *Genes Dev* 22:866–871.
20. Chong-Kopera, H., Inoki, K., Li, Y., Zhu, T., Garcia-Gonzalo, F.R., Rosa, J.L., and Guan, K.L. (2006). TSC1 stabilizes TSC2 by inhibiting the interaction between TSC2 and the HERC1 ubiquitin ligase. *J Biol Chem* 281:8313–8316.
21. Rosner, M., Hanneder, M., Siegel, N., Valli, A., and Hengstschlager, M. (2008). The tuberous sclerosis gene products hamartin and tuberin are multifunctional proteins with a wide spectrum of interacting partners. *Mutat Res* 658:234–246.
22. Mason, J.M., and Arndt, K.M. (2004). Coiled coil domains: stability, specificity, and biological implications. *Chembiochem* 5:170–176.
23. Hodges, A.K., Li, S., Maynard, J., Parry, L., Braverman, R., Cheadle, J.P., DeClue, J.E., and Sampson, J.R. (2001). Pathological mutations in TSC1 and TSC2 disrupt the interaction between hamartin and tuberin. *Hum Mol Genet* 10:2899–2905.
24. Matsumoto, S., Bandyopadhyay, A., Kwiatkowski, D.J., Maitra, U., and Matsumoto, T. (2002). Role of the Tsc1–Tsc2 complex in signaling and transport across the cell membrane in the fission yeast *Schizosaccharomyces pombe*. *Genetics* 161:1053–1063.
25. Astrinidis, A., Senapedis, W., and Henske, E.P. (2006). Hamartin, the tuberous sclerosis complex 1 gene product, interacts with polo-like kinase 1 in a phosphorylation-dependent manner. *Hum Mol Genet* 15:287–297.
26. Lupas, A. (1996). Prediction and analysis of coiled-coil structures. *Methods Enzymol* 266:513–525.
27. Noonan, D.J., Lou, D., Griffith, N., and Vanaman, T.C. (2002). A calmodulin binding site in the tuberous sclerosis 2 gene product is essential for regulation of transcription events and is altered by mutations linked to tuberous sclerosis and lymphangioleiomyomatosis. *Arch Biochem Biophys* 398:132–140.
28. York, B., Lou, D., Panettieri, R.A., Jr., Krymskaya, V.P., Vanaman, T.C., and Noonan, D.J. (2005). Cross-talk between tuberin, calmodulin, and estrogen signaling pathways. *FASEB J* 19:1202–1204.
29. York, B., Lou, D., and Noonan, D.J. (2006). Tuberin nuclear localization can be regulated by phosphorylation of its carboxyl terminus. *Mol Cancer Res* 4:885–897.
30. Jin, F., Wienecke, R., Xiao, G.H., Maize, J.C., Jr., DeClue, J.E., and Yeung, R.S. (1996). Suppression of tumorigenicity by the wild-type tuberous sclerosis 2 (Tsc2) gene and its C-terminal region. *Proc Natl Acad Sci USA* 93:9154–9159.
31. Momose, S., Kobayashi, T., Mitani, H., Hirabayashi, M., Ito, K., Ueda, M., Nabeshima, Y., and Hino, O. (2002). Identification of the coding sequences responsible for Tsc2-mediated tumor suppression using a transgenic rat system. *Hum Mol Genet* 11:2997–3006.
32. Kobayashi, T., Mitani, H., Takahashi, R., Hirabayashi, M., Ueda, M., Tamura, H., and Hino, O. (1997). Transgenic rescue from embryonic lethality and renal carcinogenesis in the Eker rat model by introduction of a wild-type Tsc2 gene. *Proc Natl Acad Sci USA* 94:3990–3993.
33. Rennebeck, G., Kleymenova, E.V., Anderson, R., Yeung, R.S., Artzt, K., and Walker, C.L. (1998). Loss of function of the tuberous sclerosis 2 tumor suppressor gene results in

embryonic lethality characterized by disrupted neuroepithelial growth and development. *Proc Natl Acad Sci USA* 95:15629–15634.
34. Kobayashi, T., Minowa, O., Kuno, J., Mitani, H., Hino, O., and Noda, T. (1999). Renal carcinogenesis, hepatic hemangiomatosis, and embryonic lethality caused by a germ-line Tsc2 mutation in mice. *Cancer Res* 59:1206–1211.
35. Onda, H., Lueck, A., Marks, P.W., Warren, H.B., and Kwiatkowski, D.J. (1999). Tsc2(+/−) mice develop tumors in multiple sites that express gelsolin and are influenced by genetic background. *J Clin Invest* 104:687–695.
36. Ito, N., and Rubin, G.M. (1999). gigas, a *Drosophila* homolog of tuberous sclerosis gene product-2, regulates the cell cycle. *Cell* 96:529–539.
37. Tapon, N., Ito, N., Dickson, B.J., Treisman, J.E., and Hariharan, I.K. (2001). The *Drosophila* tuberous sclerosis complex gene homologs restrict cell growth and cell proliferation. *Cell* 105:345–355.
38. Potter, C.J., Pedraza, L.G., and Xu, T. (2002). Akt regulates growth by directly phosphorylating Tsc2. *Nat Cell Biol* 4:658–665.
39. Inoki, K., Li, Y., Zhu, T., Wu, J., and Guan, K.L. (2002). TSC2 is phosphorylated and inhibited by Akt and suppresses mTOR signalling. *Nat Cell Biol* 4:648–657.
40. Goncharova, E.A., Goncharov, D.A., Eszterhas, A., Hunter, D.S., Glassberg, M.K., Yeung, R.S., Walker, C.L., Noonan, D., Kwiatkowski, D.J., Chou, M.M., et al. (2002). Tuberin regulates p70 S6 kinase activation and ribosomal protein S6 phosphorylation. A role for the TSC2 tumor suppressor gene in pulmonary lymphangioleiomyomatosis (LAM). *J Biol Chem* 277:30958–30967.
41. Manning, B.D., Tee, A.R., Logsdon, M.N., Blenis, J., and Cantley, L.C. (2002). Identification of the tuberous sclerosis complex-2 tumor suppressor gene product tuberin as a target of the phosphoinositide 3-kinase/akt pathway. *Mol Cell* 10:151–162.
42. Tee, A.R., Fingar, D.C., Manning, B.D., Kwiatkowski, D.J., Cantley, L.C., and Blenis, J. (2002). Tuberous sclerosis complex-1 and -2 gene products function together to inhibit mammalian target of rapamycin (mTOR)-mediated downstream signaling. *Proc Natl Acad Sci USA* 99:13571–13576.
43. Yamagata, K., Sanders, L.K., Kaufmann, W.E., Yee, W., Barnes, C.A., Nathans, D., and Worley, P.F. (1994). rheb, a growth factor- and synaptic activity-regulated gene, encodes a novel Ras-related protein. *J Biol Chem* 269:16333–16339.
44. Patel, P.H., Thapar, N., Guo, L., Martinez, M., Maris, J., Gau, C.L., Lengyel, J.A., and Tamanoi, F. (2003). *Drosophila* Rheb GTPase is required for cell cycle progression and cell growth. *J Cell Sci* 116:3601–3610.
45. Saucedo, L.J., Gao, X., Chiarelli, D.A., Li, L., Pan, D., and Edgar, B.A. (2003). Rheb promotes cell growth as a component of the insulin/TOR signalling network. *Nat Cell Biol* 5:566–571.
46. Stocker, H., Radimerski, T., Schindelholz, B., Wittwer, F., Belawat, P., Daram, P., Breuer, S., Thomas, G., and Hafen, E. (2003). Rheb is an essential regulator of S6K in controlling cell growth in *Drosophila*. *Nat Cell Biol* 5:559–565.
47. Zhang, Y., Gao, X., Saucedo, L.J., Ru, B., Edgar, B.A., and Pan, D. (2003). Rheb is a direct target of the tuberous sclerosis tumour suppressor proteins. *Nat Cell Biol* 5:578–581.
48. Inoki, K., Li, Y., Xu, T., and Guan, K.L. (2003). Rheb GTPase is a direct target of TSC2 GAP activity and regulates mTOR signaling. *Genes Dev* 17:1829–1834.
49. Tee, A.R., Manning, B.D., Roux, P.P., Cantley, L.C., and Blenis, J. (2003). Tuberous sclerosis complex gene products, Tuberin and Hamartin, control mTOR signaling by acting as a GTPase-activating protein complex toward Rheb. *Curr Biol* 13:1259–1268.

50. Castro, A.F., Rebhun, J.F., Clark, G.J., and Quilliam, L.A. (2003). Rheb binds tuberous sclerosis complex 2 (TSC2) and promotes S6 kinase activation in a rapamycin- and farnesylation-dependent manner. *J Biol Chem* 278:32493–32496.
51. Garami, A., Zwartkruis, F.J., Nobukuni, T., Joaquin, M., Roccio, M., Stocker, H., Kozma, S.C., Hafen, E., Bos, J.L., and Thomas, G. (2003). Insulin activation of Rheb, a mediator of mTOR/S6K/4E-BP signaling, is inhibited by TSC1 and 2. *Mol Cell* 11:1457–1466.
52. Sancak, Y., Thoreen, C.C., Peterson, T.R., Lindquist, R.A., Kang, S.A., Spooner, E., Carr, S.A., and Sabatini, D.M. (2007). PRAS40 is an insulin-regulated inhibitor of the mTORC1 protein kinase. *Mol Cell* 25:903–915.
53. Sato, T., Nakashima, A., Guo, L., and Tamanoi, F. (2009). Specific activation of mTORC1 by Rheb G-protein in vitro involves enhanced recruitment of its substrate protein. *J Biol Chem* 284:12783–12791.
54. Tabancay, A.P., Jr., Gau, C.L., Machado, I.M., Uhlmann, E.J., Gutmann, D.H., Guo, L., and Tamanoi, F. (2003). Identification of dominant negative mutants of Rheb GTPase and their use to implicate the involvement of human Rheb in the activation of p70S6K. *J Biol Chem* 278:39921–39930.
55. Tee, A.R., Blenis, J., and Proud, C.G. (2005). Analysis of mTOR signaling by the small G-proteins, Rheb and RhebL1. *FEBS Lett* 579:4763–4768.
56. Im, E., von Lintig, F.C., Chen, J., Zhuang, S., Qui, W., Chowdhury, S., Worley, P.F., Boss, G.R., and Pilz, R.B. (2002). Rheb is in a high activation state and inhibits B-Raf kinase in mammalian cells. *Oncogene* 21:6356–6365.
57. Marshall, C.B., Ho, J., Buerger, C., Plevin, M.J., Li, G.Y., Li, Z., Ikura, M., and Stambolic, V. (2009). Characterization of the intrinsic and TSC2-GAP-regulated GTPase activity of Rheb by real-time NMR. *Sci Signal* 2:ra3.
58. Wienecke, R., Konig, A., and DeClue, J.E. (1995). Identification of tuberin, the tuberous sclerosis-2 product. Tuberin possesses specific Rap1GAP activity. *J Biol Chem* 270:16409–16414.
59. Xiao, G.H., Shoarinejad, F., Jin, F., Golemis, E.A., and Yeung, R.S. (1997). The tuberous sclerosis 2 gene product, tuberin, functions as a Rab5 GTPase activating protein (GAP) in modulating endocytosis. *J Biol Chem* 272:6097–6100.
60. Ahmadian, M.R., Stege, P., Scheffzek, K., and Wittinghofer, A. (1997). Confirmation of the arginine-finger hypothesis for the GAP-stimulated GTP-hydrolysis reaction of Ras. *Nat Struct Biol* 4:686–689.
61. Rittinger, K., Walker, P.A., Eccleston, J.F., Smerdon, S.J., and Gamblin, S.J. (1997). Structure at 1.65 A of RhoA and its GTPase-activating protein in complex with a transition-state analogue. *Nature* 389:758–762.
62. Brinkmann, T., Daumke, O., Herbrand, U., Kuhlmann, D., Stege, P., Ahmadian, M.R., and Wittinghofer, A. (2002). Rap-specific GTPase activating protein follows an alternative mechanism. *J Biol Chem* 277:12525–12531.
63. Daumke, O., Weyand, M., Chakrabarti, P.P., Vetter, I.R., and Wittinghofer, A. (2004). The GTPase-activating protein Rap1GAP uses a catalytic asparagine. *Nature* 429:197–201.
64. Scrima, A., Thomas, C., Deaconescu, D., and Wittinghofer, A. (2008). The Rap-RapGAP complex: GTP hydrolysis without catalytic glutamine and arginine residues. *EMBO J* 27:1145–1153.
65. Li, Y., Inoki, K., and Guan, K.L. (2004). Biochemical and functional characterizations of small GTPase Rheb and TSC2 GAP activity. *Mol Cell Biol* 24:7965–7975.
66. Gromov, P.S., Madsen, P., Tomerup, N., and Celis, J.E. (1995). A novel approach for expression cloning of small GTPases: identification, tissue distribution and chromosome mapping of the human homolog of rheb. *FEBS Lett* 377:221–226.

67. van Slegtenhorst, M., Carr, E., Stoyanova, R., Kruger, W.D., and Henske, E.P. (2004). Tsc1+ and tsc2+ regulate arginine uptake and metabolism in *Schizosaccharomyces pombe*. *J Biol Chem* 279:12706–12713.
68. Urano, J., Comiso, M.J., Guo, L., Aspuria, P.J., Deniskin, R., Tabancay, A.P., Jr., Kato-Stankiewicz, J., and Tamanoi, F. (2005). Identification of novel single amino acid changes that result in hyperactivation of the unique GTPase, Rheb, in fission yeast. *Mol Microbiol* 58:1074–1086.
69. Aspuria, P.J., Sato, T., and Tamanoi, F. (2007). The TSC/Rheb/TOR signaling pathway in fission yeast and mammalian cells: temperature sensitive and constitutive active mutants of TOR. *Cell Cycle (Georgetown, Tex)* 6:1692–1695.
70. Urano, J., Tabancay, A.P., Yang, W., and Tamanoi, F. (2000). The *Saccharomyces cerevisiae* Rheb G-protein is involved in regulating canavanine resistance and arginine uptake. *J Biol Chem* 275:11198–11206.
71. Aspuria, P.J., and Tamanoi, F. (2008). The Tsc/Rheb signaling pathway controls basic amino acid uptake via the Cat1 permease in fission yeast. *Mol Genet Genomics* 279:441–450.
72. Aspuria, P.J., and Tamanoi, F. (2004). The Rheb family of GTP-binding proteins. *Cell Signal* 16:1105–1112.
73. Yang, W., Tabancay, A.P., Jr., Urano, J., and Tamanoi, F. (2001). Failure to farnesylate Rheb protein contributes to the enrichment of G0/G1 phase cells in the *Schizosaccharomyces pombe* farnesyltransferase mutant. *Mol Microbiol* 41:1339–1347.
74. Sarbassov, D.D., Guertin, D.A., Ali, S.M., and Sabatini, D.M. (2005). Phosphorylation and regulation of Akt/PKB by the rictor-mTOR complex. *Science (New York, NY)* 307:1098–1101.
75. Garcia-Martinez, J.M., and Alessi, D.R. (2008). mTOR complex 2 (mTORC2) controls hydrophobic motif phosphorylation and activation of serum- and glucocorticoid-induced protein kinase 1 (SGK1). *Biochem J* 416:375–385.
76. Facchinetti, V., Ouyang, W., Wei, H., Soto, N., Lazorchak, A., Gould, C., Lowry, C., Newton, A.C., Mao, Y., Miao, R.Q., et al. (2008). The mammalian target of rapamycin complex 2 controls folding and stability of Akt and protein kinase C. *EMBO J* 27:1932–1943.
77. Ikenoue, T., Inoki, K., Yang, Q., Zhou, X., and Guan, K.L. (2008). Essential function of TORC2 in PKC and Akt turn motif phosphorylation, maturation and signalling. *EMBO J* 27:1919–1931.
78. Jacinto, E., and Lorberg, A. (2008). TOR regulation of AGC kinases in yeast and mammals. *Biochem J* 410:19–37.
79. Huang, J., Dibble, C.C., Matsuzaki, M., and Manning, B.D. (2008). The TSC1–TSC2 complex is required for proper activation of mTOR complex 2. *Mol Cell Biol* 28:4104–4115.
80. Yang, Q., Inoki, K., Kim, E., and Guan, K.L. (2006). TSC1/TSC2 and Rheb have different effects on TORC1 and TORC2 activity. *Proc Natl Acad Sci USA* 103:6811–6816.
81. Huang, J., Wu, S., Wu, C.L., and Manning, B.D. (2009). Signaling events downstream of mammalian target of rapamycin complex 2 are attenuated in cells and tumors deficient for the tuberous sclerosis complex tumor suppressors. *Cancer Res* 69:6107–6114.
82. Shah, O.J., Wang, Z., and Hunter, T. (2004). Inappropriate activation of the TSC/Rheb/mTOR/S6K cassette induces IRS1/2 depletion, insulin resistance, and cell survival deficiencies. *Curr Biol* 14:1650–1656.
83. Harrington, L.S., Findlay, G.M., Gray, A., Tolkacheva, T., Wigfield, S., Rebholz, H., Barnett, J., Leslie, N.R., Cheng, S., Shepherd, P.R., et al. (2004). The TSC1-2 tumor suppressor controls insulin-PI3K signaling via regulation of IRS proteins. *J Cell Biol* 166:213–223.

84. Dibble, C.C., Asara, J.M., and Manning, B.D. (2009). Characterization of Rictor phosphorylation sites reveals direct regulation of mTOR complex 2 by S6K1. *Mol Cell Biol* 29:5657–5670.
85. Huang, J., and Manning, B.D. (2008). The TSC1–TSC2 complex: a molecular switchboard controlling cell growth. *Biochem J* 412:179–190.
86. Manning, B.D., and Cantley, L.C. (2007). AKT/PKB signaling: navigating downstream. *Cell* 129:1261–1274.
87. Cai, S.L., Tee, A.R., Short, J.D., Bergeron, J.M., Kim, J., Shen, J., Guo, R., Johnson, C.L., Kiguchi, K., and Walker, C.L. (2006). Activity of TSC2 is inhibited by AKT-mediated phosphorylation and membrane partitioning. *J Cell Biol* 173:279–289.
88. Zhang, H.H., Huang, J., Duvel, K., Boback, B., Wu, S., Squillace, R.M., Wu, C.L., and Manning, B.D. (2009). Insulin stimulates adipogenesis through the Akt-TSC2-mTORC1 pathway. *PLoS ONE* 4:e6189.
89. Dong, J., and Pan, D. (2004). Tsc2 is not a critical target of Akt during normal *Drosophila* development. *Genes Dev* 18:2479–2484.
90. Schleich, S., and Teleman, A.A. (2009). Akt phosphorylates both Tsc1 and Tsc2 in *Drosophila*, but neither phosphorylation is required for normal animal growth. *PLoS ONE* 4:e6305.
91. Dan, H.C., Sun, M., Yang, L., Feldman, R.I., Sui, X.M., Ou, C.C., Nellist, M., Yeung, R.S., Halley, D.J., Nicosia, S.V., et al. (2002). Phosphatidylinositol 3-kinase/Akt pathway regulates tuberous sclerosis tumor suppressor complex by phosphorylation of tuberin. *J Biol Chem* 277:35364–35370.
92. Plas, D.R., and Thompson, C.B. (2003). Akt activation promotes degradation of tuberin and FOXO3a via the proteasome. *J Biol Chem* 278:12361–12366.
93. DeYoung, M.P., Horak, P., Sofer, A., Sgroi, D., and Ellisen, L.W. (2008). Hypoxia regulates TSC1/2-mTOR signaling and tumor suppression through REDD1-mediated 14-3-3 shuttling. *Genes Dev* 22:239–251.
94. Nellist, M., Goedbloed, M.A., de Winter, C., Verhaaf, B., Jankie, A., Reuser, A.J., van den Ouweland, A.M., van der Sluijs, P., and Halley, D.J. (2002). Identification and characterization of the interaction between tuberin and 14-3-3zeta. *J Biol Chem* 277:39417–39424.
95. Shumway, S.D., Li, Y., and Xiong, Y. (2003). 14-3-3beta binds to and negatively regulates the tuberous sclerosis complex 2 (TSC2) tumor suppressor gene product, tuberin. *J Biol Chem* 278:2089–2092.
96. Anjum, R., and Blenis, J. (2008). The RSK family of kinases: emerging roles in cellular signalling. *Nat Rev* 9:747–758.
97. Tee, A.R., Anjum, R., and Blenis, J. (2003). Inactivation of the tuberous sclerosis complex-1 and -2 gene products occurs by phosphoinositide 3-kinase/Akt-dependent and -independent phosphorylation of tuberin. *J Biol Chem* 278:37288–37296.
98. Roux, P.P., Ballif, B.A., Anjum, R., Gygi, S.P., and Blenis, J. (2004). Tumor-promoting phorbol esters and activated Ras inactivate the tuberous sclerosis tumor suppressor complex via p90 ribosomal S6 kinase. *Proc Natl Acad Sci USA* 101:13489–13494.
99. Ballif, B.A., Roux, P.P., Gerber, S.A., MacKeigan, J.P., Blenis, J., and Gygi, S.P. (2005). Quantitative phosphorylation profiling of the ERK/p90 ribosomal S6 kinase-signaling cassette and its targets, the tuberous sclerosis tumor suppressors. *Proc Natl Acad Sci USA* 102:667–672.
100. Ma, L., Chen, Z., Erdjument-Bromage, H., Tempst, P., and Pandolfi, P.P. (2005). Phosphorylation and functional inactivation of TSC2 by Erk implications for tuberous sclerosis and cancer pathogenesis. *Cell* 121:179–193.

101. Lee, D.F., Kuo, H.P., Chen, C.T., Hsu, J.M., Chou, C.K., Wei, Y., Sun, H.L., Li, L.Y., Ping, B., Huang, W.C., et al. (2007). IKK beta suppression of TSC1 links inflammation and tumor angiogenesis via the mTOR pathway. *Cell* 130:440–455.
102. Zacharek, S.J., Xiong, Y., and Shumway, S.D. (2005). Negative regulation of TSC1–TSC2 by mammalian D-type cyclins. *Cancer Res* 65:11354–11360.
103. Astrinidis, A., Senapedis, W., Coleman, T.R., and Henske, E.P. (2003). Cell cycle-regulated phosphorylation of hamartin, the product of the tuberous sclerosis complex 1 gene, by cyclin-dependent kinase 1/cyclin B. *J Biol Chem* 278:51372–51379.
104. Catania, M.G., Mischel, P.S., and Vinters, H.V. (2001). Hamartin and tuberin interaction with the G2/M cyclin-dependent kinase CDK1 and its regulatory cyclins A and B. *J Neuropathol Exp Neurol* 60:711–723.
105. Bonneau, A.M., and Sonenberg, N. (1987). Involvement of the 24-kDa cap-binding protein in regulation of protein synthesis in mitosis. *J Biol Chem* 262:11134–11139.
106. Pyronnet, S., Dostie, J., and Sonenberg, N. (2001). Suppression of cap-dependent translation in mitosis. *Genes Dev* 15:2083–2093.
107. Dennis, P.B., Jaeschke, A., Saitoh, M., Fowler, B., Kozma, S.C., and Thomas, G. (2001). Mammalian TOR: a homeostatic ATP sensor. *Science (New York, NY)* 294:1102–1105.
108. Hardie, D.G. (2005). New roles for the LKB1->AMPK pathway. *Curr Opin Cell Biol* 17:167–173.
109. Shaw, R.J., Bardeesy, N., Manning, B.D., Lopez, L., Kosmatka, M., DePinho, R.A., and Cantley, L.C. (2004). The LKB1 tumor suppressor negatively regulates mTOR signaling. *Cancer Cell* 6:91–99.
110. Gwinn, D.M., Shackelford, D.B., Egan, D.F., Mihaylova, M.M., Mery, A., Vasquez, D.S., Turk, B.E., and Shaw, R.J. (2008). AMPK phosphorylation of raptor mediates a metabolic checkpoint. *Mol Cell* 30:214–226.
111. MacDonald, B.T., Tamai, K., and He, X. (2009). Wnt/beta-catenin signaling: components, mechanisms, and diseases. *Dev Cell* 17:9–26.
112. Inoki, K., Ouyang, H., Zhu, T., Lindvall, C., Wang, Y., Zhang, X., Yang, Q., Bennett, C., Harada, Y., Stankunas, K., et al. (2006). TSC2 integrates Wnt and energy signals via a coordinated phosphorylation by AMPK and GSK3 to regulate cell growth. *Cell* 126:955–968.
113. Mak, B.C., Kenerson, H.L., Aicher, L.D., Barnes, E.A., and Yeung, R.S. (2005). Aberrant beta-catenin signaling in tuberous sclerosis. *Am J Pathol* 167:107–116.
114. Mak, B.C., Takemaru, K., Kenerson, H.L., Moon, R.T., and Yeung, R.S. (2003). The tuberin–hamartin complex negatively regulates beta-catenin signaling activity. *J Biol Chem* 278:5947–5951.
115. Arsham, A.M., Howell, J.J., and Simon, M.C. (2003). A novel hypoxia-inducible factor-independent hypoxic response regulating mammalian target of rapamycin and its targets. *J Biol Chem* 278:29655–29660.
116. Liu, L., Cash, T.P., Jones, R.G., Keith, B., Thompson, C.B., and Simon, M.C. (2006). Hypoxia-induced energy stress regulates mRNA translation and cell growth. *Mol Cell* 21:521–531.
117. Reiling, J.H., and Hafen, E. (2004). The hypoxia-induced paralogs Scylla and Charybdis inhibit growth by down-regulating S6K activity upstream of TSC in Drosophila. *Genes Dev* 18:2879–2892.
118. Brugarolas, J., Lei, K., Hurley, R.L., Manning, B.D., Reiling, J.H., Hafen, E., Witters, L.A., Ellisen, L.W., and Kaelin, W.G., Jr. (2004). Regulation of mTOR function in response to hypoxia by REDD1 and the TSC1/TSC2 tumor suppressor complex. *Genes Dev* 18:2893–2904.

119. Sofer, A., Lei, K., Johannessen, C.M., and Ellisen, L.W. (2005). Regulation of mTOR and cell growth in response to energy stress by REDD1. *Mol Cell Biol* 25:5834–5845.
120. Li, Y., Inoki, K., Vacratsis, P., and Guan, K.L. (2003). The p38 and MK2 kinase cascade phosphorylates tuberin, the tuberous sclerosis 2 gene product, and enhances its interaction with 14-3-3. *J Biol Chem* 278:13663–13671.
121. Munton, R.P., Tweedie-Cullen, R., Livingstone-Zatchej, M., Weinandy, F., Waidelich, M., Longo, D., Gehrig, P., Potthast, F., Rutishauser, D., Gerrits, B., et al. (2007). Qualitative and quantitative analyses of protein phosphorylation in naive and stimulated mouse synaptosomal preparations. *Mol Cell Proteomics* 6:283–293.
122. Nellist, M., Burgers, P.C., van den Ouweland, A.M., Halley, D.J., and Luider, T.M. (2005). Phosphorylation and binding partner analysis of the TSC1–TSC2 complex. *Biochem Biophys Res Commun* 333:818–826.
123. Villen, J., Beausoleil, S.A., Gerber, S.A., and Gygi, S.P. (2007). Large-scale phosphorylation analysis of mouse liver. *Proc Natl Acad Sci USA* 104:1488–1493.
124. Zahedi, R.P., Lewandrowski, U., Wiesner, J., Wortelkamp, S., Moebius, J., Schutz, C., Walter, U., Gambaryan, S., and Sickmann, A. (2008). Phosphoproteome of resting human platelets. *J Proteome Res* 7:526–534.
125. Dephoure, N., Zhou, C., Villen, J., Beausoleil, S.A., Bakalarski, C.E., Elledge, S.J., and Gygi, S.P. (2008). A quantitative atlas of mitotic phosphorylation. *Proc Natl Acad Sci USA* 105:10762–10767.
126. Zanivan, S., Gnad, F., Wickstrom, S.A., Geiger, T., Macek, B., Cox, J., Fassler, R., and Mann, M. (2008). Solid tumor proteome and phosphoproteome analysis by high resolution mass spectrometry. *J Proteome Res* 7:5314–5326.
127. Trinidad, J.C., Thalhammer, A., Specht, C.G., Lynn, A.J., Baker, P.R., Schoepfer, R., and Burlingame, A.L. (2008). Quantitative analysis of synaptic phosphorylation and protein expression. *Mol Cell Proteomics* 7:684–696.
128. Daub, H., Olsen, J.V., Bairlein, M., Gnad, F., Oppermann, F.S., Korner, R., Greff, Z., Keri, G., Stemmann, O., and Mann, M. (2008). Kinase-selective enrichment enables quantitative phosphoproteomics of the kinome across the cell cycle. *Mol Cell* 31:438–448.
129. Cantin, G.T., Yi, W., Lu, B., Park, S.K., Xu, T., Lee, J.D., and Yates, J.R., III (2008). Combining protein-based IMAC, peptide-based IMAC, and MudPIT for efficient phosphoproteomic analysis. *J Proteome Res* 7:1346–1351.
130. Tsai, C.F., Wang, Y.T., Chen, Y.R., Lai, C.Y., Lin, P.Y., Pan, K.T., Chen, J.Y., Khoo, K.H., and Chen, Y.J. (2008). Immobilized metal affinity chromatography revisited: pH/acid control toward high selectivity in phosphoproteomics. *J Proteome Res* 7:4058–4069.
131. Olsen, J.V., Blagoev, B., Gnad, F., Macek, B., Kumar, C., Mortensen, P., and Mann, M. (2006). Global, in vivo, and site-specific phosphorylation dynamics in signaling networks. *Cell* 127:635–648.
132. Menon, S., and Manning, B.D. (2009). Common corruption of the mTOR signaling network in human tumors. *Oncogene* 27:S43–S51.
133. Manning, B.D., Logsdon, M.N., Lipovsky, A.I., Abbott, D., Kwiatkowski, D.J., and Cantley, L.C. (2005). Feedback inhibition of Akt signaling limits the growth of tumors lacking Tsc2. *Genes Dev* 19:1773–1778.
134. Inoki, K., Zhu, T., and Guan, K.L. (2003). TSC2 mediates cellular energy response to control cell growth and survival. *Cell* 115:577–590.

3

AMPK Control of mTOR Signaling and Growth

DANA M. GWINN • REUBEN J. SHAW

Molecular and Cell Biology Laboratory
Howard Hughes Medical Institute
The Salk Institute for Biological Studies
La Jolla, California, USA

I. Abstract

The AMP-activated serine/threonine protein kinase (AMPK) is a sensor of cellular energy status found in all eukaryotes that is activated under conditions of low intracellular ATP following cellular stress. Work from a large number of laboratories in the past decade has revealed that one of the major downstream signaling pathways regulated by AMPK is the mammalian target-of-rapamycin (mTOR pathway). Interestingly, like AMPK, the mTOR serine/threonine kinase plays key roles not only in growth control and cell proliferation but also in metabolism. Recent work has revealed that across eukaryotes mTOR orthologs are found in two biochemically distinct complexes and only one of those complexes (mTORC1 in mammals) is acutely sensitive to rapamycin and regulated by nutrients and AMPK. Many details of the molecular mechanism by which AMPK inhibits mTORC1 signaling have also been decoded in the past 5 years. AMPK directly phosphorylates at least two proteins to induce rapid inhibition of mTORC1 activity, the TSC2 tumor suppressor and the mTOR binding partner raptor (*r*egulatory *a*ssociated *p*rotein of m*TOR*). Here, we explore the molecular connections between AMPK and mTOR signaling pathways and examine the physiological processes in which AMPK regulation of

ISSN NO: 1874-6047
DOI: 10.1016/S1874-6047(10)28003-4

mTOR is critical for growth or metabolic control. The functional conservation of AMPK and TOR in all eukaryotes, and the sequence conservation around the AMPK phosphorylation sites in raptor examined suggest that this represents a fundamental cell growth module connecting nutrient status to the cell growth machinery across all eukaryotes. These findings have broad implications for the control of cell growth by nutrients in a number of cellular and organismal contexts.

II. AMPK is an Energy Sensing Kinase

The AMP-activated protein kinase (AMPK) is a heterotrimeric kinase complex composed of a catalytic (α) subunit and two regulatory (β and γ) subunits. AMPK is activated under conditions of energy stress, when intracellular ATP levels decline and intracellular AMP increases, as occurs during nutrient deprivation or hypoxia [1]. Upon energy stress, AMP directly binds to tandem repeats of crystathionine-β-synthase (CBS) domains in the AMPK γ subunit. Binding of AMP is thought to prevent dephosphorylation of the critical activation loop threonine in the α subunit [2]. The phosphorylation of the activation loop threonine is absolutely required for AMPK activation. Biochemical and genetic analyses in worms, flies, and mice have revealed that the serine/threonine kinase LKB1 represents the major kinase phosphorylating the AMPK activation loop under conditions of energy stress across metazoans [3]. In *Saccharomyces cerevisiae*, three related upstream kinases phosphorylate SNF1, the AMPK ortholog, and LKB1 is the human kinase sharing the greatest homology to all three [4, 5].

A fundamental requirement of all cells is that they couple the availability of nutrients to signals emanating from growth factors to drive proliferation only when nutrients are in sufficient abundance to guarantee successful cell division. Upon activation under low ATP conditions, AMPK acts as a metabolic checkpoint in the cell, halting cell growth and suppressing ATP-consuming biosynthetic processes while stimulating ATP-generating processes to repair the initiating loss of ATP [6]. In addition to its widespread cell-autonomous role as an energy checkpoint, AMPK also plays key roles in glucose and lipid metabolism in specialized metabolic tissues in mammals and higher eukaryotes such as liver, muscle, and adipose [1]. Thus, AMPK not only governs cellular energetics, but indeed overall organismal bioenergetics by coordinating the response between tissues to nutritional input.

Consistent with a role for AMPK in growth control, its major upstream kinase LKB1 is a human tumor suppressor. LKB1 was originally identified as the gene mutated in Peutz–Jeghers syndrome (PJS), an autosomal dominant inherited cancer disorder [7]. In addition, somatic LKB1 mutations are prevalent in a large percentage (30–40%) of sporadic non-small-cell lung cancers (NSCLC) [8, 9]. Reintroduction of LKB1 into some LKB1-deficient cancer cell lines results in growth arrest and reversion of tumorigenecity [10]. Moreover, genetically engineered mice bearing a conditionally inactivated allele of LKB1 have been deleted in a number of tissues revealing that loss of LKB1 in prostate, skin, uterus, gut, and pancreas is sufficient to initiate hyperplasia and tumorigenesis in some of these tissues [10].

In addition to phosphorylating the activation loop of AMPK, LKB1 also phosphorylates the activation loops of a family of 12 kinases related to AMPK, also resulting in their activation [11, 12]. Importantly, of these 14 LKB1-dependent kinases, only AMPKα1 and AMPKα2 are activated under low ATP conditions, probably due to the fact that only they interact with AMPKγ which contains the AMP-binding domains [13]. However, two other family members, SNARK/Nuak2 and SIK2, have been reported to be activated under low energy conditions [14–16]. Further work is needed to examine the regulation of the 12 AMPK-related kinases in additional cell types, and to better examine whether they also can bind the AMP-responsive gamma subunits of AMPK under some conditions.

Unlike many kinases, LKB1 is not regulated by phosphorylation of its activation loop; instead, its active conformation is achieved through obligate interactions with the pseudokinase STRAD, and the scaffolding protein MO25, which stabilizes the activation loop of LKB1 [17].

Little is currently known about what stimuli direct LKB1 toward any of these AMPK-related kinases and current evidence suggests that LKB1 is constitutively active and these other kinases may be regulated through phosphorylation at other sites outside of their activation loops. Several of these family members have been shown to be transcriptionally upregulated following specific stimuli, including SIK1 following isoproterenol in a CREB-dependent manner [18], and SNARK/Nuak2 following FAS ligand in an NF-κB-dependent manner [19]. The reported overlap in downstream substrates for several of these kinases suggests that their transcriptional upregulation following specific stimuli would result in the temporally controlled production of active kinases owing to their constitutive activation by LKB1. Whether the dephosphorylation of the activation loop threonine is an acutely regulated step for any of the kinases other than AMPKα1 and AMPKα2 also remains to be determined. Finally, thus far there is no substantial mutational data from human tumors to directly support any of the downstream kinases, including AMPK, as being the critical target for LKB1 in tumorigenesis. One confounding issue with the lack of mutations found in these downstream kinases is that there is a great deal of redundancy and functional overlap among them, suggesting that loss of any one of them may be compensated for by other family members, unlike the case for LKB1 where no other specific kinase has been shown to functionally compensate for its regulation of these kinases *in vivo*.

Despite the lack of direct mutational evidence implicating AMPK in cancer, a number of studies in the past 5 years have revealed key growth suppressive roles for AMPK and the identification of a number of direct substrates for AMPK with well-established roles in growth control is consistent with the concept of AMPK as a metabolic checkpoint. The tumor suppressor p53 [20] and the CDK inhibitor protein p27 [21] have been reported to be direct substrates of AMPK, though neither of the reported AMPK phosphorylation sites in these proteins exactly conform to the established optimal substrate motif for AMPK, perhaps suggesting they may be regulated indirectly. Activation of p53-dependent transcription has been previously been shown to occur downstream of LKB1 and AMPK activation [22–24].

Notably, PJS shares a number of clinical features with Cowden's disease, which is caused by inactivating mutations in the PTEN tumor suppressor. This phenotypic overlap suggested that LKB1-dependent signaling might negatively regulate some aspect of PI3-kinase signaling, analogous to PTEN function. However, while classic PI3K/Akt signaling is not elevated in LKB1-deficient cells, we and others discovered that mTOR signaling is uniquely hyperactivated in LKB1-deficient murine embryonic fibroblasts (MEFs) and liver [25–27]. Consistent with this data from noncancerous settings, tumors arising in LKB1-heterozygous mice and human lung cancer lines lacking LKB1 similarly exhibit hyperactivation of mTOR [28]. Given that mTOR is one of the key downstream targets of PI3-kinase signaling, the observation that mTOR is also elevated in many LKB1-deficient cells and tumors suggested a common biochemical basis for the clinical overlap observed in PJS and Cowden's disease.

III. mTOR is a Central Conserved Regulator of Growth and Metabolism

mTOR, a serine/threonine kinase highly conserved in all eukaryotes, is a central regulator of cell growth [29]. Whereas AMPK is active under nutrient-poor conditions and inactive under nutrient-rich conditions, mTOR is activated in the inverse pattern. In higher eukaryotes, mTOR activation requires positive signals from both nutrients (glucose, amino acids) and growth factors. mTOR, like its budding yeast orthologs, is found in two biochemically and functionally discrete signaling complexes [29].

In mammals, the mTORC1 complex is composed of four known subunits: raptor, a negative regulatory subunit PRAS40, mLST8, and mTOR. Raptor acts as a scaffold to recruit substrates to the mTORC1 complex [30, 31]. The mTORC2 complex contains rictor (*r*apamycin *i*nsensitive *c*ompanion of m*TOR*), mSIN1, PRR5/Protor, mLST8, and mTOR [32]. Signaling from mTOR complex 1 (mTORC1) is nutrient-sensitive, acutely inhibited by the bacterial macrolide rapamycin, and controls cell growth, angiogenesis, and metabolism. In contrast, mTORC2 is not sensitive to nutrients, nor acutely inhibited by rapamycin, and its known substrates include the hydrophobic motif phosphorylation ("PDK2") sites in AGC kinases including Akt, SGK, and PKC family members [32, 33].

Downstream of the raptor-mTOR (mTORC1) complex are its two well-characterized substrates: 4EBP1 and the p70 ribosomal S6 kinase. Phosphorylation of 4EBP1 by mTORC1 suppresses its ability to bind and inhibit the translation initiation factor eIF4E. mTORC1 mediates phosphorylation of S6K at a Thr residue in a hydrophobic motif at the C-terminus of the kinase domain. A specific motif (TOS motif) found in both 4EBP1 and S6K was shown to mediate direct binding of these proteins to raptor allowing them to be phosphorylated in the mTORC1 complex. Mechanistic details of how mTORC1 regulates the assembly of translational initiation complexes via a number of ordered phosphorylation events were recently discovered [34]. mTORC1-dependent translation is known to control a number of specific cell growth regulators, including cyclin D1, the HIF-1α transcriptional factor, and c-myc, which in turn promote processes including cell cycle progression, cell growth, glycolysis, and angiogenesis, all contributing to enhanced tumorigenesis [32]. Interestingly, HIF-1α is independently negatively regulated by the VHL tumor suppressor, providing another link between the mTOR pathway and cancer. Furthermore, eIF4E is an oncogene upregulated in a number of cancers [35], and eIF4E overexpression has been shown to be sufficient to promote rapamycin-resistant growth of tumors [36] and promote increased cyclin D1 [37].

Upstream components of the mTORC1 complex were initially discovered through classical cancer genetics. The TSC2 tumor suppressor tuberin and its obligate binding partner hamartin (TSC1) are mutated in a familial tumor syndrome called tuberous sclerosis complex (TSC). TSC patients are predisposed to widespread benign tumors termed hamartomas in kidney, lung, brain, and skin. Genetic studies in *Drosophila* and mammalian cells identified the TSC tumor suppressors as critical upstream inhibitors of the mTORC1 complex. The TSC gene products form a complex that acts as a GTPase activation protein (GAP) for the small Ras-like GTPase Rheb, which in its GTP-bound state has been shown to directly activate mTORC1 [38]. Inactivation or loss of TSC1 or TSC2 therefore leads to elevated GTP-loading of Rheb and consequent hyperactivation of mTORC1. Phosphorylation of TSC1 and TSC2 serves as an integration point for a wide variety of environmental signals that regulate mTORC1 [39]. One of the key activators of the mTORC1 pathway is PI3-kinase, which plays a key role in promoting cell growth and insulin-mediated effects on metabolism. PI3-kinase via PDK1 activates the serine/threonine kinase Akt, which directly phosphorylates and inactivates TSC2 through a poorly understood mechanism. In addition to TSC2, Akt also directly phosphorylates the mTORC1 inhibitory subunit, PRAS40, causing its dissociation from the complex [38, 40]. In addition to being directly phosphorylated by Akt, TSC2 is reportedly phosphorylated and inactivated by Erk as well as its downstream kinase Rsk [39].

IV. AMPK Inhibits mTORC1 Through Phosphorylation of TSC2 and Raptor

In addition to these growth stimulatory cues that activate mTORC1, the complex is rapidly inactivated by a wide variety of cell stresses, thereby ensuring that cells do not continue to grow under unfavorable conditions. One of the unique aspects of the mTORC1 complex is that, unlike many of the aforementioned growth factor activated kinases, it is dependent on nutrient availability for its kinase activity. Withdrawal of glucose, amino acids, or oxygen leads to rapid suppression of mTORC1 activity [39]. Upon LKB1- and AMP-dependent activation of AMPK by energy stress, AMPK directly phosphorylates the TSC2 tumor suppressor on conserved serine sites distinct from those targeted by other kinases, which stimulates GAP activity toward Rheb and constitutes one mechanism by which glucose and oxygen control mTORC1 activation [26, 27, 41, 42]. Ser1387 of human TSC2 (Ser1345 in the rat cDNA) is phosphorylated by AMPK *in vivo*, though there may be multiple additional sites. Phosphorylation of Ser1387 was also shown to serve as a priming site for additional phosphorylation by GSK-3 at Ser1383 and 1379, which is well established to target serines four residues away from a priming phosphorylation site [43]. This phosphorylation of TSC2 by GSK-3 is inhibited by the Wnt signaling pathway, creating a signal integration at these sites depending on both the activation state of AMPK and of GSK-3 that dictates the amount of active TSC2 and hence downstream mTORC1 signaling.

While TSC2 is clearly a central receiver of a wide variety of positive and negative inputs that regulate mTORC1, cells lacking TSC2 still partially suppress mTORC1 following AMPK activation (albeit less so than wild-type cells) suggesting that additional AMPK substrates may directly or indirectly modulate mTORC1 activity [44, 45]. An additional reason to suspect additional AMPK substrates that regulate mTOR is revealed by the fact that glucose inactivation of AMPK orthologs and stimulation of TOR orthologs is conserved across all eukaryotes, including several that lack TSC2 orthologs such as *Caenorhabditis elegans* and *S. cerevisiae*. This suggests that either additional mechanisms exist to coordinate the kinase activity of these two master regulators of cell growth and metabolism, or AMPK must target additional conserved components of the pathway. Using an unbiased peptide library screening methodology [46], we recently identified an optimal substrate motif for AMPK which matches well with previous optimal AMPK peptide substrate analyses [47]. Using the AMPK optimal substrate motif, we performed bioinformatics searches to identify highly conserved substrates of AMPK that may be involved in growth control. We discovered the critical mTOR binding partner raptor as a direct substrate of AMPK, and demonstrated that phosphorylation of raptor by AMPK at two highly conserved serines—Ser722 and Ser792—induces their direct binding to 14-3-3, which leads to a suppression of mTORC1 kinase activity toward its downstream substrates [45]. Like previous observations in LKB1-deficient MEFs, AMPK-deficient MEFs are also unable to downregulate mTORC1 following energy stress [45].

Using cells reconstituted with a nonphosphorylatable raptor cDNA or wild-type raptor cDNA, we found that phosphorylation of these two sites in raptor is required for suppression of mTORC1 activity and cell cycle arrest by AMPK agonists including AICAR and phenformin in MEFs. While the p53 tumor suppressor is required for a G1 phase cell cycle arrest following treatment of MEFs with AMPK agonists [20], cells lacking both p53 and TSC2 still maintain a G2/M cell cycle arrest following AICAR treatment. We demonstrated that this G2/M arrest is ablated in cells reconstituted with raptor cDNA mutated at the two AMPK phosphorylation sites [45]. These data demonstrate for the first time that downregulation of mTOR is required for efficient cell cycle arrest following AMPK activation. These results also suggest that mTORC1 regulates some step of the G2/M cell cycle progression.

Taken together with previous studies, our findings indicate that energy stress results in LKB-dependent activation of AMPK, which directly phosphorylates both TSC2 and raptor to inhibit mTORC1 activity by a dual-pronged mechanism (see Figure 3.1). While the data from MEFs suggests that the phosphorylation of TSC2 and raptor are required for AMPK to suppress mTORC1, it remains possible that additional substrates of AMPK also contribute to suppression of mTORC1. Interestingly, mTOR itself has been reported to serve as an AMPK substrate [48], and while the reported site does not conform to the established AMPK substrate consensus, it remains possible there are other bona fide sites for AMPK in mTOR.

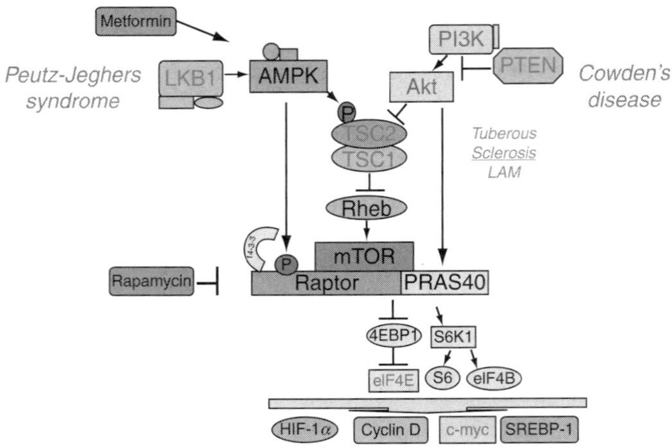

FIG. 3.1. Multiple hamartoma disorders converge on the mTORC1 complex.

Another stress context where AMPK suppresses mTORC1 has recently been uncovered. p53-dependent transcription has been known to result in mTORC1 inhibition in a number of cell types, though the mechanism remained obscure [49]. Recently, the critical transcriptional targets of p53 that mediate the suppression of mTORC1 were identified as Sestrin1 and Sestrin2 [50]. Overexpression of Sestrin1 or Sestrin2 led to increases in AMPK activation and suppression of mTORC1 signaling, whereas mice lacking Sestrin2 failed to downregulate mTORC1 following the carcinogen DEN. The molecular mechanism by which sestrins activate AMPK remains to be elucidated and whether sestrins themselves behave as tumor suppressors and in what tissues they are rate-limiting for suppression of mTORC1 following genotoxic stress remains to be fully explored.

Importantly, we expect that the relative contribution of TSC2 as compared to raptor in suppressing mTORC1 downstream of AMPK will vary from tissue to tissue depending on TSC2 and raptor expression patterns and cross-talk on these two proteins from other signaling pathways in individual tissues. Future studies will be needed to define the relative contribution of AMPK-mediated phosphorylation of raptor and TSC2 in various tissue settings to define the physiological contexts in which each plays a significant role in controlling endogenous mTORC1 activity. We explore below some of the tissue settings where AMPK and mTOR have been shown to play specialized roles.

V. AMPK and mTOR Have Opposing Roles in Specialized Metabolic Tissues in Mammals

In addition to broad roles in controlling cell growth in all mammalian cell types, mTOR and AMPK play key roles in a number of "professional" metabolic tissues in mammals. Here, we briefly examine a few tissue settings where AMPK and mTOR have been shown to demonstrate opposing control of metabolic function.

In the hypothalamus, food intake is controlled by neurons in the arcuate nucleus. Hypothalamic AMPK is activated in response to low glucose, endocannibinoids, AgRP, or the gastric hormone ghrelin, all of which are increased during fasting. Conversely, AMPK activity is decreased upon refeeding or administration of insulin or leptin. Consistent with its suppressive effect on AMPK, leptin induces mTORC1 activity in the neurons of the arcuate nucleus [51]. Mice bearing disruptions in AMPK or core mTORC1 components exhibit a variety of defects in food intake and organismal energy metabolism [52, 53].

Skeletal and cardiac muscles are additional tissues where AMPK and mTOR play key roles in glucose metabolism, hypertrophy, and the response to exercise. In skeletal muscle, AMPK activation has been shown to promote mitochondrial biogenesis at least in part through transcription effects downstream of PGC-1α and PPARδ [54, 55]. Consistently, mice lacking AMPK function in muscle, either from expression of a dominant-negative AMPK or deletion of LKB1, exhibit loss of mitochondrial mass and a dramatic reduction in voluntary exercise [56, 57]. Resistance exercise in humans has been shown to decrease mTORC1-dependent phosphorylation of 4EBP1 coincident with maximal activation of AMPK [58]. Whether AMPK-deficient mice show elevations in mTOR within subtypes of muscle following exercise and whether mTORC1 plays any role in the metabolic reprogramming of muscle fiber type remains to be examined. The picture in skeletal muscle is likely to be complex, as both AMPK and mTOR have been reported to stimulate PGC-1α-dependent mitochondrial biogenesis in this tissue, albeit via distinct mechanisms [54, 59]. Indeed, previous observation suggests isoform-specific AMPK activation in individual muscle types [60], suggesting a thorough analysis of all fiber types and muscle groups in the individual AMPKα1 and AMPKα2 knockout mice and ultimately a skeletal muscle tissue-specific deletion of both, will prove necessary to define where and when AMPK is most rate-limiting for mTOR suppression and PGC-1α regulation following specific stimuli. Notably, mTORC1 signaling following insulin or leucine or electrical stimulation is suppressed by AICAR pretreatment in EDL, gastrocnemius, and extensor digitorum longus muscles, respectively [61–63]. Importantly, utilizing muscle from knockout mice lacking AMPKα2 or AMPKγ3, it was recently demonstrated that each of these AMPK isoforms are required for AICAR to suppress mTORC1 activity [61]. Indeed AMPK activation by metformin or AICAR or by overexpression of activated LKB1 inhibits protein synthesis and hypertrophy in neonatal rat cardiac myocytes coincident with suppression of mTORC1 signaling [64, 65]. Consistent with these findings, induction of hypertrophy by angiotensin II is accompanied by inhibition of AMPK and activation of mTORC1 [66], whereas loss of AMPKα1 accelerates overloading-induced hypertrophy and results in elevated mTORC1 activity [67]. In critical genetic tests of the involvement of AMPK in cardiac hypertrophy, two independent studies found that following isoproterenol [68], or transverse aortic constriction [69], increased hypertrophy was observed in AMPKα2-deficient mice, which correlated with dramatic increases in mTORC1 signaling in the AMPKα2-deficient hearts [69]. Recently, cardiac myocyte-specific deletion of LKB1 was shown to cause developmental defects in the atria and ventricles accompanied by increased left ventricular hypertrophy with reduced AMPK and elevated mTORC1 activity [70]. These defects resulted in cardiac dysfunction and atrial fibrillation with the mice dying within 6 months of age [70]. Although much remains to be elucidated in these models and the molecular interplay between AMPK and mTOR signaling in skeletal and cardiac muscle, it is clear that AMPK modulation of mTOR may play a central role in cardiac hypertrophy.

In liver, AMPK plays key roles in glucose and lipid metabolism. Hormones that activate AMPK in liver including glucagon [71] and adiponectin [72] have been reported to suppress mTORC1 signaling. AMPK-dependent effects on hepatic gluconeogenesis are mediated through direct phosphorylation of transcription factors and coactivators that control transcription of gluconeogenic enzymes [73–76]. Hence regulation of gluconeogenesis by AMPK is likely independent of effects on mTOR. In contrast, lipogenesis is controlled in part by mTORC1-dependent signals. One key regulator of lipogenesis is the SREBP-1 transcription factor. SREBP-1 is a sterol-sensing transcription factor that drives lipogenesis not only in liver, but also in a large variety of mammalian cells. Recently, mTORC1 signaling was shown to be required for nuclear accumulation of SREBP-1 and the induction of SREBP-1 target genes [77]. Similar to rapamycin, treatment with AMPK agonists including AICAR and 2DG resulted in suppression of nuclear SREBP-1 accumulation [77]. Similarly, nuclear SREBP-1 is suppressed in the liver of mice treated with metformin [78]. Metformin treatment or overexpression of an activated allele of AMPK was found to be sufficient to reduce triglyceride content in insulin-resistant HepG2 cells [79]. Mice lacking hepatic AMPK function due to liver-specific LKB1 deletion show elevated SREBP-1 and SREBP-1 target genes resulting in lipid accumulation and hepatic steatosis [25]. Metformin treatment of mice leads to robust phosphorylation of raptor Ser792 in murine liver, which is ablated in the LKB1-liver-specific knockout mice further illustrating that this molecular event may be relevant in the context of AMPK-mediated control of lipid metabolism [45]. Beyond hepatic lipogenesis, SREBP-1 has been shown to be critical for cell growth in both *Drosophila* and mammalian cells [77] suggesting that it may be a critical target of AMPK and mTOR signaling not only in the context of metabolic disease, but also in tumorigenesis. Consistent with this idea, expression the SREBP-1 transcriptional target fatty acid synthase (FASN) has been linked to breast cancer proliferation and FASN inhibitors are beginning to be explored clinically as anticancer agents [80]. In future studies, it will be important to define how much of the lipid-reducing effects of AMPK are due to direct phosphorylation of lipogenic enzymes such as acetyl-CoA carboxylase (ACC), and how much are due to effects on SREBP-1-dependent transcription through effects of AMPK on mTORC1.

Deletion of LKB1 in adult β-cells results in an mTORC1-dependent hypertrophy of β-cells as well as altering their polarity [81, 82]. Upon glucose stimulation, LKB1 deleted β-cells secrete more insulin than their wild-type counterparts, and mice with loss of LKB1 in β-cells exhibit elevated glucose tolerance. After 16 weeks on a high fat diet, control mice had deficient glucose tolerance, whereas mice lacking LKB1 in β-cells displayed glucose tolerance similar to mice on a normal diet [81, 82].

One final metabolic function that AMPK downregulation of mTORC1 may help explain is the effect of AMPK activators as an insulin sensitizer. The ability of metformin and other AMPK-activating drugs to act as insulin sensitizers is well documented [83], and one explanation for that effect is that by AMPK lowering hepatic gluconeogenesis independently from insulin, it relieves the amount of insulin required to be made by the pancreas to reduce circulating blood glucose. Another, cell-autonomously molecular explanation for increased insulin resistance observed in the metabolic syndrome has also been decoded in the past 5 years. Upon activation by insulin, the insulin receptor dimerizes and transphorylates the IRS1 and IRS2 scaffolding proteins which bind which then serve as docking sites for the regulatory subunit of PI3-kinase, and are responsible for insulin-dependent PI3-kinase activation [84]. A large number of laboratories in the past 5 years have discovered that one effect of hyperglycemia, hyperlipidemia, and hyperinsulinemia in the circulation is chronic hyperactivation of the mTORC1 complex, and that mTORC1 and its downstream effector S6K1 both directly phosphorylate the IRS family of scaffolds leading to their targeted degradation [85, 86]. This negative feedback loop suppressing of IRS-PI3-kinase activation can explain effects of overnutrition in the organism as well as in cultured cells [87]. Since mTORC1 activity is dictating the extent of the feedback and suppression of PI3-kinase activity, AMPK activation actually serves to attenuate the feedback and promotes cell-autonomous restoration of IRS protein levels and IRS signaling to PI3-kinase. The net effect is most clearly observed on IRS protein levels and Akt activation. When mTORC1 activity is high, IRS proteins are low and Akt is inhibited. When AMPK is activated, it suppresses mTORC1, restoring IRS protein levels and Akt activation.

VI. AMPK and TOR Function in Model Organisms to Control Growth, Metabolism, Autophagy, and Aging

The central role of AMPK in growth and metabolic regulation is widely conserved across eukaryotes. First studied in budding yeast, the AMPK ortholog Snf1 coordinates the response to altered carbon sources as well as diverse environmental stresses including salt stress, nutrient deprivation, and hypoxia [88]. In a manner parallel to its effects on mammalian physiology, under low glucose conditions Snf1 is activated and promotes a reprogramming of glucose metabolism via regulation of a number of transcription factors. Snf1 is also activated upon nitrogen starvation, paralleling inactivation of TOR [89]. Further emphasizing the convergence of these pathways across evolution, Snf1- and TOR-dependent signals have been shown to converge in the transcriptional regulation of stress-responsive transcription factors Gln3 and Msn2 [90, 91]. In addition, deletions in AMPK α or γ subunit orthologs result in shortened lifespan in response to nutrient deprivation in yeast [92]. Opposing the effects seen for AMPK, depletion of TOR or its effector S6K ortholog Sch9 results in lifespan extension in budding yeast [93]. The lifespan extension effects of TOR in yeast have been further linked to suppression of protein translation [94, 95].

Similarly to yeast, AMPK orthologs in plants (KIN10/KIN11) also play a key role in the response to nutrient deprivation, salt, and herbicide stress [96]. Depletion of AMPK results in shortened lifespan in response to environmental stresses including constant darkness in plants [96, 97]. The TOR pathway is also conserved in photosynthetic eukaryotes including the model plant *Arabidopsis thaliana* and the green algae *Chlamydomonas reinhardtii*, and studies using genetic TOR mutants have revealed that TOR similarly controls cell growth [98]. Many components of the TORC1 pathway are conserved in plants, and the AMPK phosphorylation sites are found in the *Arabidopsis* ortholog of raptor.

The strongest support for opposing roles of AMPK and TOR in growth control, physiology, and aging among lower eukaryotes come from studies in the nematode *C. elegans*. AMPK (aak-2) and LKB1 (par-4) orthologs are required for the extended cell cycle arrest of germ cells in dauer worms [99] as well as the arrest of L1 stage V lineage cells under starvation conditions [100]. In both lineages, AMPK or LKB1 loss causes inappropriate proliferation under nutrient-poor conditions. Furthermore, dauer worms lacking AMPK or LKB1 expend their lipid stores resulting in premature death [99]. In addition, AMPK activation is required in *C. elegans* for lifespan extension by daf-2, heat shock, and glycolytic inhibitors [101–103]. Conversely, suppression of TOR promotes lifespan extension in *C. elegans* [104–106]. Lifespan extension has also been observed with depletion of the TORC1 downstream effectors S6K and eIF4E [107–109], as well as with TOR-dependent autophagy components and the Pha-4 Foxa transcription factor [110, 111], which was previously found as a key lifespan extension gene required for caloric restriction to extend lifespan in *C. elegans* [112].

It is likely that conserved and opposing roles for AMPK and TOR will be uncovered in other model organisms as well. TOR has a central role in cell growth in all eukaryotes examined to date and recently AMPK activation has been shown to suppress cell proliferation in both *Drosophila* and *Dictyostelium* [113, 114].

In addition to effects on protein translation that impact cell growth, metabolism, and aging, TOR also has an independent role in the suppression of autophagy under nutrient-rich conditions. Autophagy is the process by which cells consume their own proteins and organelles to maintain levels of essential building blocks and promote survival under nutrient-poor conditions [115, 116]. The conserved roles of AMPK to promote autophagy from yeast to fly to mammals [117–119] and of TOR to suppress autophagy across those species [120–122] suggest that these two central metabolic integrators play key roles in the molecular regulation of this process. A conserved role for TOR in autophagy control is well established from budding yeast studies demonstrating that TOR controls the first committed step of autophagy through regulation of the ATG1 serine/threonine kinase [123]. mTOR is also known to control mammalian autophagy, and the ATG1 orthologs ULK1 and ULK2 also have reported roles in autophagy in different mammalian cell types [124] though much further work is needed to dissect the molecular details in higher eukaryotes. The ability of AMPK to promote cell survival under conditions of environmental stress may lie in part in its ability to stimulate autophagy. This cell survival mechanism is likely to play a significant role in tumor cells facing a variety of energetic and environmental stresses as well as in the control of stress resistance and lifespan of the aforementioned species. Notably, autophagy has been tied to lifespan extension as well [110, 125].

Given this plethora of conserved functions for AMPK and TOR, it will be interesting to determine if the predicted AMPK phosphorylation sites in raptor similarly dictate nutrient-dependent responses governing cell growth, aging, and stress response in lower organisms.

VII. Therapeutic Implications

The suppression of mTORC1 signaling by AMPK has two immediate therapeutic implications for the treatment of human cancer using existing FDA-approved agents:

1. Tumors in Peutz–Jeghers patients or sporadic NSCLC lacking LKB1 may be sensitive to mTORC1 inhibitors including rapamycin analogs or mTOR kinase inhibitors
2. Any tumor exhibiting hyperactivation of mTORC1 may be sensitive to growth suppression by AMPK agonists

A. Rapamycin as a Therapeutic for Hamartomas and Other LKB1-Deficient Tumors

Aberrant activation of the mTORC1 pathway has been observed in spontaneously arising tumors in mice genetically engineered for loss of the tumor suppressors Pten, Nf1, Tsc2, or Lkb1 [39]. Mutations in these genes are responsible for the inherited cancer syndromes Cowden's disease, Neurofibromatosis type I, TSC, and PJS, collectively referred to as phakomatoses and all sharing overlapping clinical features including the development of hamartomas and pigmentation defects. As illustrated in Figure 3.1, biochemical and cell biological studies from the past decade have revealed that these tumor suppressors all are direct components of the mTOR signaling pathway that serve to inhibit mTORC1 activity.

The underlying hypothesis is that mutational inactivation of these tumor suppressors in individual cells lead to cell-autonomous hyperactivation of mTORC1, promoting cell growth and ultimately resulting in tumors that are subsequently reliant on mTORC1 signaling for tumor maintenance. Consistent with this possibility, rapamycin analogs have been examined for their therapeutic efficacy in the suppression of tumors that arise in a number of the aforementioned mouse models. The $Pten^{+/-}$, $Nf1^{+/-}$, $Tsc^{+/-}$, and activated Akt transgenic mouse models have also proven to be responsive to rapamycin analogs. These drugs have been proven to effectively inhibit mTORC1 *in vivo* and reduce tumor burden through mTORC1-dependent mechanisms, including suppression of cyclin D, Mcl-1, or HIF-1α and its targets [36, 126–128].

In recent clinical trials, rapamycin analogs were shown to have palliative success on patients with PTEN-deficient glioblastomas and metastatic renal cell carcinoma [129, 130]. Furthermore, in a pair of phase II clinical trials involving TSC and lymphangioleiomyomatosis (LAM) patients, partial responses to the rapamycin analogs were observed, including regression of angiomyoliomas [131, 132]. Combined with data from mouse models, these clinical data suggest that hamartoma syndromes with hyperactivation of mTORC1 may be particularly responsive to rapamycin analogs as a single agent. To date there are no therapies to treat PJS and the only course of treatment is resection of arising gastrointestinal hamartomatous polyps. We and others have recently completed a preclinical trial using rapamycin to treat the $Lkb1^{+/-}$ mouse model of PJS which displayed great efficacy [133]. And in a model of invasive endometrial cancer caused by loss of LKB1, rapamycin therapy was highly effective in slowing disease progression and shrinking preexisting tumors [134]. While these results are encouraging for the use of rapamycin analogs as therapeutics for PJS, like the recent Phase II clinical trial findings with TSC patients, removal of the drug may result in rapid return of the initial tumor due to the largely cytostatic nature of the response [131]. Perhaps new, targeted inhibitors directed at the kinase domain of mTOR will produce greater therapeutic response with targeted cytotoxicity, or perhaps kinase inhibitors that dually inactivate mTOR and PI3K would be even more effective, as PI3K provides a survival signal in most epithelial cell types. As observed in most cancers studied to date, combinations of targeted therapeutics, or of targeted and traditional chemotherapeutics may find the ultimate utility in the treatment of this disease. Importantly, it is worth noting here that rapamycin treatment may not only be therapeutically useful for the hamartomas that arise in Peutz–Jeghers patients, but also in preventing and reducing any secondary malignancies that arise in these patients at additional sites (breast, pancreas, ovary).

In terms of human disease however, the number of patients with PJS is dwarfed by the number of people with sporadic lung tumors containing LKB1 mutations. Recent studies have revealed that LKB1 mutations are among the most common in smoking-associated lung tumors, which are still the most common form of cancer in the United States and worldwide. LKB1 mutations are found in non-small-cell lung carcinomas, which represent 85% of lung cancer, and mutation rate estimates vary between 20% and 40% depending on the study and the methodology employed for detecting LKB1 alterations [9, 10]. Analysis of human NSCLC cell lines and primary tumors have revealed loss of AMPK and hyperactivation of mTORC1 [28], indicating that the AMPK/mTOR pathway is likely a relevant target of LKB1 for the suppression of tumorigenesis in the setting of the lung. Interestingly, the K-ras oncogene is frequently comutated with LKB1 in NSCLC [135]. To model these tumors, mice bearing a conditional activated allele of K-ras were crossed to those bearing a conditional inactivating allele of LKB1 and then mice were treated with an aerosolized adenovirus inducing deletion and hence simultaneous activation of K-ras and loss of LKB1 in isolated lung cells [9]. The K-ras, LKB1 double mutant mice showed a dramatic increase in their tumor incidence and metastasis resulting in rapid acceleration of death (25 weeks for K-ras alone vs. 10 weeks for K-ras and LKB1 mutant). Immunohistochemical analysis of these tumors revealed that the LKB1-deficient tumors showed loss of AMPK signaling and even further increases in mTORC1 signaling beyond what was observed with K-ras alone [9]. Interestingly, c-myc expression in lung results in long latency tumors that frequently carry K-Ras or LKB1 mutations [136]. Combined with the human data, these findings in mouse models reinforce that the AMPK/mTORC1 pathway may be relevant for this tumorigenesis setting. Whether mTORC1 inhibitors as a single agent would show any utility against these tumors is unclear, although given that the tumors are not only mutant for LKB1, but also have all the K-ras-dependent mitogenic signaling pathways stimulated, one might imagine a combination of mTORC1 and Erk, or mTORC1 and Akt inhibitors might demonstrate greater clinical efficacy.

Beyond PJS and NSCLC patients, the fact that LKB1 is rate-limiting for tumor formation in a number of tissue settings in the mouse suggests that LKB1, and hence AMPK, may have a wider role in human cancers than currently appreciated. In particular, LKB1 deficiency in prostate, keratinocytes, uterine epithelium, pancreas, and bone all give rise to tumor-prone phenotypes [10]. Whether these tumors all possess elevations of mTORC1 remains to be determined. As AMPK likely has multiple substrates involved growth control, and LKB1 regulates 12 kinases in addition to AMPK, the relative importance of mTORC1 hyperactivation to the tumor phenotype in each of these contexts is an open question that remains to be answered.

Finally, therapeutics aimed at AMPK activation also hold potential for cancer. AMPK activation by metformin or AICAR inhibits the growth of tumor cell in culture as well as in xenograft models [137]. In endometrial cancer cell lines, metformin treatment inhibits proliferation and reduces levels of hTERT [138]. In glioblastomas with activating mutations in EGFR, AICAR treatment inhibits tumor cell proliferation better than rapamycin treatment primarily through inhibition of lipogenesis [139]. Moreover, metformin treatment suppresses naturally arising tumors in transgenic and carcinogen-treated rodent cancer models [140–142].

Given the known pharmacokinetics and widespread long-term clinical use of metformin, its utility as a potential chemotherapeutic modality deserves further attention. In addition, the biguanide phenformin which was removed from clinical use for type 2 diabetes due to incidence of fatalities from lactic acidosis may find modern utility as a anticancer agent as the dosing and duration of its use would be quite distinct from that used for diabetes. Recently, the first study to directly compare the antitumoral efficacy of metformin, phenformin, and the small molecule compound Abbott A769662 was reported [142]. The authors examined the ability of each of these AMPK activators to suppress tumors in Pten+/− mice, which are prone to spontaneously arising lymphomas. While all three compounds resulted in delayed tumor onset, phenformin and Abbott A769662 suppressed tumors with much greater efficacy, which correlated with their ability to activate AMPK and suppress mTORC1 in a wide number of tissues *in vivo*, unlike metformin, which caused more transient and less robust changes on the signaling pathways in most tissues examined. Perhaps key to the success observed here is the fact that the tumors initiated from loss of Pten and hence activation of PI3-kinase, making mTORC1 hyperactivation of the biochemical initiating events for this tumor type and increasing the odds that mTORC1 signaling is required for the tumor. These data also suggest a possible therapeutic window for the use of AMPK agonists to treat tumors arising in patients with TSC or for tumors exhibiting hyperactivation of mTOR by other genetic lesions. The fact that the targeted Abbott compound also performed well suggests that AMPK is in fact the key target of the biguanides in tumor reduction, though off-target effects of the Abbott compound have also been reported [143].

Interestingly, the antifolate drug pemexetred was approved by the FDA as an anticancer drug and was recently shown to activate AMPK and inhibit mTOR [144], suggesting that like other AMPK activators, pemexetred may exhibit clinical efficacy in tumors with aberrant mTOR activity. Notably, pemexetred is used clinically in non-small-cell lung carcinoma, 30% of which bear LKB1 mutations, thereby introducing the possibility that pemexetred may work selectively on either LKB1-positive or LKB1-negative tumors. Future studies should help illuminate this exciting new possibility.

Importantly, compounds that activate AMPK will not only inhibit tumorigenesis via suppression of mTORC1 and lipogenic targets such as ACC, but perhaps also through alterations in organismal metabolism such as reducing blood glucose and insulin resistance, leading to lowered systemic blood insulin levels as well. Given the number of type 2 diabetics worldwide taking metformin daily (>100 million), epidemiologists have begun examining cancer incidence of those taking metformin. Initial studies have revealed that diabetic patients taking metformin exhibit a statistical reduction of tumor incidence [145, 146]. Many additional epidemiological studies are required to determine whether there is indeed a clear effect of prolonged use of these compounds. Most important will be determining if any particular tumor types and any specific tumor genotypes best predict therapeutic efficacy of these compounds.

In fact, tumor cells lacking LKB1 are hypersensitive to apoptosis in culture following treatment with energy stress inducing agents, presumably originating from an inability to restore ATP levels due to AMPK deficiency [6, 28, 147]. However, in cells with functional AMPK signaling, the ability of AMPK to restore ATP levels results in cell survival effects in a number of cell types, which might result in increased tumor cell survival even if it caused a growth arrest of cells within the tumor. However, the magnitude of ATP depletion and effective level of energy stress different cell populations face *in vivo* induced by each combination of drugs, combined with the status of apoptotic and prosurvival signals in these cells is something that can only be modeled accurately *in vivo*, reinforcing the need for better mouse models of spontaneously arising tumors in their natural tissue setting.

VIII. Future Perspectives

The existence of a nutrient-regulated tumor suppressor pathway that couples cell growth to glucose and lipid metabolism results in a number of intriguing predictions and unanswered questions. For example, do environmental factors such as diet and exercise that contribute to physiological AMPK activation modulate cancer risk through inhibition of mTORC1? It is clear from a large number of epidemiology studies that cancer risk is elevated among individuals with metabolic syndrome, or obesity, or type 2 diabetes alone [148]. While a number of metabolic factors are deregulated in these individuals, not the least of which are increased insulin levels and commensurate changes in the insulin/IGF-1 signaling axis, these individuals also show hyperactivation of mTORC1 in many peripheral cell types which may contribute to increased cell proliferation of certain lineages. The identity of the cell types most sensitive to growth suppression effects of AMPK and LKB1 may reveal those lineages in which cell growth is most tightly coupled to dietary conditions. Conversely, exercise and caloric restriction, which both activate AMPK in some lineages, can lower overall cancer risk and improve cancer prognosis [149]. The mammalian cell types in which exercise and caloric restriction suppress cell growth and cancer risk also remain to be delineated. Though much remains to be done to examine whether AMPK mediates some of the beneficial effects of exercise and caloric restriction on cancer risk, a recent study revealed that AMPK was activated, and mTORC1 signaling was suppressed, in some rodent tissues in a dose-dependent manner by increasing amounts of dietary restriction [150]. Conversely, high fat diet was observed to increase mTOR and decrease AMPK activity in some mouse tissues [151].

Future studies are needed to further delineate the full set of tissue types and physiological conditions in which AMPK plays a prominent role in the regulation of overall mTOR activity. The investigation of mice deficient in AMPKα1 or AMPKα2, as well as mice bearing tissue-specific deletions of both will be critical for these studies, and will also be useful for defining tissues where loss of AMPK is sufficient to promote tumorigenesis. While AMPK regulation of mTOR is most likely to play a role in tissues that experience physiological or pathological energy stress, in other tissues and conditions, the activation of mTOR by nutrients and growth factors may play a more prominent role in dictating the level of mTOR activity.

In addition to cell proliferation, cell survival, and autophagy, inhibition of mTORC1 by AMPK will likely play a physiological role in other biological processes that mTORC1 is known to regulate including angiogenesis, mitochondrial metabolism, viral infection, and specific transcriptional responses. The existence of diverse biological and metabolic processes regulated by AMPK for which there are few known direct effectors suggests that there are many critical substrates yet to be identified. It will also be interesting to determine which of these will turn out to involve inhibition of mTORC1 as the as-of-yet unappreciated molecular mechanism.

In addition to clear roles in cell growth, the central role of AMPK and mTOR in autophagic control may be key to many differentiated nondividing adult cells throughout the body in response to dietary and hormonal homeostasis fluctuations. The critical role that autophagy is beginning to be appreciated to play in the pathology of cancer, neurological diseases, and metabolic diseases, is just one more indication that this basic signaling pathway connecting AMPK to mTORC1 is likely to lie at the heart of many human maladies [115]. As the response to a shortage of environmental nutrients and resultant loss in cellular energy represents one of the most fundamental pathological events in all organisms, we anticipate that further investigation of the downstream targets of AMPK will provide great insight into the emerging nexus of cancer and metabolic disease controlled by this ancestral signaling pathway.

Acknowledgments

Research in the author's laboratory is generously supported by grants from the NIH (P01 CA120964, R01 DK080425), The Howard Hughes Medical Institute, American Cancer Society, American Diabetes Association, V Foundation for Cancer Research, and Adler Foundation.

References

1. Kahn, B.B., Alquier, T., Carling, D., and Hardie, D.G. (2005). AMP-activated protein kinase: ancient energy gauge provides clues to modern understanding of metabolism. *Cell Metab* 1:15–25.
2. Sanders, M.J., Grondin, P.O., Hegarty, B.D., Snowden, M.A., and Carling, D. (2007). Investigating the mechanism for AMP activation of the AMP-activated protein kinase cascade. *Biochem J* 403:139–148.
3. Hardie, D.G. (2007). AMP-activated/SNF1 protein kinases: conserved guardians of cellular energy. *Nat Rev Mol Cell Biol* 8:774–785.
4. Hong, S.P., Leiper, F.C., Woods, A., Carling, D., and Carlson, M. (2003). Activation of yeast Snf1 and mammalian AMP-activated protein kinase by upstream kinases. *Proc Natl Acad Sci USA* 100:8839–8843.
5. Sutherland, C.M., et al. (2003). Elm1p is one of three upstream kinases for the *Saccharomyces cerevisiae* SNF1 complex. *Curr Biol* 13:1299–1305.
6. Shaw, R.J., et al. (2004). The tumor suppressor LKB1 kinase directly activates AMP-activated kinase and regulates apoptosis in response to energy stress. *Proc Natl Acad Sci USA* 101:3329–3335.
7. Hemminki, A., et al. (1998). A serine/threonine kinase gene defective in Peutz–Jeghers syndrome. *Nature* 391:184–187.
8. Sanchez-Cespedes, M., et al. (2002). Inactivation of LKB1/STK11 is a common event in adenocarcinomas of the lung. *Cancer Res* 62:3659–3662.
9. Ji, H., et al. (2007). LKB1 modulates lung cancer differentiation and metastasis. *Nature* 448:807–810.
10. Hezel, A.F., and Bardeesy, N. (2008). LKB1; linking cell structure and tumor suppression. *Oncogene* 27:6908–6919.
11. Lizcano, J.M., et al. (2004). LKB1 is a master kinase that activates 13 kinases of the AMPK subfamily, including MARK/PAR-1. *EMBO J* 23:833–843.
12. Jaleel, M., et al. (2005). Identification of the sucrose non-fermenting related kinase SNRK, as a novel LKB1 substrate. *FEBS Lett* 579:1417–1423.
13. Al-Hakim, A.K., et al. (2005). 14-3-3 cooperates with LKB1 to regulate the activity and localization of QSK and SIK. *J Cell Sci* 118:5661–5673.
14. Suzuki, A., Kusakai, G., Kishimoto, A., Minegichi, Y., Ogura, T., and Esumi, H. (2003). Induction of cell–cell detachment during glucose starvation through F-actin conversion by SNARK, the fourth member of the AMP-activated protein kinase catalytic subunit family. *Biochem Biophys Res Commun* 311:156–161.
15. Lefebvre, D.L., and Rosen, C.F. (2005). Regulation of SNARK activity in response to cellular stresses. *Biochim Biophys Acta* 1724:71–85.
16. Du, J., Chen, Q., Takemori, H., and Xu, H. (2008). SIK2 can be activated by deprivation of nutrition and it inhibits expression of lipogenic genes in adipocytes. *Obesity (Silver Spring)* 16:531–538.
17. Zeqiraj, E., Filippi, B.M., Deak, M., Alessi, D.R., and van Aalten, D.M. (2009). Structure of the LKB1-STRAD-MO25 complex reveals an allosteric mechanism of kinase activation. *Science* 326:1707–1711.
18. Berdeaux, R., et al. (2007). SIK1 is a class II HDAC kinase that promotes survival of skeletal myocytes. *Nat Med* 13:597–603.
19. Legembre, P., Schickel, R., Barnhart, B.C., and Peter, M.E. (2004). Identification of SNF1/AMP kinase-related kinase as an NF-kappaB-regulated anti-apoptotic kinase involved in CD95-induced motility and invasiveness. *J Biol Chem* 279:46742–46747.

20. Jones, R.G., et al. (2005). AMP-activated protein kinase induces a p53-dependent metabolic checkpoint. *Mol Cell* 18:283–293.
21. Liang, J., et al. (2007). The energy sensing LKB1–AMPK pathway regulates p27(kip1) phosphorylation mediating the decision to enter autophagy or apoptosis. *Nat Cell Biol* 9:218–224.
22. Imamura, K., Ogura, T., Kishimoto, A., Kaminishi, M., and Esumi, H. (2001). Cell cycle regulation via p53 phosphorylation by a 5′-AMP activated protein kinase activator, 5-aminoimidazole-4-carboxamide-1-beta-D-ribofuranoside, in a human hepatocellular carcinoma cell line. *Biochem Biophys Res Commun* 287:562–567.
23. Karuman, P., et al. (2001). The Peutz–Jegher gene product LKB1 is a mediator of p53-dependent cell death. *Mol Cell* 7:1307–1319.
24. Tiainen, M., Vaahtomeri, K., Ylikorkala, A., and Makela, T.P. (2002). Growth arrest by the LKB1 tumor suppressor: induction of p21(WAF1/CIP1). *Hum Mol Genet* 11:1497–1504.
25. Shaw, R.J., et al. (2005). The kinase LKB1 mediates glucose homeostasis in liver and therapeutic effects of metformin. *Science* 310:1642–1646.
26. Corradetti, M.N., Inoki, K., Bardeesy, N., DePinho, R.A., and Guan, K.L. (2004). Regulation of the TSC pathway by LKB1: evidence of a molecular link between tuberous sclerosis complex and Peutz–Jeghers syndrome. *Genes Dev* 18:1533–1538.
27. Shaw, R.J., et al. (2004). The LKB1 tumor suppressor negatively regulates mTOR signaling. *Cancer Cell* 6:91–99.
28. Carretero, J., et al. (2007). Dysfunctional AMPK activity, signalling through mTOR and survival in response to energetic stress in LKB1-deficient lung cancer. *Oncogene* 26:1616–1625.
29. Wullschleger, S., Loewith, R., and Hall, M.N. (2006). TOR signaling in growth and metabolism. *Cell* 124:471–484.
30. Nojima, H., et al. (2003). The mammalian target of rapamycin (mTOR) partner, raptor, binds the mTOR substrates p70 S6 kinase and 4E-BP1 through their TOR signaling (TOS) motif. *J Biol Chem* 278:15461–15464.
31. Schalm, S.S., Fingar, D.C., Sabatini, D.M., and Blenis, J. (2003). TOS motif-mediated raptor binding regulates 4E-BP1 multisite phosphorylation and function. *Curr Biol* 13:797–806.
32. Guertin, D.A., and Sabatini, D.M. (2007). Defining the role of mTOR in cancer. *Cancer Cell* 12:9–22.
33. Garcia-Martinez, J.M., and Alessi, D.R. (2008). mTOR complex 2 (mTORC2) controls hydrophobic motif phosphorylation and activation of serum- and glucocorticoid-induced protein kinase 1 (SGK1). *Biochem J* 416:375–385.
34. Holz, M.K., Ballif, B.A., Gygi, S.P., and Blenis, J. (2005). mTOR and S6K1 mediate assembly of the translation preinitiation complex through dynamic protein interchange and ordered phosphorylation events. *Cell* 123:569–580.
35. Ruggero, D., and Pandolfi, P.P. (2003). Does the ribosome translate cancer? *Nat Rev Cancer* 3:179–192.
36. Wendel, H.G., et al. (2006). Determinants of sensitivity and resistance to rapamycin-chemotherapy drug combinations in vivo. *Cancer Res* 66:7639–7646.
37. Averous, J., Fonseca, B.D., and Proud, C.G. (2008). Regulation of cyclin D1 expression by mTORC1 signaling requires eukaryotic initiation factor 4E-binding protein 1. *Oncogene* 27:1106–1113.
38. Sancak, Y., et al. (2007). PRAS40 is an insulin-regulated inhibitor of the mTORC1 protein kinase. *Mol Cell* 25:903–915.

39. Shaw, R.J., and Cantley, L.C. (2006). Ras, PI(3)K and mTOR signalling controls tumour cell growth. *Nature* 441:424–430.
40. Vander Haar, E., Lee, S.I., Bandhakavi, S., Griffin, T.J., and Kim, D.H. (2007). Insulin signalling to mTOR mediated by the Akt/PKB substrate PRAS40. *Nat Cell Biol* 9:316–323.
41. Inoki, K., Zhu, T., and Guan, K.L. (2003). TSC2 mediates cellular energy response to control cell growth and survival. *Cell* 115:577–590.
42. Liu, L., Cash, T.P., Jones, R.G., Keith, B., Thompson, C.B., and Simon, M.C. (2006). Hypoxia-induced energy stress regulates mRNA translation and cell growth. *Mol Cell* 21:521–531.
43. Inoki, K., et al. (2006). TSC2 integrates Wnt and energy signals via a coordinated phosphorylation by AMPK and GSK3 to regulate cell growth. *Cell* 126:955–968.
44. Hahn-Windgassen, A., Nogueira, V., Chen, C.C., Skeen, J.E., Sonenberg, N., and Hay, N. (2005). Akt activates the mammalian target of rapamycin by regulating cellular ATP level and AMPK activity. *J Biol Chem* 280:32081–32089.
45. Gwinn, D.M., et al. (2008). AMPK phosphorylation of raptor mediates a metabolic checkpoint. *Mol Cell* 30:214–226.
46. Hutti, J.E., et al. (2004). A rapid method for determining protein kinase phosphorylation specificity. *Nat Methods* 1:27–29.
47. Scott, J.W., Norman, D.G., Hawley, S.A., Kontogiannis, L., and Hardie, D.G. (2002). Protein kinase substrate recognition studied using the recombinant catalytic domain of AMP-activated protein kinase and a model substrate. *J Mol Biol* 317:309–323.
48. Cheng, S.W., Fryer, L.G., Carling, D., and Shepherd, P.R. (2004). Thr2446 is a novel mammalian target of rapamycin (mTOR) phosphorylation site regulated by nutrient status. *J Biol Chem* 279:15719–15722.
49. Levine, A.J., Feng, Z., Mak, T.W., You, H., and Jin, S. (2006). Coordination and communication between the p53 and IGF-1-AKT-TOR signal transduction pathways. *Genes Dev* 20:267–275.
50. Budanov, A.V., and Karin, M. (2008). p53 target genes sestrin1 and sestrin2 connect genotoxic stress and mTOR signaling. *Cell* 134:451–460.
51. Woods, S.C., Seeley, R.J., and Cota, D. (2008). Regulation of food intake through hypothalamic signaling networks involving mTOR. *Annu Rev Nutr* 28:295–311.
52. Cota, D., et al. (2006). Hypothalamic mTOR signaling regulates food intake. *Science* 312:927–930.
53. Claret, M., et al. (2007). AMPK is essential for energy homeostasis regulation and glucose sensing by POMC and AgRP neurons. *J Clin Invest* 117:2325–2336.
54. Jager, S., Handschin, C., St-Pierre, J., and Spiegelman, B.M. (2007). AMP-activated protein kinase (AMPK) action in skeletal muscle via direct phosphorylation of PGC-1alpha. *Proc Natl Acad Sci USA* 104:12017–12022.
55. Narkar, V.A., et al. (2008). AMPK and PPARdelta agonists are exercise mimetics. *Cell* 134:405–415.
56. Mu, J., Brozinick, J.T., Jr., Valladares, O., Bucan, M., and Birnbaum, M.J. (2001). A role for AMP-activated protein kinase in contraction- and hypoxia-regulated glucose transport in skeletal muscle. *Mol Cell* 7:1085–1094.
57. Thomson, D.M., Porter, B.B., Tall, J.H., Kim, H.J., Barrow, J.R., and Winder, W.W. (2007). Skeletal muscle and heart LKB1 deficiency causes decreased voluntary running and reduced muscle mitochondrial marker enzyme expression in mice. *Am J Physiol Endocrinol Metab* 292:E196–E202.
58. Dreyer, H.C., Fujita, S., Cadenas, J.G., Chinkes, D.L., Volpi, E., and Rasmussen, B.B. (2006). Resistance exercise increases AMPK activity and reduces 4E-BP1 phosphorylation and protein synthesis in human skeletal muscle. *J Physiol* 576:613–624.

59. Cunningham, J.T., Rodgers, J.T., Arlow, D.H., Vazquez, F., Mootha, V.K., and Puigserver, P. (2007). mTOR controls mitochondrial oxidative function through a YY1-PGC-1alpha transcriptional complex. *Nature* 450:736–740.
60. McGee, S.L., Mustard, K.J., Hardie, D.G., and Baar, K. (2008). Normal hypertrophy accompanied by phosphoryation and activation of AMP-activated protein kinase alpha1 following overload in LKB1 knockout mice. *J Physiol* 586:1731–1741.
61. Deshmukh, A.S., Treebak, J.T., Long, Y.C., Viollet, B., Wojtaszewski, J.F., and Zierath, J.R. (2008). Role of adenosine 5′-monophosphate-activated protein kinase subunits in skeletal muscle mammalian target of rapamycin signaling. *Mol Endocrinol* 22:1105–1112.
62. Pruznak, A.M., Kazi, A.A., Frost, R.A., Vary, T.C., and Lang, C.H. (2008). Activation of AMP-activated protein kinase by 5-aminoimidazole-4-carboxamide-1-beta-D-ribonucleoside prevents leucine-stimulated protein synthesis in rat skeletal muscle. *J Nutr* 138:1887–1894.
63. Thomson, D.M., Fick, C.A., and Gordon, S.E. (2008). AMPK activation attenuates S6K1, 4E-BP1, and eEF2 signaling responses to high-frequency electrically stimulated skeletal muscle contractions. *J Appl Physiol* 104:625–632.
64. Chan, A.Y., Soltys, C.L., Young, M.E., Proud, C.G., and Dyck, J.R. (2004). Activation of AMP-activated protein kinase inhibits protein synthesis associated with hypertrophy in the cardiac myocyte. *J Biol Chem* 279:32771–32779.
65. Noga, A.A., Soltys, C.L., Barr, A.J., Kovacic, S., Lopaschuk, G.D., and Dyck, J.R. (2007). Expression of an active LKB1 complex in cardiac myocytes results in decreased protein synthesis associated with phenylephrine-induced hypertrophy. *Am J Physiol Heart Circ Physiol* 292:H1460–H1469.
66. Stuck, B.J., Lenski, M., Bohm, M., and Laufs, U. (2008). Metabolic switch and hypertrophy of cardiomyocytes following treatment with angiotensin II are prevented by AMP-activated protein kinase. *J Biol Chem* 283:32562–32569.
67. Mounier, R., et al. (2009). Important role for AMPKalpha1 in limiting skeletal muscle cell hypertrophy. *FASEB J* 23:2264–2273.
68. Zarrinpashneh, E., et al. (2008). AMPKalpha2 counteracts the development of cardiac hypertrophy induced by isoproterenol. *Biochem Biophys Res Commun* 376:677–681.
69. Zhang, P., et al. (2008). AMP activated protein kinase-alpha2 deficiency exacerbates pressure-overload-induced left ventricular hypertrophy and dysfunction in mice. *Hypertension* 52:918–924.
70. Ikeda, Y., et al. (2009). Cardiac-specific deletion of LKB1 leads to hypertrophy and dysfunction. *J Biol Chem* 284:35839–35849.
71. Kimball, S.R., Siegfried, B.A., and Jefferson, L.S. (2004). Glucagon represses signaling through the mammalian target of rapamycin in rat liver by activating AMP-activated protein kinase. *J Biol Chem* 279:54103–54109.
72. Wang, C., et al. (2007). Adiponectin sensitizes insulin signaling by reducing p70 S6 kinase-mediated serine phosphorylation of IRS-1. *J Biol Chem* 282:7991–7996.
73. Yang, W., Hong, Y.H., Shen, X.Q., Frankowski, C., Camp, H.S., and Leff, T. (2001). Regulation of transcription by AMP-activated protein kinase: phosphorylation of p300 blocks its interaction with nuclear receptors. *J Biol Chem* 276:38341–38344.
74. Hong, Y.H., Varanasi, U.S., Yang, W., and Leff, T. (2003). AMP-activated protein kinase regulates HNF4alpha transcriptional activity by inhibiting dimer formation and decreasing protein stability. *J Biol Chem* 278:27495–27501.
75. Koo, S.H., et al. (2005). The CREB coactivator TORC2 is a key regulator of fasting glucose metabolism. *Nature* 437:1109–1111.
76. Inoue, E., and Yamauchi, J. (2006). AMP-activated protein kinase regulates PEPCK gene expression by direct phosphorylation of a novel zinc finger transcription factor. *Biochem Biophys Res Commun* 351:793–799.

77. Porstmann, T., et al. (2008). SREBP activity is regulated by mTORC1 and contributes to Akt-dependent cell growth. *Cell Metab* 8:224–236.
78. Zhou, G., et al. (2001). Role of AMP-activated protein kinase in mechanism of metformin action. *J Clin Invest* 108:1167–1174.
79. Zang, M., et al. (2004). AMP-activated protein kinase is required for the lipid-lowering effect of metformin in insulin-resistant human HepG2 cells. *J Biol Chem* 279:47898–47905.
80. Menendez, J.A., and Lupu, R. (2007). Fatty acid synthase and the lipogenic phenotype in cancer pathogenesis. *Nat Rev Cancer* 7:763–777.
81. Fu, A., et al. (2009). Loss of Lkb1 in adult beta cells increases beta cell mass and enhances glucose tolerance in mice. *Cell Metab* 10:285–295.
82. Granot, Z., et al. (2009). LKB1 regulates pancreatic beta cell size, polarity, and function. *Cell Metab* 10:296–308.
83. Towler, M.C., and Hardie, D.G. (2007). AMP-activated protein kinase in metabolic control and insulin signaling. *Circ Res* 100:328–341.
84. Manning, B.D. (2004). Balancing Akt with S6K: implications for both metabolic diseases and tumorigenesis. *J Cell Biol* 167:399–403.
85. Harrington, L.S., et al. (2004). The TSC1-2 tumor suppressor controls insulin-PI3K signaling via regulation of IRS proteins. *J Cell Biol* 166:213–223.
86. Shah, O.J., Wang, Z., and Hunter, T. (2004). Inappropriate activation of the TSC/Rheb/mTOR/S6K cassette induces IRS1/2 depletion, insulin resistance, and cell survival deficiencies. *Curr Biol* 14:1650–1656.
87. Um, S.H., et al. (2004). Absence of S6K1 protects against age- and diet-induced obesity while enhancing insulin sensitivity. *Nature* 431:200–205.
88. Hong, S.P., and Carlson, M. (2007). Regulation of snf1 protein kinase in response to environmental stress. *J Biol Chem* 282:16838–16845.
89. Orlova, M., Kanter, E., Krakovich, D., and Kuchin, S. (2006). Nitrogen availability and TOR regulate the Snf1 protein kinase in *Saccharomyces cerevisiae*. *Eukaryot Cell* 5:1831–1837.
90. Bertram, P.G., Choi, J.H., Carvalho, J., Chan, T.F., Ai, W., and Zheng, X.F. (2002). Convergence of TOR-nitrogen and Snf1-glucose signaling pathways onto Gln3. *Mol Cell Biol* 22:1246–1252.
91. Mayordomo, I., Estruch, F., and Sanz, P. (2002). Convergence of the target of rapamycin and the Snf1 protein kinase pathways in the regulation of the subcellular localization of Msn2, a transcriptional activator of STRE (Stress Response Element)-regulated genes. *J Biol Chem* 277:35650–35656.
92. Ashrafi, K., Lin, S.S., Manchester, J.K., and Gordon, J.I. (2000). Sip2p and its partner snf1p kinase affect aging in *S. cerevisiae*. *Genes Dev* 14:1872–1885.
93. Kaeberlein, M., et al. (2005). Regulation of yeast replicative life span by TOR and Sch9 in response to nutrients. *Science* 310:1193–1196.
94. Steffen, K.K., et al. (2008). Yeast life span extension by depletion of 60s ribosomal subunits is mediated by Gcn4. *Cell* 133:292–302.
95. Kaeberlein, M., and Kennedy, B.K. (2008). Protein translation, 2008. *Aging Cell* 7:777–782.
96. Baena-Gonzalez, E., Rolland, F., Thevelein, J.M., and Sheen, J. (2007). A central integrator of transcription networks in plant stress and energy signalling. *Nature* 448:938–942.
97. Thelander, M., Olsson, T., and Ronne, H. (2004). Snf1-related protein kinase 1 is needed for growth in a normal day–night light cycle. *EMBO J* 23:1900–1910.
98. Diaz-Troya, S., Perez-Perez, M.E., Florencio, F.J., and Crespo, J.L. (2008). The role of TOR in autophagy regulation from yeast to plants and mammals. *Autophagy* 4:851–865.

99. Narbonne, P., and Roy, R. (2006). Inhibition of germline proliferation during *C. elegans* dauer development requires PTEN, LKB1 and AMPK signalling. *Development* 133:611–619.
100. Baugh, L.R., and Sternberg, P.W. (2006). DAF-16/FOXO regulates transcription of cki-1/Cip/Kip and repression of lin-4 during *C. elegans* L1 arrest. *Curr Biol* 16:780–785.
101. Apfeld, J., O'Connor, G., McDonagh, T., DiStefano, P.S., and Curtis, R. (2004). The AMP-activated protein kinase AAK-2 links energy levels and insulin-like signals to lifespan in *C. elegans*. *Genes Dev* 18:3004–3009.
102. Curtis, R., O'Connor, G., and DiStefano, P.S. (2006). Aging networks in *Caenorhabditis elegans*: AMP-activated protein kinase (aak-2) links multiple aging and metabolism pathways. *Aging Cell* 5:119–126.
103. Schulz, T.J., Zarse, K., Voigt, A., Urban, N., Birringer, M., and Ristow, M. (2007). Glucose restriction extends *Caenorhabditis elegans* life span by inducing mitochondrial respiration and increasing oxidative stress. *Cell Metab* 6:280–293.
104. Vellai, T., Takacs-Vellai, K., Zhang, Y., Kovacs, A.L., Orosz, L., and Muller, F. (2003). Genetics: influence of TOR kinase on lifespan in *C. elegans*. *Nature* 426:620.
105. Jia, K., Chen, D., and Riddle, D.L. (2004). The TOR pathway interacts with the insulin signaling pathway to regulate *C. elegans* larval development, metabolism and life span. *Development* 131:3897–3906.
106. Kapahi, P., Zid, B.M., Harper, T., Koslover, D., Sapin, V., and Benzer, S. (2004). Regulation of lifespan in *Drosophila* by modulation of genes in the TOR signaling pathway. *Curr Biol* 14:885–890.
107. Hansen, M., Taubert, S., Crawford, D., Libina, N., Lee, S.J., and Kenyon, C. (2007). Lifespan extension by conditions that inhibit translation in *Caenorhabditis elegans*. *Aging Cell* 6:95–110.
108. Pan, K.Z., et al. (2007). Inhibition of mRNA translation extends lifespan in *Caenorhabditis elegans*. *Aging Cell* 6:111–119.
109. Syntichaki, P., Troulinaki, K., and Tavernarakis, N. (2007). eIF4E function in somatic cells modulates ageing in *Caenorhabditis elegans*. *Nature* 445:922–926.
110. Hansen, M., Chandra, A., Mitic, L.L., Onken, B., Driscoll, M., and Kenyon, C. (2008). A role for autophagy in the extension of lifespan by dietary restriction in *C. elegans*. *PLoS Genet* 4:e24.
111. Sheaffer, K.L., Updike, D.L., and Mango, S.E. (2008). The target of rapamycin pathway antagonizes pha-4/FoxA to control development and aging. *Curr Biol* 18:1355–1364.
112. Panowski, S.H., Wolff, S., Aguilaniu, H., Durieux, J., and Dillin, A. (2007). PHA-4/Foxa mediates diet-restriction-induced longevity of *C. elegans*. *Nature* 447:550–555.
113. Mandal, S., Guptan, P., Owusu-Ansah, E., and Banerjee, U. (2005). Mitochondrial regulation of cell cycle progression during development as revealed by the tenured mutation in *Drosophila*. *Dev Cell* 9:843–854.
114. Bokko, P.B., et al. (2007). Diverse cytopathologies in mitochondrial disease are caused by AMP-activated protein kinase signaling. *Mol Biol Cell* 18:1874–1886.
115. Levine, B., and Kroemer, G. (2008). Autophagy in the pathogenesis of disease. *Cell* 132:27–42.
116. Mizushima, N., Levine, B., Cuervo, A.M., and Klionsky, D.J. (2008). Autophagy fights disease through cellular self-digestion. *Nature* 451:1069–1075.
117. Meijer, A.J., and Codogno, P. (2007). AMP-activated protein kinase and autophagy. *Autophagy* 3:238–240.
118. Lippai, M., Csikos, G., Maroy, P., Lukacsovich, T., Juhasz, G., and Sass, M. (2008). SNF4Agamma, the *Drosophila* AMPK gamma subunit is required for regulation of developmental and stress-induced autophagy. *Autophagy* 4:476–486.

119. Hoyer-Hansen, M., and Jaattela, M. (2007). AMP-activated protein kinase: a universal regulator of autophagy? *Autophagy* 3:381–383.
120. Noda, T., and Ohsumi, Y. (1998). Tor, a phosphatidylinositol kinase homologue, controls autophagy in yeast. *J Biol Chem* 273:3963–3966.
121. Arsham, A.M., and Neufeld, T.P. (2006). Thinking globally and acting locally with TOR. *Curr Opin Cell Biol* 18:589–597.
122. Dubouloz, F., Deloche, O., Wanke, V., Cameroni, E., and De Virgilio, C. (2005). The TOR and EGO protein complexes orchestrate microautophagy in yeast. *Mol Cell* 19:15–26.
123. Kamada, Y., Sekito, T., and Ohsumi, Y. (2004). Autophagy in yeast: a TOR-mediated response to nutrient starvation. *Curr Top Microbiol Immunol* 279:73–84.
124. Stephan, J.S., and Herman, P.K. (2006). The regulation of autophagy in eukaryotic cells: do all roads pass through Atg1? *Autophagy* 2:146–148.
125. Jia, K., and Levine, B. (2007). Autophagy is required for dietary restriction-mediated life span extension in *C. elegans*. *Autophagy* 3:597–599.
126. Podsypanina, K., et al. (2001). An inhibitor of mTOR reduces neoplasia and normalizes p70/S6 kinase activity in $Pten^{+/-}$ mice. *Proc Natl Acad Sci USA* 98:10320–10325.
127. Majumder, P.K., et al. (2004). mTOR inhibition reverses Akt-dependent prostate intraepithelial neoplasia through regulation of apoptotic and HIF-1-dependent pathways. *Nat Med* 10:594–601.
128. Johannessen, C.M., et al. (2008). TORC1 is essential for NF1-associated malignancies. *Curr Biol* 18:56–62.
129. Hudes, G., et al. (2007). Temsirolimus, interferon alfa, or both for advanced renal-cell carcinoma. *N Engl J Med* 356:2271–2281.
130. Cloughesy, T.F., et al. (2008). Antitumor activity of rapamycin in a Phase I trial for patients with recurrent PTEN-deficient glioblastoma. *PLoS Med* 5:e8.
131. Bissler, J.J., et al. (2008). Sirolimus for angiomyolipoma in tuberous sclerosis complex or lymphangioleiomyomatosis. *N Engl J Med* 358:140–151.
132. Davies, D.M., et al. (2008). Sirolimus therapy in tuberous sclerosis or sporadic lymphangioleiomyomatosis. *N Engl J Med* 358:200–203.
133. Wei, C., et al. (2008). Suppression of Peutz–Jeghers polyposis by targeting mammalian target of rapamycin signaling. *Clin Cancer Res* 14:1167–1171.
134. Contreras, C.M., et al. (2010). Lkb1 inactivation is sufficient to drive endometrial cancers that are aggressive yet highly responsive to mTOR inhibitor monotherapy. *Dis Model Mech* 3:181–193.
135. Makowski, L., and Hayes, D.N. (2008). Role of LKB1 in lung cancer development. *Br J Cancer* 99:683–688.
136. Rapp, U.R., et al. (2009). MYC is a metastasis gene for non-small-cell lung cancer. *PLoS One* 4:e6029.
137. Buzzai, M., et al. (2007). Systemic treatment with the antidiabetic drug metformin selectively impairs p53-deficient tumor cell growth. *Cancer Res* 67:6745–6752.
138. Cantrell, L.A., Zhou, C., Mendivil, A., Malloy, K.M., Gehrig, P.A., and Bae-Jump, V.L. (2010). Metformin is a potent inhibitor of endometrial cancer cell proliferation—implications for a novel treatment strategy. *Gynecol Oncol* 116:92–98.
139. Guo, D., et al. (2009). The AMPK agonist AICAR inhibits the growth of EGFRvIII-expressing glioblastomas by inhibiting lipogenesis. *Proc Natl Acad Sci USA* 106:12932–12937.
140. Schneider, M.B., et al. (2001). Prevention of pancreatic cancer induction in hamsters by metformin. *Gastroenterology* 120:1263–1270.

141. Anisimov, V.N., et al. (2005). Effect of metformin on life span and on the development of spontaneous mammary tumors in HER-2/neu transgenic mice. *Exp Gerontol* 40:685–693.
142. Huang, X., et al. (2008). Important role of the LKB1–AMPK pathway in suppressing tumorigenesis in PTEN-deficient mice. *Biochem J* 412:211–221.
143. Moreno, D., Knecht, E., Viollet, B., and Sanz, P. (2008). A769662, a novel activator of AMP-activated protein kinase, inhibits non-proteolytic components of the 26S proteasome by an AMPK-independent mechanism. *FEBS Lett* 582:2650–2654.
144. Racanelli, A.C., Rothbart, S.B., Heyer, C.L., and Moran, R.G. (2009). Therapeutics by cytotoxic metabolite accumulation: pemetrexed causes ZMP accumulation, AMPK activation, and mammalian target of rapamycin inhibition. *Cancer Res* 69:5467–5474.
145. Evans, J.M., Donnelly, L.A., Emslie-Smith, A.M., Alessi, D.R., and Morris, A.D. (2005). Metformin and reduced risk of cancer in diabetic patients. *BMJ* 330:1304–1305.
146. Bowker, S.L., Majumdar, S.R., Veugelers, P., and Johnson, J.A. (2006). Increased cancer-related mortality for patients with type 2 diabetes who use sulfonylureas or insulin: response to Farooki and Schneider. *Diabetes Care* 29:1990–1991.
147. Memmott, R.M., et al. (2008). Phosphatidylinositol ether lipid analogues induce AMP-activated protein kinase-dependent death in LKB1-mutant non small cell lung cancer cells. *Cancer Res* 68:580–588.
148. Martinez, M.E., Marshall, J.R., and Giovannucci, E. (2008). Diet and cancer prevention: the roles of observation and experimentation. *Nat Rev Cancer* 8:694–703.
149. McTiernan, A. (2008). Mechanisms linking physical activity with cancer. *Nat Rev Cancer* 8:205–211.
150. Jiang, W., Zhu, Z., and Thompson, H.J. (2008). Dietary energy restriction modulates the activity of AMP-activated protein kinase, Akt, and mammalian target of rapamycin in mammary carcinomas, mammary gland, and liver. *Cancer Res* 68:5492–5499.
151. Moore, T., et al. (2008). Dietary energy balance modulates signaling through the Akt/mammalian target of rapamycin pathways in multiple epithelial tissues. *Cancer Prev Res (Phila Pa)* 1:65–76.

4

mTOR Signaling by Amino Acid Nutrients: Involvement of MAP4K3

LIJUN YAN[a,b] • RICHARD F. LAMB[a]

[a]*Department of Oncology*
University of Alberta Edmonton
Alberta, Canada

[b]*The Center for Life Sciences and Environmental Sciences*
Harbin University of Commerce
Harbin, P.R. China

I. Abstract

The mTOR (mammalian target of rapamycin) signaling pathway plays a major role in cell growth and metabolism in eukaryotes. In metazoans, the mTOR kinase exists in at least two distinct complexes, mTORC1 and mTORC2. mTORC1 integrates nutrient-derived signals in coordination with other signals, such as growth factor and energy status, to regulate cell growth. As a result of the pivotal role of mTOR in these processes, mTOR dysregulation has been implicated in various human diseases such as cancer, obesity, and diabetes. In this chapter, we will discuss some recent advances in understanding the mechanisms by which amino acid nutrients signal to mTORC1, with an emphasis on the regulation of the protein kinase MAP4K3 in mTOR signaling.

ISSN NO: 1874-6047
DOI: 10.1016/S1874-6047(10)28004-6

II. Nutrient Signaling to mTOR: Introduction

The target of rapamycin (TOR) was originally described in 1991 as a target protein of the antifungal and immunosuppressant agent rapamycin [1]. In the budding yeast *Saccharomyces cerevisiae*, there are two TOR genes—TOR1 and TOR2 [2]. TOR1 and TOR2 depletion or their inactivation by rapamycin/FKBP12 (FK506 Binding Protein 12) mimics the effects of nutrient deprivation resulting in an inhibition of protein synthesis, upregulation of autophagy, and induction of several stress-responsive transcription factors [3]. This led to the suggestion that TOR functionally mediates physiological processes in response to nutrient cues. In *S. cerevisiae*, TOR kinases exist in two complexes termed TOR Complex 1 (TORC1, that contains mainly TOR1 and a small amount of TOR2) and TOR Complex 2 (TORC2, that contains only TOR2). TORC1 and TORC2 have distinct cellular functions [4, 5] and locations [6], and both are essential for viability in *S. cerevisiae*. In *S. cerevisiae*, only TORC1 is sensitive to rapamycin and regulates temporal processes of growth in response to nutrient availability [7]. TORC2, unlike TORC1, is thought to be rapamycin-insensitive [5, 8]. However, at least in mammalian cells long-term exposure of cells in culture to rapamycin leads to inhibition of mTORC2 assembly and reduced mTORC2 activity in certain cell types [9]. In *S. cerevisiae*, TORC2 functions in spatial control of cell growth by promoting actin assembly and polarization [7], and although evidence for a conservation of this function for mTORC2 in mammals has been provided [8], the exact mechanism of how mTORC2 controls the actin cytoskeleton in metazoans is not clear. During development, mTOR is likely to primarily control growth, whereas in the adult, where there is relatively little further growth, mTOR may control aspects of nutrient-related physiology and, through an unknown process involving S6K1 [10, 11], restrict lifespan.

Unlike yeast, other eukaryotes contain only one TOR gene. The core mTORC1 proteins are composed of mTOR, the WD-40 repeat protein raptor and GβL [5, 12–14], while the core mTORC2 proteins are mTOR, rictor, Sin1, and GβL [5, 8, 15–17]. In yeast, TOR activity is regulated by the availability of nutrients such as a source of nitrogen (with the preferred nitrogen source being the amino acid glutamine) and a carbon source [3, 4].Similarly, mTORC1 activity and the phosphorylation and activity of its well-studied downstream effectors p70 S6 kinase (S6K1) and eukaryotic initiation factor eIF4E-binding protein 1 (4E-BP1) are regulated by amino acid levels in mammalian cells [18]. In addition, mitogens and hormones, such as insulin, or agents such as bacterial lipopolysaccharide (LPS, Virginie Mieulet and R. F. L., unpublished data) are insufficient to drive S6K1 activation in the absence of amino acids [18], indicating that mTORC1 integrates inputs from ambient amino acid nutrients and other stimuli. Insulin and other growth factors promote mTORC1 through mechanisms that converge on inhibiting the tuberous sclerosis heterodimeric complex (TSC1/TSC2), thereby stimulating GTP loading of the small GTPase Rheb (Ras homolog enriched in brain) [19–21]. In contrast, amino acid nutrients act on mTORC1 signaling via a distinct signaling pathway [22, 23]. TORC1 in *S. cerevisiae* responds to nutrients [24] despite the absence of TSC orthologs.

Although the sensing of amino acid nutrients appears key in single-celled organisms, whether there is any role for regulated changes in levels of amino acids exerting physiological consequences in metazoans is unclear. However, a recent study indicated that one potential pathological context for altered levels of particular amino acids may be in obesity. In obese humans a few amino acids, including the most critical mTORC1 activators BCAA (branched-chain amino acids; leucine, isoleucine, and valine) and arginine, were found to be elevated in blood plasma. In rats subject to a diet high in fat to mimic obesity, supplementation of BCAAs activated mTORC1 signaling in skeletal muscle resulting in reduced insulin sensitivity [25], potentially via the previously described negative feedback loop from mTORC1 to PI3K [26]. Thus an elevation of BCAA, at least in the context of a high-fat diet, may lead to hyperactive mTORC1 signaling. Aside from the BCAAs, altered levels of the other amino acid critical for mTORC1 activation, arginine (see Section III), has been of interest in a quite different setting, the innate immune response. In mice fasted of all food, only three amino acids were found to decrease significantly in plasma; arginine, alanine, and aspartic acid. Fasting was also found to inhibit activation of both the MAPK and mTORC1 pathway in liver macrophages (Kupffer cells) stimulated with LPS, a potent activator of TLR-mediated signaling to MAPKs and mTORC1, resulting in a blunted production of TNFα [27]. Restoration of arginine to fasted mice was found to stimulate both MAPK activation in Kupffer cells, and TNFα induction and mTORC1 activation (Virginie Mieulet and R. F. L., unpublished observations). Thus, decreased circulating levels of arginine that occur during periods of malnutrition may lead to a defective response of inflammatory cells to ligands that promote TLR signaling.

III. The Sensing of Amino Acid Nutrients

The mechanism by which amino acids regulate mTORC1 still remains an open question, despite extensive research efforts. The pioneering study by Avruch and colleagues first indicated that removal of particular amino acids from cell culture media differentially inhibited S6K1 activity [18]. Thus, removal of either leucine or arginine caused the greatest decline in S6K1 activity compared to removal of several other amino acids. Surprisingly however, in isolation, arginine and leucine were found to be unable to support S6K1 activity. Today, the mechanism underlying this necessary but insufficient requirement for both arginine and leucine remains unsolved. In considering this dual requirement, an important point to note is that the effects on mTORC1 signaling of depleting or supplementing these amino acids from the growth media are rapid. Removal of either leucine or arginine results in S6K1 T389 dephosphorylation within 5–10 min in HEK293T cells, and S6K1 is rephosphorylated within a similar time frame upon supplementation of the omitted amino acid (L. Y. and R. F. L, unpublished observations). A previous study in *Xenopus* oocytes indicated a further important mechanistic consideration. mTORC1 signaling to S6K1 was found to be increased by microinjection of leucine into oocytes that normally lack the system L leucine transporter [28], indicating that the effect of leucine is probably due to its intracellular abundance, rather than via a specific signaling event associated with transport into the cell. Thus, two conclusions can be made: (1) if the effect on mTORC1 is not due to the specific amino acid itself, but rather an amino acid metabolite, such a metabolite must change in abundance within a relatively short time frame; and (2) the sensing mechanism is likely to detect intracellular amino acid abundance. So how might this dual sensing of arginine and leucine occur? A relatively trivial explanation that cannot presently be discounted is that only one amino acid is actually sensed, and that when either leucine or arginine is removed, the other amino acid is not effectively transported into cells, thus lowering its intracellular abundance (Figure 4.1). Alternatively, one could invoke either independent sensors for arginine and leucine, or a common sensor able to detect the coincidence of both amino acids, with signaling to mTORC1 only occurring when both are present (Figure 4.1). In this latter scenario, one would require a putative sensor that is simultaneously able to bind both leucine and arginine.

Regarding a potential leucine sensor, a possible candidate has recently emerged. UBR1 and UBR2 are E3 ubiquitin ligases that recognize the identity of N-terminal residues and contribute to selective destabilization of target proteins. Both ligases were identified as leucine-binding proteins using a leucine-immobilized affinity approach [29]. Their overexpression was found to result in a reduction in mTOR-dependent S6K1 phosphorylation, whereas their depletion increased S6K1 phosphorylation in amino acid-starved HEK293T cells. *In vitro* leucine was found to bind to the substrate-recognition domain of UBR2 and inhibit degradation of N-end rule substrates. Thus, leucine might activate this pathway in part through inhibition of UBR1/2 ubiquitin ligase activity, although the identity of proteins targeted by these proteins to regulate mTORC1 activation remains unknown.

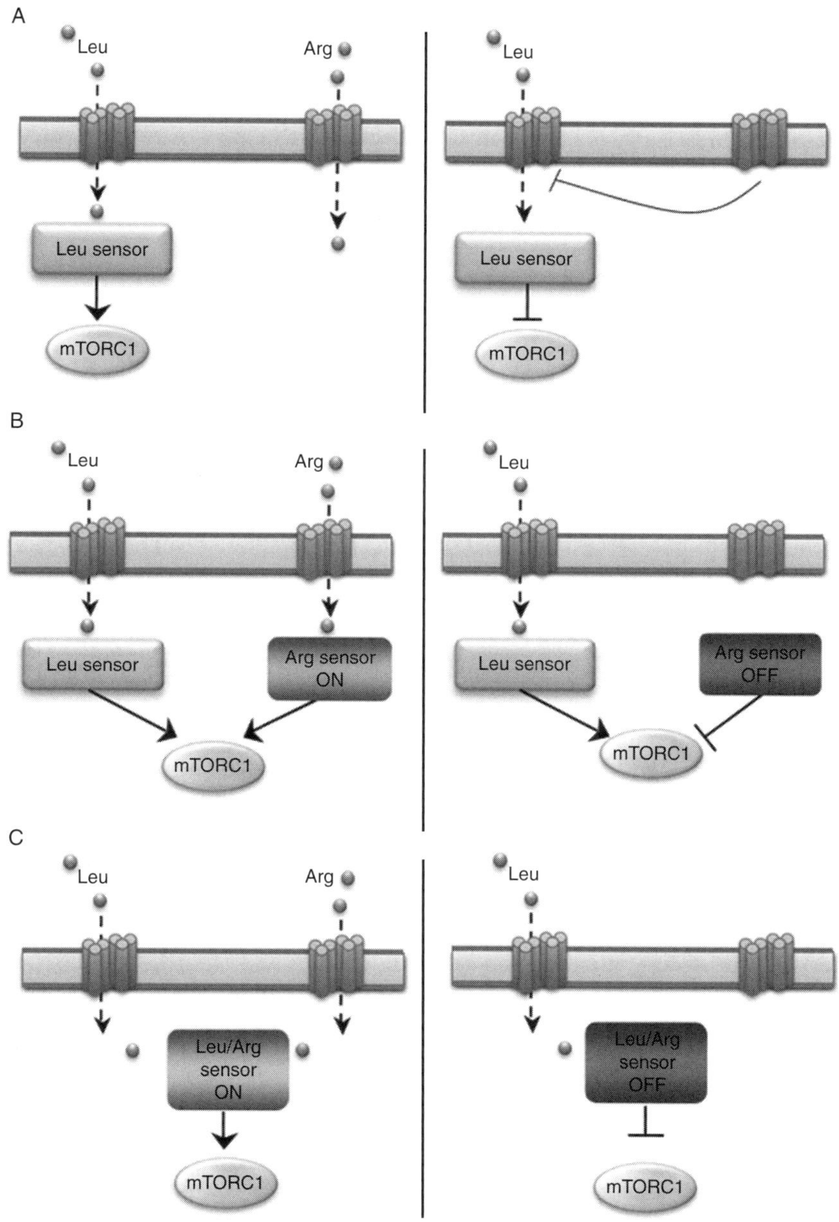

FIG. 4.1. Models to explain the dual requirement of arginine and leucine in activation of mTORC1. In each model, sufficiency of both leucine and arginine is depicted on the left, while deficiency of one amino acid, arginine (Arg) depicted on the right. For simplicity, in each model, deficiency of arginine is shown, although each model would apply equally with

Regarding why arginine and leucine in isolation from other amino acids are unable to fully support S6K1 T389 phosphorylation, there is currently no obvious explanation. However, one simple possibility is that once all amino acids are removed from tissue culture media, transport of leucine and/or arginine is prevented. This could occur at the level of the amino acid transporters, as at least one antiporter has been identified that imports leucine in exchange for intracellular glutamine [30].

IV. Amino Acid Transporters and mTOR Signaling

Amino acid deprivation results in a rapid decrease in the intracellular levels of essential amino acids (EAA), particularly the BCAA [31], indicating that intracellular pools of free amino acids are likely to be critical in activating mTORC1. Several different amino acid transporters are expressed in the plasma membrane of most cells [32], and a variety of evidence, including loss of function mutations in *Drosophila*, have linked them to mTORC1 ([33], Figure 4.2). The system L transporters, which transport amino acids into the cell in a Na+-independent manner [34], are the primary route for cellular entry of BCAAs and large aromatic side chain neutral amino acids (i.e., phenylalanine and tyrosine) in many mammalian cell types. The system L transporters are composed of heterodimers of an amino acid permeases (L-type amino acid transporter 1, 2, 3, and 4) and the 4F2hc/CD98 glycoprotein [35–38]. Among amino acids, changes in leucine levels are sufficient to regulate the phosphorylation state and activity of both downstream components of the mTORC1 pathway [18, 39], suggesting that import of leucine into the cytosol can be a critical determining factor. A recent study indicated that the uptake of leucine to regulate mTORC1 in some cells is via one such L amino acid transporter termed SLC7A5-SLC3A2, a bidirectional transporter that regulates the simultaneous efflux of L-glutamine and influx of leucine (and other EAA) into cells [30]. Although intriguing, the generality of this mode of regulating mTORC1 by glutamine efflux/leucine influx *in vivo* remains to be determined.

---

deficiency in leucine, and a reciprocal effect on the arginine transporter or sensor. (A) Only a single sensor for leucine is present acting upstream of mTORC1. Deficiency in arginine would then exert an inhibitory effect on mTORC1 by inhibiting leucine transport at the level of the leucine transporter. A reduction in intracellular free leucine would then be sensed, inhibiting a positive signal to mTORC1. (B) Independent sensors for leucine and arginine are depicted. Deficiency in arginine would exert an inhibitory effect on mTORC1 because both sensors are required to be engaged to activate mTORC1. (C) A single dual sensor for leucine and arginine is depicted that must simultaneously be engaged by both amino acids in order to activate mTORC1.

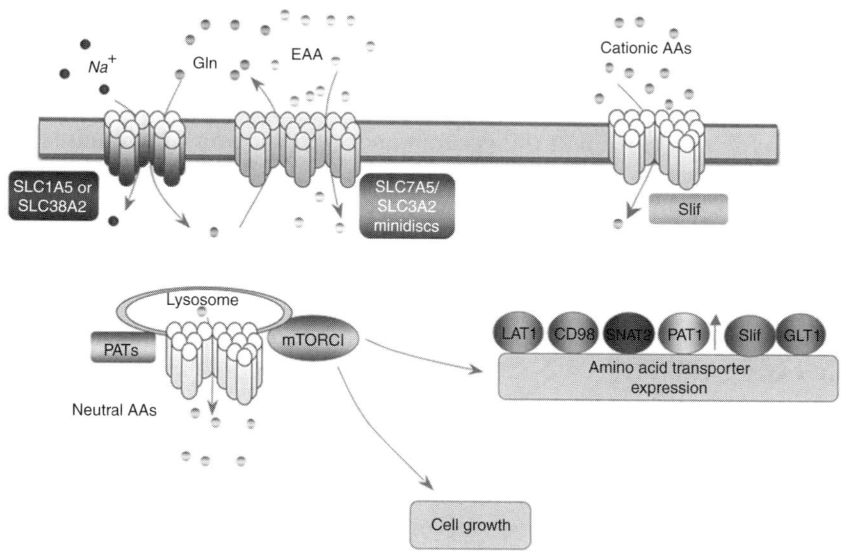

FIG. 4.2. Amino acid transporters implicated in TOR activation. Depicted are amino acid transporters implicated in activation of mTORC1 and/or growth. Plasmamembrane transporters for glutamine SLC1A5 (or SLC38A2/SNAT2), leucine/EAA (SLC7A5/SLC3A2, Minidiscs), and cationic amino acids (Slif), as well as lysosomal transporters for neutral amino acids (PATs) are shown acting upstream of mTORC1. For simplicity, active mTORC1 is depicted associated with lysosomes in accord with the model described by Sancak and colleagues [41]. Downstream of mTOR amino acid transporters shown to be altered in expression are depicted, LAT1, CD98, SNAT2, PAT1, Slif, and the glutamate transporter GLT1 [56, 57, 92].

The recent suggestion that amino acid sufficiency may require the relocalization of mTORC1 to lysosomal membranes [40, 41] (see Section V) has implicated lysosomes and their yeast counterparts, vacuoles [42] in mTORC1 activation. The PATs (proton-assisted amino acid transporters) were initially identified on the surface of mammalian lysosomes [43, 44], and later also found at the plasma membrane and in endosomal compartments [44, 45]. Two PAT-related transporters in *Drosophila*, CG3424 and CG1139/*Path*, have been shown to mediate growth in flies [46]. CG1139 specifically transports alanine, glycine, and proline, amino acids that do not appear to significantly promote mTORC1 signaling, at least in cell culture [18], while *Path*, although it has poor transport properties, was found able to activate mTORC1 in a heterologous *Xenopus* oocyte system in response to alanine microinjection. Although the mechanistic basis for these effects on mTORC1 signaling remain to be established, both transporters interact genetically with dTOR and other InR signaling components in flies [46], while two PAT transporters have been found to be able to promote mTORC1 signaling in cultured human cells [47], indicating that they may in some manner modulate mTORC1. PATs are related to AVTs (amino acid vacuolar transporters), which can transport amino acids out of the vacuole [48]. Interestingly, the *S. cerevisiae* orthologs of Rags, Gtr1p and Gtr2p [49], are also found localized at the vacuole and were shown to be required for proper shuttling of the amino acid permease Gap1p to the plasmamembrane [50, 51], although this has recently been called into question [24]. The presence of PATs on the surface of lysosomes/vacuoles, together with their relatively poor transport properties, has suggested to some that they may act as amino acid sensors (transceptors), rather than transporters [51]. In *S. cerevisiae*, an amino acid permease-like membrane protein called Ssy1p and two other proteins provide a precedent for this possibility as they are known to form a complex that senses external amino acid concentration and transmit an intracellular signal leading to transcriptional regulation [52]. However, to our knowledge no functionally analogous sensing complex has been identified in metazoans. It would be interesting therefore to determine whether by localizing mTORC1 to lysosomes—a manipulation that activates mTOR in the absence of extracellular amino acids (see Section V.C [41])—lysosomal PAT transporters are providing the signal to activate mTORC1.

Two additional amino acid transporters have been shown to regulate growth and/or mTOR signaling in flies (Figure 4.2). *Minidiscs* (a homologue of the CD98 light chains LAT1 and -2) is involved in leucine transport [53, 54]), while *Slif* (Slimfast) is a cationic amino acid transporter that may have important functions in regulating overall body size by regulating mTORC1 signaling in the *Drosophila* fat body [55]. These transporters are likely to be involved in transporting the two critical amino acids involved in activating mTORC1, leucine and arginine.

In addition to promoting mTORC1 activation, there is some evidence that amino acid transporters are also downstream targets of the mTOR signaling pathway. The surface expression of several transporters associated with mTORC1 signaling (LAT1, CD98, SNAT2, and PAT1) is rapidly and transiently upregulated following EAA ingestion in humans [56], while in *Drosophila* dTOR has been shown to promote surface expression of *Slif* [57]. A decrease in the intracellular amino acid pool over a relatively long time frame activates the General Control Nonrepressed (GCN2) pathway, resulting in an increased expression of the activating transcription factor 4 (ATF4) [58]. ARF4 has been found to also upregulate amino acid transporters such as SNAT2 [59, 60] apparently to enable amino acid efflux during conditions associated with muscle catabolism [61, 62], indicating that these changes may serve as a cellular adaptation to improve amino acid transport.

V. Evidence That Intracellular Signaling Molecules Relay the Presence of Amino Acid Sufficiency to mTORC1

A. Rheb GTPase

The dominant positive regulator of mTORC1 is the GTP-charged form of the Ras-like GTPase Rheb. In a model suggested by Sancak and coworkers, amino acids regulate mTORC1 by controlling the ability of Rheb-GTP to activate mTORC1 [63]. Overexpressed Rheb binds directly to mTOR at the carboxyterminal lobe of the mTOR catalytic domain and has been shown to somewhat stimulate its weak kinase activity measured *in vitro* [64, 65] and promote substrate recruitment via raptor [66]. However, whether Rheb at endogenous levels is able to bind to mTOR and stimulate its activity is open to question. Thus far an interaction between Rheb and mTOR has not been demonstrated at endogenous levels during stimulation of mTORC1 by growth factor signaling. However, although amino acid withdrawal can lead to a decrease in Rheb-GTP levels in some cell types [67], Rheb-GTP levels do not drop in TSC1- or TSC2-null cells following amino acid withdrawal despite S6K1 T389 dephosphorylation, indicating that a separate pathway promoted by amino acid sufficiency activates mTORC1 independently of Rheb-GTP levels [22, 67].

B. hVps34

The Class III PI 3-kinase, hVps34 (Vacuolar Protein Sorting 34), has also been suggested to regulate mTOR, S6K1, and 4E-BP1, and to be a critical component of the nutrient sensing apparatus [22, 68]. This conclusion was derived from experiments in which siRNA knockdown of hVps34 blocked amino acid- and insulin-induced S6K1 activation, but had no effect on PKB activation [22, 68]. hVps34 did not appear to be a part of the insulin input to S6K1, as its activity was not stimulated by insulin, and inhibition of hVps34 had no effect on phosphorylation of Akt or TSC2 in insulin-stimulated cells [68]. In addition, stimulation of cells with amino acids led to an increase in the concomitant production of PI3P (phosphatidylinositol-3-phosphate), the product of hVps34 [22, 68]. PI3P levels were found to decrease in cells after amino acid deprivation, whereas readdition of amino acids to amino acid-starved cells led to increased PI3P levels [22]. In addition, hVps34 is also inhibited by activation of the AMP-activated kinase, which inhibits mTOR/S6K1 in glucose-starved cells, indicating hVps34 activity was also regulated by energy as well as by amino acids [68]. However, a general role of hVps34 in mTORC1 activation by amino acids in metazoans has been cast in doubt by the finding that a loss of function of *Drosophila* dVps34 does not apparently lead to inhibition of mTORC1 signaling [69].

C. Rag GTPase Heterodimers

Two independent teams have shown that the Rag proteins, a family of four related small GTPases in mammals, promote mTORC1 activation by amino acids [70, 71]. The Rag GTPases are highly conserved from yeast to mammals. Similar to the mammalian Rag GTPases, the yeast Rag GTPase homolog Gtr1, a component of the vacuolar-membrane-associated EGO complex (EGOC), interacts with and activates TORC1 in an amino acid-sensitive manner. Expression of a constitutively active (GTP-bound) $Gtr1^{GTP}$, which interacts strongly with TORC1, rendered TORC1 partially resistant to leucine deprivation, whereas expression of a growth inhibitory, GDP-bound $Gtr1^{GDP}$, caused constitutively low TORC1 activity. Vam6 (Vacuolar Morphogenesis protein 6) was also identified as a conserved guanine nucleotide exchange factor (GEF) for Gtr1 and controls TORC1 function by promoting GTP loading of the Gtr1 subunit of the EGOC [24]. The mammalian Rag GTPases interact with mTORC1 and are proposed to activate it in response to amino acids by promoting mTORC1 translocation to a membrane-bound compartment that contains the mTORC1 activator, Rheb [71]. The noticeable difference between yeast and mammalian cells is that amino acids apparently do not alter the subcellular localization of Tor1 and Tco89 (a TORC1 component) in yeast. The subcellular localization of Gtr1 and Gtr2 is not affected by amino acids, suggesting the proposed mechanism of TORC1 activation via the amino acid-induced subcellular localization of TORC1 is not conserved [72]. However, how the GTP loading of Rag proteins are regulated in other eukaryotes outwith *S. cerevisiae* is not currently known.

Two recent studies indicate that amino acid sufficiency may induce the movement of mTORC1 to lysosomal membranes [40, 41], where the Rag proteins reside [41]. A complex encoded by MAPKSP1 (MP1), ROBLD3 (p14), and c11orf59 (p18) (termed the "Ragulator") was found to interact with the Rag GTPases, recruiting them to lysosomes. These proteins had previously been implicated in signaling from late endosomes/lysosomes, particularly to the MAPK pathway [73–75], and more generally in membrane trafficking, particularly the biogenesis of lysosomes [76], and trafficking of the EGF receptor [74]. Given these prior results, it is presently difficult to discount the possibility that the signaling defect in amino acid-induced mTORC1 activation when Ragulator components are deleted might reflect a more general defect in membrane trafficking events that impact upon multiple signaling pathways, including activation of mTORC1. However, it appears that at least one signaling pathway, serum-induced activation of Akt/PKB is normal in p14 and p18 null cells [41]. Further supporting the notion that the lysosome is a critical location for activation of mTORC1 however, constitutive targeting of mTORC1 to the lysosomal surface, but not the plasmamembrane, by fusing raptor to the C-terminal region of Rheb, also appears sufficient to render the mTORC1 pathway amino acid insensitive and independent of Rag and Ragulator, but not Rheb [41]. Thus, Rag–Ragulator-mediated translocation of mTORC1 to lysosomal membranes has been argued to be an important event in amino acid signaling to mTORC1 [41]. However, what is less clear is whether targeting to lysosomes is permissive for mTORC1 activation as a consequence of locally concentrating mTORC1 on an internal endomembrane, or whether there is some specific characteristic of lysosomal membranes that promote mTORC1 activation. Similarly, Rheb—the proposed activator of mTORC1 on lysosomes—exists predominantly in the soluble (nonmembraneous) fraction at least in HEK293T cells [77]. Thus, why Rheb might be able to activate mTORC1 only when mTORC1 is associated with the lysosomal membrane is currently unclear.

VI. MAP4K3 Participates in Amino Acid Signaling and Maintenance of Cell Size

Recently, the finding that MAP4K3 (mitogen-activated protein kinase kinase kinase kinase 3, also known as GLK, germinal centre-like kinase) mediates amino acid signaling to mTORC1 has provided some new insights into the amino acid regulation of mTORC1 [23, 77, 78]. MAP4K3 was identified as a regulator of mTORC1 through an RNA interference screen of *Drosophila* protein kinases required for signaling to dS6K. BLAST homology searches revealed that the closest human ortholog of *Drosophila CG7097* is MAP4K3. Subsequent siRNA and overexpression experiments revealed that MAP4K3 is required for maximal mTORC1-dependent S6K/4E-BP1 phosphorylation and a regulator of cell growth, at least in cell culture [23]. MAP4K3 is a member of the *Ste20*-related family, named after Sterile 20 involved in pheromone signaling in *S. cerevisiae* [79]. There are about 30 Ste20-related kinases in mammals [80] in addition to homologs in *Drosophila*, *Caenorhabditis elegans*, and other organisms. They are characterized by the presence of a conserved kinase domain together with a noncatalytic region that enables these kinases to interact with various signaling molecules and regulatory proteins [81]. Phylogenetic analysis reveals that the *Ste20*-related family are classified into two major subgroups including germinal center kinases (GCKs) and p21-activated kinases (PAKs), and MAP4K3 lies within the GCK-1 subfamily where its closest relatives are GCK and HPK1 (hematopoietic progenitor kinase 1) [81]. However, the ubiquitous tissue distribution of MAP4K3 differs from that of GCK and HPK1 [82].

Although the relevant substrates of MAP4K3 are not currently known, MAP4K3 activity was found to be positively regulated by amino acid sufficiency, suggesting that it might be involved in amino acid sensing processes to control cell growth [23]. Further studies identified a phosphorylation site at Ser170 in the kinase activation loop of MAP4K3 [77]. The phosphorylation of Ser170 paralleled MAP4K3 activation and mutation of Ser170 to Ala abolished MAP4K3 activity and impaired mTORC1-mediated phosphorylation of S6K1. Importantly, MAP4K3 Ser170 phosphorylation was rapidly inhibited by removal of all amino acids from the growth medium and was stimulated by their readdition, but was unaffected by stimulation of insulin. Phosphorylation of MAP4K3 Ser170 likely proceeds via a transautophosphorylation event catalyzed by MAP4K3 itself, as an inactive DFG-AFG mutant of the kinase is only poorly phosphorylated at Ser170, presumably because endogenous MAP4K3 is only poorly able to phosphorylate the inactive kinase. Although the presence of all amino acids promotes autophosphorylation and therefore promotes MAP4K3 activity, it is not known whether amino acids critical for activation of mTORC1 signaling, arginine or leucine, similarly affect MAP4K3 activity.

In addition to an N-terminal kinase domain and a C-terminal citron homology (CNH) domain, MAP4K3 contains tandem polyproline motifs in its central region (Figure 4.3). The multiple proline-rich domains, which are putative SH3 binding sites, are likely to be involved in protein–protein interaction, and indeed at least one SH3-containing protein, endophilin, has been shown to bind to MAP4K3 [83]. However, the functional significance of this binding in terms of activation of mTORC1 is not known. Since Rag

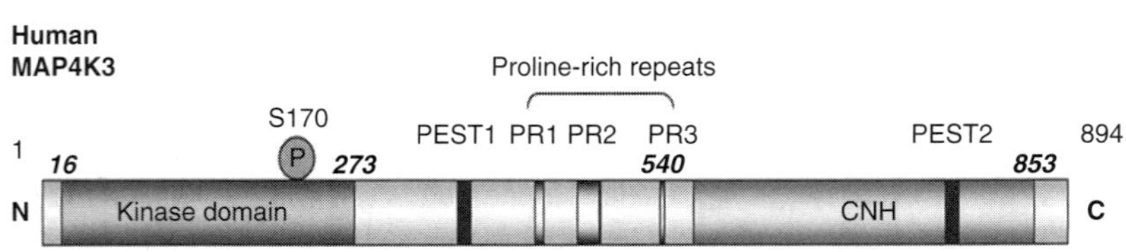

FIG. 4.3. Structural profile of human MAP4K3. MAP4K3 contains an N-terminal serine/threonine kinase domain (aa 16–273), a C-terminal citron homology domain (CNH, aa 540–853, pfam00780), three proline-rich repeats (PR1, PR2, and PR3) that are the likely endophilin I-binding site (aa 475–480 [83]), and also the likely PP2AT61ε interaction area (amino acids 431–540 [77]). Indicated are two putative PEST (proline, glutamic acid, serine, and threonine) sequences that likely act as a signal for MAP4K3 degradation, and Ser170 in the catalytic domain which when dephosphorylated is involved in MAP4K3 activity and activation of mTORC1 [77]. The amino acid position numbers of human MAP4K3 that define each domain are indicated.

GTPases are also involved in amino acid signaling to mTORC1 [70, 71], the relationship between Rags and MAP4K3 has been investigated. Although it is not known whether MAP4K3 plays a role in organ growth or body size in higher organisms, a recent study indicates that MAP4K3 plays an important role in growth in *Drosophila* [83]. A P-element insertion was identified in flies that significantly inhibited dMAP4K3 expression. The resulting dMAP4K3 mutant exhibited several defects reminiscent of dTOR mutants, including reduced lipid levels, larval growth rate and cell size in the wing, along with reduced signaling to dS6K and d4E-BP. Thus, dMAP4K3 appears to be a bone fide regulator of signaling to TORC1 in *Drosophila*. In *Drosophila*, expression of active RagA (Q61L) in a MAP4K3 knockdown mutant background was still able to induce tissue overgrowth [84], while suppression of Rag C and Rag D in human cells was found to impair the ability of MAP4K3 to activate mTORC1 [77]. Thus, Rag proteins appear to act genetically downstream of MAP4K3. In *Drosophila*, dMAP4K3 interacts with dRagC, however this does not appear to be the case with the human proteins, based upon the failure of MAP4K3 to coimmunoprecipitate human RagC. Similarly, in mammalian cells, Rag A, B, or C does not appear to be direct substrates of MAP4K3 [77]. Therefore, at present, the exact relationship between MAP4K3 and Rag GTPases in activation of mTORC1 remains unknown.

Although the mechanism by which amino acids signal to mTORC1 via MAP4K3 is not clear, an investigation of how MAP4K3 is regulated by amino acid sufficiency led to the identification of an interesting new regulator of cell growth. MAP4K3 was found to interact with a targeting subunit of the PP2A phosphatase, PR61ε, by mass spectrometry [77]. PR61ε protein localizes to the cytoplasm [85] and can be associated with PP2A core dimers, consisting of a structural subunit (PP2A-A, PR65) and a catalytic subunit (PP2A-C). The regulatory subunit of PP2A is thought to be critical for specific substrate recognition [86], and indeed suppression of PR61ε led to defective dephosphorylation of phosphorylated Ser170 during amino acid withdrawal. Further study showed that overexpression of PR61ε also inhibited both MAP4K3 Ser170 phosphorylation and the ability of MAP4K3 to activate mTORC1-S6K1. Finally, suppression of PR61ε also impaired mTORC1 inhibition (dephosphorylation of S6 at Ser240/244) after amino acid withdrawal, supporting the idea that PP2AT61ε is a negative modulator of mTORC1 via dephosphorylation of MAP4K3 at Ser170. This view was strengthened by the finding that suppression of PR61ε in HEK293 cells caused an increase in cell size. How are PP2AT61ε and MAP4K3 however involved in regulation of amino acids sufficiency upstream of mTORC1? PP2AT61ε was found to preferentially interact with MAP4K3 after amino acid withdrawal leading to more rapid Ser170 dephosphorylation and mTORC1-mediated dephosphorylation of S6K1.

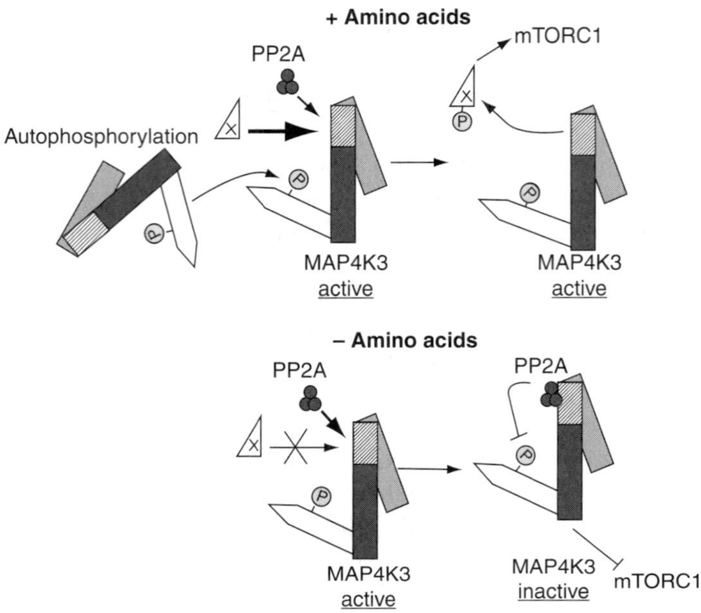

FIG. 4.4 Model of amino acid regulation of MAP4K3 and activation of mTORC1. In the presence of amino acids (+Amino acids), MAP4K3 is transautophosphorylated at Ser170 in the kinase activation segment, by a presumed transient interaction between MAP4K3 polypeptides, causing MAP4K3 activation. When active, MAP4K3 is able to phosphorylate a putative substrate involved in activation of mTORC1 (depicted as a triangle, X-P). The interaction of this substrate with MAP4K3 is favored (upper, thick arrow) over binding to an overlapping region of MAP4K3 by PP2A (upper, thin arrow). When amino acids are depleted from tissue culture media (–Amino acids), the putative substrate is no longer able to bind and be phosphorylated by MAP4K3 (lower, thin arrow), allowing PP2A to efficiently bind MAP4K3 (lower, thick arrow) and dephosphorylate Ser170, leading to MAP4K3 inhibition. The dephosphorylated substrate is then predicted to be incapable of supporting mTORC1 activation [77].

A truncation (MAP4K3 aa 1–431) exhibiting the least binding to PP2AT61ε was not inhibited by amino acid withdrawal but was impaired in activation of mTORC1 signaling. Thus, the C-terminal region of MAP4K3 is important both for binding PP2A and activation of mTORC1 (Figure 4.4).

VII. MAP4K3 Promotes Apoptosis via Regulation of the BH3-Only Proteins

An important additional function for MAP4K3 may be to induce apoptosis [87, 88]. Apoptosis can be initiated either by activation of death receptors on the cell surface membranes (extrinsic pathway) or through a series of cellular events primarily processed at mitochondria (intrinsic pathway) [89]. Apoptosis induced by MAP4K3 intersects both the intrinsic pathway and extrinsic cell death pathways.

Lam et al. first demonstrated that MAP4K3 is a potent apoptosis inducer via a screen to identify novel modulators of the intrinsic pathway of apoptosis. MAP4K3 was known to be activated by exposure of cells to UV irradiation [82], an activator of the intrinsic pathway of apoptosis. The proapoptotic BH3-only proteins, which are members of the BCL-2 family, trigger the intrinsic pathway as sensors of apoptotic signals. PUMA, BAD, BIM, and BID, which comprise a subset of BH3-only proteins, lead to the activation of proapoptotic BCL2 family members, BAX and BAK, that are thought to be critical for the permeabilization of the outer mitochondrial membrane and induction of downstream apoptotic execution routines.

The levels of active BAX were found to increase upon overexpression of MAP4K3, while the levels of active BAX induced following UV-mediated induction of cell death were reduced upon MAP4K3 suppression. This study further showed that overexpression of MAP4K3 increases expression of the BH3-only proteins PUMA and BAD, an upregulation that seems to require MAP4K3 kinase activity and was rapamycin-sensitive [87]. Thus, the mTORC1 pathway may act downstream of MAP4K3 to promote upregulation of PUMA and BAD. This view was strengthened by the finding that overexpression of MAP4K3 while suppressing *Raptor* partially suppressed the induction of PUMA and BAD and resulted in a rescue of MAP4K3-induced cell death. Notably, MAP4K3 signaling through mTORC1 resulted in increased expression of PUMA and BAD through a posttranscriptional mechanism rather than at transcription level.

In parallel, overexpression of MAP4K3 resulted in a significant increase in the level of phospho-BIM (Ser69) that was partially suppressed by inhibition of JNKs (c-Jun N-terminal kinases), suggesting that JNKs might signal through MAP4K3 leading to the phosphorylation and activation of BIM. The direct activation of BAX by BIM, together with the neutralization of BCL-XL (a member of proapoptotic BCL2 family) by PUMA or BAD, may then be capable of promoting the activation of the intrinsic apoptosis pathway [89].

MAP4K3 can also be activated by the proinflammatory cytokine TNFα [82], which can activate the death receptor (extrinsic) pathway of apoptosis. The extrinsic pathway of apoptosis intersects the intrinsic pathway through the caspase-8-mediated proteolytic processing of the BH3-only protein BID to tBID [90, 91]. Subsequently, tBID is capable of translocating to mitochondria and enhancing executioner caspase activation through the intrinsic pathway.

Overexpression of MAP4K3 also results in an increase in the levels of BID protein, which appears to require MAP4K3 kinase activity but is rapamycin-insensitive [88], suggesting that MAP4K3 may play a role in the regulation of BID in an mTORC1-independent manner. In conclusion, it is clear that under some circumstances MAP4K3 can regulate cell death; however, the physiological significance of this regulation relative to the regulation of cell growth by MAP4K3 remains to be established.

Acknowledgments

The authors thank Virginie Mieulet for discussion and acknowledge grant support from the Alberta Cancer Foundation, the Tuberous Sclerosis Association (UK), and Cancer Research (UK).

References

1. Heitman, J., Movva, N.R., and Hall, M.N. (1991). Targets for cell cycle arrest by the immunosuppressant rapamycin in yeast. *Science* 253:905–909.
2. Helliwell, S.B., Wagner, P., Kunz, J., Deuter-Reinhard, M., Henriquez, R., and Hall, M.N. (1994). TOR1 and TOR2 are structurally and functionally similar but not identical phosphatidylinositol kinase homologues in yeast. *Mol Biol Cell* 5:105–118.
3. Wullschleger, S., Loewith, R., and Hall, M.N. (2006). TOR signaling in growth and metabolism. *Cell* 124:471–484.
4. Helliwell, S.B., Howald, I., Barbet, N., and Hall, M.N. (1998). TOR2 is part of two related signaling pathways coordinating cell growth in *Saccharomyces cerevisiae. Genetics* 148:99–112.
5. Loewith, R., et al. (2002). Two TOR complexes, only one of which is rapamycin sensitive, have distinct roles in cell growth control. *Mol Cell* 10:457–468.
6. Sturgill, T.W., Cohen, A., Diefenbacher, M., Trautwein, M., Martin, D.E., and Hall, M.N. (2008). TOR1 and TOR2 have distinct locations in live cells. *Eukaryot Cell* 7:1819–1830.
7. Martin, D.E., and Hall, M.N. (2005). The expanding TOR signaling network. *Curr Opin Cell Biol* 17:158–166.
8. Jacinto, E., Loewith, R., Schmidt, A., Lin, S., Ruegg, M.A., Hall, A., and Hall, M.N. (2004). Mammalian TOR complex 2 controls the actin cytoskeleton and is rapamycin insensitive. *Nat Cell Biol* 6:1122–1128.
9. Zeng, Z., et al. (2007). Rapamycin derivatives reduce mTORC2 signaling and inhibit AKT activation in AML. *Blood* 109:3509–3512.
10. Harrison, D.E., et al. (2009). Rapamycin fed late in life extends lifespan in genetically heterogeneous mice. *Nature* 460:392–395.
11. Selman, C., et al. (2009). Ribosomal protein S6 kinase 1 signaling regulates mammalian life span. *Science* 326:140–144.
12. Hara, K., et al. (2002). Raptor, a binding partner of target of rapamycin (TOR), mediates TOR action. *Cell* 110:177–189.
13. Kim, D.H., Sarbassov, D.D., Ali, S.M., King, J.E., Latek, R.R., Erdjument-Bromage, H., Tempst, P., and Sabatini, D.M. (2002). mTOR interacts with raptor to form a nutrient-sensitive complex that signals to the cell growth machinery. *Cell* 110:163–175.

14. Kim, D.H., Sarbassov, D.D., Ali, S.M., Latek, R.R., Guntur, K.V., Erdjument-Bromage, H., Tempst, P., and Sabatini, D.M. (2003). GbetaL, a positive regulator of the rapamycin-sensitive pathway required for the nutrient-sensitive interaction between raptor and mTOR. *Mol Cell* 11:895–904.
15. Sarbassov, D.D., Ali, S.M., Kim, D.H., Guertin, D.A., Latek, R.R., Erdjument-Bromage, H., Tempst, P., and Sabatini, D.M. (2004). Rictor, a novel binding partner of mTOR, defines a rapamycin-insensitive and raptor-independent pathway that regulates the cytoskeleton. *Curr Biol* 14:1296–1302.
16. Yang, Q., Inoki, K., Ikenoue, T., and Guan, K.L. (2006). Identification of Sin1 as an essential TORC2 component required for complex formation and kinase activity. *Genes Dev* 20:2820–2832.
17. Jacinto, E., et al. (2006). SIN1/MIP1 maintains rictor–mTOR complex integrity and regulates Akt phosphorylation and substrate specificity. *Cell* 127:125–137.
18. Hara, K., Yonezawa, K., Weng, Q.P., Kozlowski, M.T., Belham, C., and Avruch, J. (1998). Amino acid sufficiency and mTOR regulate p70 S6 kinase and eIF-4E BP1 through a common effector mechanism. *J Biol Chem* 273:14484–14494.
19. Crino, P.B., Nathanson, K.L., and Henske, E.P. (2006). The tuberous sclerosis complex. *N Engl J Med* 355:1345–1356.
20. Inoki, K., Corradetti, M.N., and Guan, K.L. (2005). Dysregulation of the TSC–mTOR pathway in human disease. *Nat Genet* 37:19–24.
21. Kwiatkowski, D.J., and Manning, B.D. (2005). Tuberous sclerosis: a GAP at the crossroads of multiple signaling pathways. *Hum Mol Genet* 14(Spec. No. 2)**:**R251–R258.
22. Nobukuni, T., et al. (2005). Amino acids mediate mTOR/raptor signaling through activation of class 3 phosphatidylinositol 3OH-kinase. *Proc Natl Acad Sci USA* 102:14238–14243.
23. Findlay, G.M., Yan, L., Procter, J., Mieulet, V., and Lamb, R.F. (2007). A MAP4 kinase related to Ste20 is a nutrient-sensitive regulator of mTOR signalling. *Biochem J* 403:13–20.
24. Binda, M., Peli-Gulli, M.P., Bonfils, G., Panchaud, N., Urban, J., Sturgill, T.W., Loewith, R., and De Virgilio, C. (2009). The Vam6 GEF controls TORC1 by activating the EGO complex. *Mol Cell* 35:563–573.
25. Newgard, C.B., et al. (2009). A branched-chain amino acid-related metabolic signature that differentiates obese and lean humans and contributes to insulin resistance. *Cell Metab* 9:311–326.
26. Harrington, L.S., Findlay, G.M., and Lamb, R.F. (2005). Restraining PI3K: mTOR signalling goes back to the membrane. *Trends Biochem Sci* 30:35–42.
27. Mieulet, V., Yan, L., Choisy, C., Sully, K., Procter, J., et al. (2010). TPL-2 activation of MAPK downstream of TLR4 is coupled to arginine availibility. *Sci Signal* 3, ra61.
28. Christie, G.R., Hajduch, E., Hundal, H.S., Proud, C.G., and Taylor, P.M. (2002). Intracellular sensing of amino acids in *Xenopus laevis* oocytes stimulates p70 S6 kinase in a target of rapamycin-dependent manner. *J Biol Chem* 277:9952–9957.
29. Kume, K., Iizumi, Y., Shimada, M., Ito, Y., Kishi, T., Yamaguchi, Y., and Handa, H. (2010). Role of N-end rule ubiquitin ligases UBR1 and UBR2 in regulating the leucine-mTOR signaling pathway. *Genes Cells* 15:339–349.
30. Nicklin, P., et al. (2009). Bidirectional transport of amino acids regulates mTOR and autophagy. *Cell* 136:521–534.
31. Dennis, P.B., Jaeschke, A., Saitoh, M., Fowler, B., Kozma, S.C., and Thomas, G. (2001). Mammalian TOR: a homeostatic ATP sensor. *Science* 294:1102–1105.
32. Nishimura, M., and Naito, S. (2008). Tissue-specific mRNA expression profiles of human solute carrier transporter superfamilies. *Drug Metab Pharmacokinet* 23:22–44.

33. Taylor, P.M. (2009). Amino acid transporters: eminences grises of nutrient signalling mechanisms? *Biochem Soc Trans* 37:237–241.
34. Christensen, H.N. (1990). Role of amino acid transport and countertransport in nutrition and metabolism. *Physiol Rev* 70:43–77.
35. Kanai, Y., Segawa, H., Miyamoto, K., Uchino, H., Takeda, E., and Endou, H. (1998). Expression cloning and characterization of a transporter for large neutral amino acids activated by the heavy chain of 4F2 antigen (CD98). *J Biol Chem* 273:23629–23632.
36. Segawa, H., Fukasawa, Y., Miyamoto, K., Takeda, E., Endou, H., and Kanai, Y. (1999). Identification and functional characterization of a Na+-independent neutral amino acid transporter with broad substrate selectivity. *J Biol Chem* 274:19745–19751.
37. Babu, E., et al. (2003). Identification of a novel system L amino acid transporter structurally distinct from heterodimeric amino acid transporters. *J Biol Chem* 278:43838–43845.
38. Bodoy, S., Martin, L., Zorzano, A., Palacin, M., Estevez, R., and Bertran, J. (2005). Identification of LAT4, a novel amino acid transporter with system L activity. *J Biol Chem* 280:12002–12011.
39. Lynch, C.J., Fox, H.L., Vary, T.C., Jefferson, L.S., and Kimball, S.R. (2000). Regulation of amino acid-sensitive TOR signaling by leucine analogues in adipocytes. *J Cell Biochem* 77:234–251.
40. Flinn, R.J., Yan, Y., Goswami, S., Parker, P.J., and Backer, J.M. (2010). The late endosome is essential for mTORC1 signaling. *Mol Biol Cell* 21:833–841.
41. Yasemin Sancak, L.B.-P., Roberto Zoncu, L.B.-P., Markhard, A.L., Shigeyuki Nada, A.L., and Sabatini, D.M. (2010). Ragulator–Rag complex targets mTORC1 to the lysosomal surface and is necessary for its activation by amino acids. *Cell* 141:290–303.
42. Li, S.C., and Kane, P.M. (2009). The yeast lysosome-like vacuole: endpoint and crossroads. *Biochim Biophys Acta* 1793:650–663.
43. Sagne, C., Agulhon, C., Ravassard, P., Darmon, M., Hamon, M., El Mestikawy, S., Gasnier, B., and Giros, B. (2001). Identification and characterization of a lysosomal transporter for small neutral amino acids. *Proc Natl Acad Sci USA* 98:7206–7211.
44. Boll, M., Foltz, M., Rubio-Aliaga, I., Kottra, G., and Daniel, H. (2002). Functional characterization of two novel mammalian electrogenic proton-dependent amino acid cotransporters. *J Biol Chem* 277:22966–22973.
45. Rubio-Aliaga, I., Boll, M., Vogt Weisenhorn, D.M., Foltz, M., Kottra, G., and Daniel, H. (2004). The proton/amino acid cotransporter PAT2 is expressed in neurons with a different subcellular localization than its paralog PAT1. *J Biol Chem* 279:2754–2760.
46. Goberdhan, D.C., Meredith, D., Boyd, C.A., and Wilson, C. (2005). PAT-related amino acid transporters regulate growth via a novel mechanism that does not require bulk transport of amino acids. *Development* 132:2365–2375.
47. Heublein, S., Kazi, S., Ogmundsdottir, M.H., Attwood, E.V., Kala, S., Boyd, C.A., Wilson, C., and Goberdhan, D.C. (2010). Proton-assisted amino-acid transporters are conserved regulators of proliferation and amino-acid-dependent mTORC1 activation. *Oncogene*. 29:4068–4079.
48. Russnak, R., Konczal, D., and McIntire, S.L. (2001). A family of yeast proteins mediating bidirectional vacuolar amino acid transport. *J Biol Chem* 276:23849–23857.
49. Nakashima, N., Noguchi, E., and Nishimoto, T. (1999). *Saccharomyces cerevisiae* putative G protein, Gtr1p, which forms complexes with itself and a novel protein designated as Gtr2p, negatively regulates the Ran/Gsp1p G protein cycle through Gtr2p. *Genetics* 152:853–867.
50. Gao, M., and Kaiser, C.A. (2006). A conserved GTPase-containing complex is required for intracellular sorting of the general amino-acid permease in yeast. *Nat Cell Biol* 8:657–667.

51. Goberdhan, D.C., Ogmundsdottir, M.H., Kazi, S., Reynolds, B., Visvalingam, S.M., Wilson, C., and Boyd, C.A. (2009). Amino acid sensing and mTOR regulation: inside or out? *Biochem Soc Trans* 37:248–252.
52. Forsberg, H., and Ljungdahl, P.O. (2001). Genetic and biochemical analysis of the yeast plasma membrane Ssy1p-Ptr3p-Ssy5p sensor of extracellular amino acids. *Mol Cell Biol* 21:814–826.
53. Martin, J.F., Hersperger, E., Simcox, A., and Shearn, A. (2000). minidiscs encodes a putative amino acid transporter subunit required non-autonomously for imaginal cell proliferation. *Mech Dev* 92:155–167.
54. Reynolds, B., Roversi, P., Laynes, R., Kazi, S., Boyd, C.A., and Goberdhan, D.C. (2009). *Drosophila* expresses a CD98 transporter with an evolutionarily conserved structure and amino acid-transport properties. *Biochem J* 420:363–372.
55. Colombani, J., Raisin, S., Pantalacci, S., Radimerski, T., Montagne, J., and Leopold, P. (2003). A nutrient sensor mechanism controls *Drosophila* growth. *Cell* 114:739–749.
56. Drummond, M.J., Glynn, E.L., Fry, C.S., Timmerman, K.L., Volpi, E., and Rasmussen, B.B. (2010). An increase in essential amino acid availability upregulates amino acid transporter expression in human skeletal muscle. *Am J Physiol Endocrinol Metab* 298:E1011–E1018.
57. Hennig, K.M., Colombani, J., and Neufeld, T.P. (2006). TOR coordinates bulk and targeted endocytosis in the *Drosophila melanogaster* fat body to regulate cell growth. *J Cell Biol* 173:963–974.
58. Ameri, K., and Harris, A.L. (2008). Activating transcription factor 4. *Int J Biochem Cell Biol* 40:14–21.
59. Harding, H.P., et al. (2003). An integrated stress response regulates amino acid metabolism and resistance to oxidative stress. *Mol Cell* 11:619–633.
60. Palii, S.S., Thiaville, M.M., Pan, Y.X., Zhong, C., and Kilberg, M.S. (2006). Characterization of the amino acid response element within the human sodium-coupled neutral amino acid transporter 2 (SNAT2) System A transporter gene. *Biochem J* 395:517–527.
61. Rennie, M.J., Ahmed, A., Khogali, S.E., Low, S.Y., Hundal, H.S., and Taylor, P.M. (1996). Glutamine metabolism and transport in skeletal muscle and heart and their clinical relevance. *J Nutr* 126:1142S–1149S.
62. Bode, B.P. (2001). Recent molecular advances in mammalian glutamine transport. *J Nutr* 131:2475S–2485Sdiscussion 2486S–2487S.
63. Avruch, J., Long, X., Ortiz-Vega, S., Rapley, J., Papageorgiou, A., and Dai, N. (2009). Amino acid regulation of TOR complex 1. *Am J Physiol Endocrinol Metab* 296:E592–E602.
64. Long, X., Lin, Y., Ortiz-Vega, S., Yonezawa, K., and Avruch, J. (2005). Rheb binds and regulates the mTOR kinase. *Curr Biol* 15:702–713.
65. Sancak, Y., Thoreen, C.C., Peterson, T.R., Lindquist, R.A., Kang, S.A., Spooner, E., Carr, S.A., and Sabatini, D.M. (2007). PRAS40 is an insulin-regulated inhibitor of the mTORC1 protein kinase. *Mol Cell* 25:903–915.
66. Sato, T., Nakashima, A., Guo, L., and Tamanoi, F. (2009). Specific activation of mTORC1 by Rheb G-protein in vitro involves enhanced recruitment of its substrate protein. *J Biol Chem* 284:12783–12791.
67. Roccio, M., Bos, J.L., and Zwartkruis, F.J. (2006). Regulation of the small GTPase Rheb by amino acids. *Oncogene* 25:657–664.
68. Byfield, M.P., Murray, J.T., and Backer, J.M. (2005). hVps34 is a nutrient-regulated lipid kinase required for activation of p70 S6 kinase. *J Biol Chem* 280:33076–33082.
69. Juhasz, G., Hill, J.H., Yan, Y., Sass, M., Baehrecke, E.H., Backer, J.M., and Neufeld, T.P. (2008). The class III PI(3)K Vps34 promotes autophagy and endocytosis but not TOR signaling in *Drosophila*. *J Cell Biol* 181:655–666.

70. Kim, E., Goraksha-Hicks, P., Li, L., Neufeld, T.P., and Guan, K.L. (2008). Regulation of TORC1 by Rag GTPases in nutrient response. *Nat Cell Biol* 10:935–945.
71. Sancak, Y., Peterson, T.R., Shaul, Y.D., Lindquist, R.A., Thoreen, C.C., Bar-Peled, L., and Sabatini, D.M. (2008). The Rag GTPases bind raptor and mediate amino acid signaling to mTORC1. *Science* 320:1496–1501.
72. Li, L., and Guan, K.L. (2009). Amino acid signaling to TOR activation: Vam6 functioning as a Gtr1 GEF. *Mol Cell* 35:543–545.
73. Teis, D., Wunderlich, W., and Huber, L.A. (2002). Localization of the MP1-MAPK scaffold complex to endosomes is mediated by p14 and required for signal transduction. *Dev Cell* 3:803–814.
74. Teis, D., et al. (2006). p14-MP1-MEK1 signaling regulates endosomal traffic and cellular proliferation during tissue homeostasis. *J Cell Biol* 175:861–868.
75. Nada, S., Hondo, A., Kasai, A., Koike, M., Saito, K., Uchiyama, Y., and Okada, M. (2009). The novel lipid raft adaptor p18 controls endosome dynamics by anchoring the MEK–ERK pathway to late endosomes. *EMBO J* 28:477–489.
76. Bohn, G., et al. (2007). A novel human primary immunodeficiency syndrome caused by deficiency of the endosomal adaptor protein p14. *Nat Med* 13:38–45.
77. Yan, L., et al. (2010). PP2A T61 epsilon is an inhibitor of MAP4K3 in nutrient signaling to mTOR. *Mol Cell* 37:633–642.
78. Yan, L., Mieulet, V., and Lamb, R.F. (2010). Nutrient regulation of mTORC1 and cell growth. *Cell Cycle* 9:2473–2474.
79. Leberer, E., Dignard, D., Harcus, D., Thomas, D.Y., and Whiteway, M. (1992). The protein kinase homologue Ste20p is required to link the yeast pheromone response G-protein beta gamma subunits to downstream signalling components. *EMBO J* 11:4815–4824.
80. Manning, G., Whyte, D.B., Martinez, R., Hunter, T., and Sudarsanam, S. (2002). The protein kinase complement of the human genome. *Science* 298:1912–1934.
81. Dan, I., Watanabe, N.M., and Kusumi, A. (2001). The Ste20 group kinases as regulators of MAP kinase cascades. *Trends Cell Biol* 11:220–230.
82. Diener, K., Wang, X.S., Chen, C., Meyer, C.F., Keesler, G., Zukowski, M., Tan, T.H., and Yao, Z. (1997). Activation of the c-Jun N-terminal kinase pathway by a novel protein kinase related to human germinal center kinase. *Proc Natl Acad Sci USA* 94:9687–9692.
83. Ramjaun, A.R., Angers, A., Legendre-Guillemin, V., Tong, X.K., and McPherson, P.S. (2001). Endophilin regulates JNK activation through its interaction with the germinal center kinase-like kinase. *J Biol Chem* 276:28913–28919.
84. Bryk, B., Hahn, K., Cohen, S.M., and Teleman, A.A. (2010). MAP4K3 regulates body size and metabolism in *Drosophila*. *Dev Biol* 344:150–157.
85. Janssens, V., and Goris, J. (2001). Protein phosphatase 2A: a highly regulated family of serine/threonine phosphatases implicated in cell growth and signalling. *Biochem J* 353:417–439.
86. Eichhorn, P.J., Creyghton, M.P., and Bernards, R. (2009). Protein phosphatase 2A regulatory subunits and cancer. *Biochim Biophys Acta* 1795:1–15.
87. Lam, D., Dickens, D., Reid, E.B., Loh, S.H., Moisoi, N., and Martins, L.M. (2009). MAP4K3 modulates cell death via the post-transcriptional regulation of BH3-only proteins. *Proc Natl Acad Sci USA* 106:11978–11983.
88. Lam, D., and Martins, L.M. (2009). MAP4K3 enhances the expression of the BH3-only protein BID. *Cell Cycle* 8:3248–3249.
89. Youle, R.J., and Strasser, A. (2008). The BCL-2 protein family: opposing activities that mediate cell death. *Nat Rev Mol Cell Biol* 9:47–59.

90. Luo, X., Budihardjo, I., Zou, H., Slaughter, C., and Wang, X. (1998). Bid, a Bcl2 interacting protein, mediates cytochrome *c* release from mitochondria in response to activation of cell surface death receptors. *Cell* 94:481–490.
91. Li, H., Zhu, H., Xu, C.J., and Yuan, J. (1998). Cleavage of BID by caspase 8 mediates the mitochondrial damage in the Fas pathway of apoptosis. *Cell* 94:491–501.
92. Wu, X., Kihara, T., Akaike, A., Niidome, T., and Sugimoto, H. (2010). PI3K/Akt/mTOR signaling regulates glutamate transporter 1 in astrocytes. *Biochem Biophys Res Commun* 393:514–518.

5

mTORC2: The Other Facet of mTOR

CAROLINE TREINS[a] • JULIAN DOWNWARD[b]

[a]*Inserm U845, Laboratory of Cell Growth Control by Nutrients*
Medical School, Paris Descartes University
Paris, France

[b]*Signal Transduction Laboratory*
Cancer Research UK London Research Institute
London, United Kingdom

I. Abstract

The *t*arget *o*f *r*apamycin (TOR) protein was first discovered in the early 1990s in the budding yeast *Saccharomyces cerevisiae* as the gene product in which mutations confer resistance to the growth inhibitory properties of the macrocyclic lactone, rapamycin [1]. Soon after, TOR (also known in mammals as mTOR, FRAP, or RAFT1) was identified and cloned in mammalian cells [2–5]. TOR is a highly conserved serine/threonine kinase, which is a major regulator of cell growth, proliferation, survival, and metabolism in all eukaryotes. Perhaps surprisingly, higher eukaryotes have only a single *TOR* gene, unlike budding and fission yeast, which contain two *TOR* genes suggesting the existence of two separate TOR signaling axes. The existence of these two signaling networks have been subsequently demonstrated by the identification of two structurally and functionally distinct multiprotein complexes in yeast (TOR Complex 1 and TOR Complex 2) and in mammals (mTORC1 and mTORC2), in which TOR (or mTOR) is the core component [6–10]. These complexes contain shared and distinct partners and control many cellular processes in response to diverse environmental cues. In this chapter, we will provide a detailed overview on the structure, function, and regulation of mTORC2 and discuss its potential as a target for drug development.

ISSN NO: 1874-6047
DOI: 10.1016/S1874-6047(10)28005-8

II. Structure of mTOR Complex 2 (mTORC2)

mTORC2 is formed by five elements: mTOR, *m*ammalian *l*ethal with *s*ec *t*hirteen protein 8 (mLST8; also known as *G*-protein β subunit-*l*ike protein, GβL), *r*apamycin-*i*nsensitive *c*ompanion of m*TOR* (Rictor; also known as mAVO3), *m*ammalian *s*tress-activated protein kinase *in*teracting protein (mSin1; also known as *M*EKK2 *i*nteracting *p*rotein 1, MIP1), and *pr*otein *o*bserved with Ric*tor* (Protor; also known as *pr*oline *r*ich *p*rotein 5, PRR5) [6, 9–17]. These mTOR interacting proteins have been identified in a wide variety of organisms ranging from yeast to mammals (Figure 5.1). The structure of each mTORC2 member is represented in Figure 5.2.

mTOR, the only component of mTORC2 which displays a known catalytic activity, operates as a protein serine/threonine kinase despite being a member of the *p*hospho*i*nositide-*3*-*k*inase-related *k*inase (PIKK) family [18–23]. mTOR is composed of 2549 amino acids and possesses in its amino-terminal region two tandemly repeated arrays of HEAT motifs (*H*untingtin, *E*longation Factor 3, regulatory *A* subunit of PP2A, *T*OR1) which have been found to be involved in protein–protein interaction, dimerization, and membrane localization [7, 24–29]. The HEAT region is directly followed by a FAT (*F*RAP, *A*TM, and *T*RRAP) domain which is relatively conserved in PIKK family members. A second small FAT domain (FATC) is located in the extreme carboxy-terminal region [30, 31]. These domains might modulate mTOR kinase activity [32]. The kinase domain lies between the FRB (*F*KBP12/*r*apamycin *b*inding) domain and the FATC domain [33, 34].

The FRB domain is a unique feature of mTOR, highly conserved among all TOR orthologs. Rapamycin (also known as sirolimus) was first discovered as a product of the soil bacterium *Streptomyces hygroscopicus* [35, 36]. On entering the cell, rapamycin binds to the prolyl-isomerase *FK*506-*b*inding *p*rotein of *12* kDa (FKBP12) and interacts with the FRB

S. Cerevisiae	*S. Pombe*	*Dictyostelium*	*C. elegans*	*D. melanogaster*	*Mammals*
TOR2	Tor1	TOR	CeTOR	dTOR	mTOR
AVO1	SIN1	PiaA	SINH1	Sin1	mSin1
AVO2	–	–	–	–	–
AVO3	Ste20	RIP3	RICT1	Rictor	Rictor
LST8	WAT1	LST8	LST8	LST8	mLST8
BIT61	BIT61	–	–	–	PRR5/Protor-1

FIG. 5.1. Conserved core components of mTORC2 in different eukaryotes.

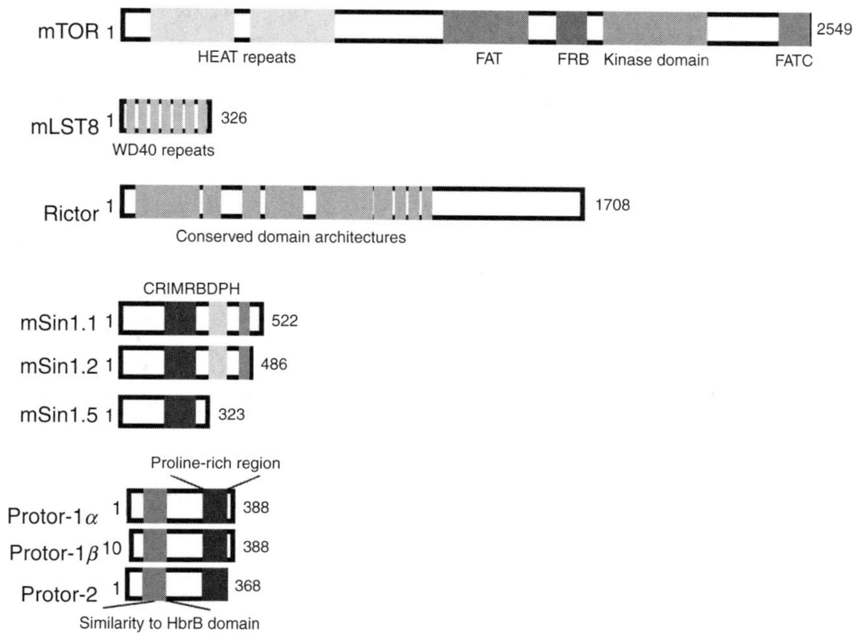

FIG. 5.2. Structure of mTORC2 subunits.

domain of mTOR leading to the inactivation of mTORC1 [33, 34]. The complex of FKBP12 with rapamycin has been shown to be unable to physically interact with, or inhibit, mTORC2 [6, 10]. On the basis of their sensitivities to rapamycin, mTORC1 and mTORC2 have been referred, respectively, as the rapamycin-sensitive and rapamycin-insensitive mTOR complexes. Indeed, despite the fact that prolonged rapamycin treatment has been described to disrupt mTORC2 assembly in a subset of cells, rapamycin does not appear to be a general inhibitor of mTORC2 [37].

Only two members of mTORC2 are shared with mTORC1, mTOR and mLST8, which was previously identified as a *G*-protein β subunit-*l*ike protein (GβL) [38]. mLST8 is a 36 kDa protein consisting almost entirely of seven WD40 repeats, which interacts with mTOR kinase domain [11]. Initially described as required for mTORC1 activity, deletion of the gene encoding mLST8 in mice reveals that it is not essential for mTORC1 function but in contrast is needed for mTORC2 activity [11, 39].

Rictor was the first mTOR partner identified as exclusively binding to mTOR in mTORC2. Rictor is a large protein of 1708 amino acids with no functional domains described, aside from some domain structures in the amino-terminal region that are relatively conserved among species [6, 10].

So far, the binding domain between Rictor and mTOR has not been mapped, but this interaction seems to be dependent of the presence of another mTORC2-specific partner mSin1 [12–14].

In mammals, mSin1 has been isolated as a MEKK2-interacting protein and has since then been shown to interact with various stress-associated kinases [40–42]. Five isoforms of mSin1 have been described, three of which have been shown to assemble into mTORC2 to generate three distinct mTORC2s [12, 43]. Most of the Sin1 orthologs contain both a Raf-like *R*as-*b*inding *d*omain (RBD) and a *p*leckstrin *h*omology (PH) domain, in addition to a highly conserved region of unknown function in the middle of the protein [43, 44]. mSin1, as well as Rictor, is absolutely required for mTORC2 activity [6, 10, 12–14]. In contrast, the physiological function of the last mTORC2 partner described, Protor, has not yet been defined [15–17]. Indeed, Protor binds to mTORC2 via Rictor but does not affect mTORC2 integrity or kinase activity. Five isoforms of Protor have been described, with three of them interacting with Rictor: Protor-1 α and β (also known as PRR5-1 and -2) and Protor-2 (also known as PRR5 Like, PRR5L) [15–17, 45]. Protor-1 contains in its amino-terminal a region with similarities to the HbrB domain, a domain found in proteins involved in hyphal growth and polarity, and a proline-rich region in the carboxy-terminal part [46]. None of these domains are implicated in the interaction with Rictor [16]. To increase the complexity, mTORC2 has been shown to behave as a multimer *in vivo*, both in yeast and in mammals [27, 28].

Aside from Protor, knock out of which has not yet been described, the deletion of each component of mTORC2 in mice is embryonic lethal [13, 14, 39, 47–49]. The strength of the phenotype obtained emphasizes the overall physiological importance of mTORC2 and highlights the key roles played by each member of this complex. Germline disruption of mTOR in mice causes the strongest phenotype with embryonic lethality at or around implantation [47, 48]. The early lethality of mTOR-null mice appears to result from a loss of mTORC1 function because mice where another mTORC1 specific member, *r*egulatory *a*ssociated *p*rotein of m*TOR* (Raptor), has been deleted are phenotypically similar [39]. The inactivation of mLST8, Rictor, or mSin1 in mouse induces embryonic lethality that is delayed relative to that observed in mTOR knockout mouse, with death occurring at E10.5–E11.5 [13, 14, 39, 49].

III. Role of mTORC2

The identification of mTORC2 opened up a new horizon in the understanding of the mTOR pathway. Prior to that, much of our knowledge on mTOR was limited to mTORC1. Indeed, mTOR targets and functions had been mainly identified using rapamycin, which inhibits only one side of the mTOR pathway, so a hunt for mTORC2-specific targets and functions was started.

A. mTORC2 Substrates

Akt (also known as *p*rotein *k*inase *B*, PKB), and more precisely the serine residue 473 which is part of Akt *h*ydrophobic *m*otif (HM), was the first direct target of mTORC2 reported [50], just a few months after the identification of mTORC2. Gene knockout studies in mice have provided genetic evidence that mTORC2 is the HM kinase for Akt, as in cells deficient for either Rictor, mSin1, or mLST8 there is no detectable phosphorylation of Ser 473 [13, 14, 39]. However, the phosphorylation of Akt at serine 473 is decreased but not abolished in models of muscle-specific deletion of Rictor as well as in one report of Rictor-null mice where residual Ser 473 phosphorylation was observed even after the deletion of Rictor [49, 51, 52]. The reason for this discrepancy is unknown but raises the possibility that mTORC2 is not the exclusive HM kinase for Akt. Indeed, several other kinases such as Ataxia *t*elangiectasia *m*utant (ATM), *DNA*-dependent *p*rotein *k*inase (DNA-PK), and *I*ntegrin-*l*ike *k*inase (ILK) have been implicated in the phosphorylation of Akt at serine 473 [53–56]. It is worth noticing that Rictor has been identified as an ILK interactor in a proteomic screen which has been suggested to define a new complex regulating Akt phosphorylation at serine 473 [57].

Akt is a serine/threonine kinase that is part of the AGC kinase family (reviewed in Ref. [58]). Its full activation has been reported to require dual phosphorylation on its activation loop within the catalytic domain at threonine 308, by PDK1 and at serine 473 within the HM in the carboxy-terminal noncatalytic region by mTORC2 [50, 59, 60]. Although earlier studies showed reduced phosphorylation of Akt at both sites, serine 473 and threonine 308, after knockdown of Rictor, analysis of mouse embryonic fibroblasts (MEFs) derived from Rictor, mSin1, or mLST8 null mice in which mTORC2 activity was ablated revealed that the residue threonine 308 could still be phosphorylated to the same extent as in wild-type cells even in the absence of phosphorylation at serine 473 [13, 14, 39, 49, 50, 61]. Furthermore, rather than being inactive, Akt without phosphorylation at serine 473 remains largely active on many substrates, with only forkhead class O transcription factor 1/3a (FOXO1/3a), *B*cl-2-*a*ssociated *d*eath promoter (BAD), and *A*kt *s*ubstrate of *160* kDa (AS160) phosphorylation level being affected by mTORC2 activity ablation [13, 14, 39, 51, 62]. Thus, mTORC2-mediated Akt phosphorylation seems to control Akt substrate specificity more than its activity.

Surprisingly, in contrast to what has been shown in MEFs, the phosphorylation at threonine 308 induced by PTEN loss has recently been described to require mTORC2 activity. The authors suggested that coregulation of Akt by mTORC2 and PDK1 is linked in cancer cells, while otherwise these two inputs are uncoupled [63].

Other AGC kinases, *p*rotein *k*inase *C* (PKC) and *s*erum- and *g*lucocorticoid-induced protein *k*inase *1* (SGK1), have been identified as mTORC2 targets [10, 39, 64–66]. SGK1 has recently been shown to be directly phosphorylated by mTORC2 at serine 422 lying within the carboxy-terminal HM. Indeed, SGK1 HM phosphorylation and kinase activity, as shown by the decrease of the phosphorylation of the SGK1 substrate *N*-myc *d*ownstream *reg*ulated *1* (NDRG1), are abolished in Rictor-, mSin1-, or mLST8-deficient MEFs [66]. With regard to the role of mTORC2 in the regulation of PKC, it seems clear that mTORC2 is required for phosphorylation of all conventional PKCs and the novel PKCε at their HMs [10, 39, 64, 65]. However, so far there is no direct evidence that PKC is a direct target of mTORC2 as no study has been able to demonstrate *in vitro* a direct phosphorylation of any PKC isoform by mTORC2.

In addition to the HMs, mTORC2 has been shown to regulate the phosphorylation levels of the *t*urn *m*otifs (TM) of PKCs and Akt, thereby modulating their stability [64, 65]. The TM of the AGC kinases is located a few amino acids before the HM. In contrast to the phosphorylation of the activation loop and the HM of Akt, which are stimulated by growth factors through *p*hosphatidyl*i*nositol-*3*-*k*inase (PI3K), TM is constitutively phosphorylated during or soon after synthesis. This phosphorylation stabilizes the active conformation of the kinase and may also protect the HM from dephosphorylation [67]. In mTORC2-deficient MEFs, Akt and PKC are not phosphorylated at TM leading to a decrease in their stability, although it is still unclear if this phosphorylation event is direct or not [64, 65].

In conclusion, mTORC2 have been described to play a key role in the control of the HM and TM phosphorylation level of a least two AGC kinases, Akt and PKC, and to regulate the phosphorylation level of the HM of SGK1. The role of mTORC2 could extend to the regulation of other AGC kinases, which is subject of intensive investigations.

B. mTORC2 Functions

The identification of mTORC2 as a major regulator of several AGC kinases emphasized the key role played by mTORC2 (Figure 5.3). Indeed, by modulating Akt, PKC and SGK1 phosphorylation levels, activity and stability, mTORC2 is involved in the control of a large number of processes

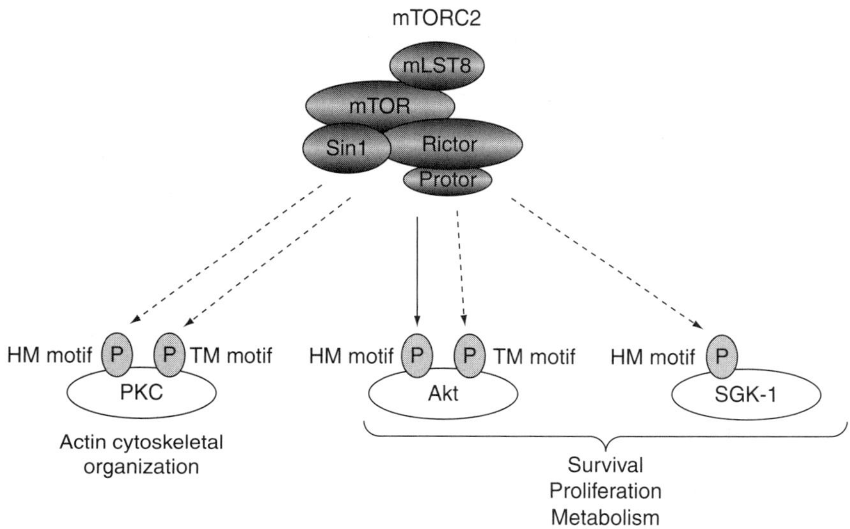

FIG. 5.3. mTORC2 substrates and functions.

within the cell, as suggested by the embryonic lethality associated with the inactivation of each single mTORC2 member.

mTORC2 has first been described to be involved in the regulation of the actin cytoskeleton, both in yeast and in mammalian cells. Indeed, modifications in the actin network have been observed after knockdown of the expression of mTORC2-specific components [6, 10, 13]. The molecular mechanisms by which mTORC2 regulates the actin organization remains unclear. However, these studies have suggested that mTORC2 controls the actin cytoskeleton by promoting PKCα phosphorylation, phosphorylation and relocalization to focal adhesions of paxillin, and the GTP loading of RhoA and Rac1 [6, 10]. In contrast with these results, no remarkable changes in actin cytoskeleton has been observed in Rictor or mLST8-null MEFs, suggesting that maintenance of cytoskeletal integrity is not a critical downstream function of mTORC2 [39, 49]. The discrepancy of these results could be due to differences in the models used. In knockout mouse models, adaptive mechanisms could have been set up to overcome the actin organization defects. More study will be needed to obtain a better view of the role of mTORC2 in the regulation of the cytoskeletal organization.

Akt has been described to play a major role in the regulation of cell survival, proliferation, and metabolism through the phosphorylation of various effectors. As discussed above, mTORC2 phosphorylates Akt, which seems to lead to a modification of its substrate specificity, thus changing the activation status of some specific Akt targets. So far, inhibition of Akt following mTORC2 depletion has been shown to reduce the phosphorylation, and therefore activates, FoxO1/3a and BAD, and more recently has been shown to reduce the phosphorylation, and therefore inhibits, AS160 [13, 14, 39, 51, 62]. In contrast, the phosphorylation state of *t*uberous *s*clerosis *c*omplex *2* (TSC2) or *g*lycogen *s*ynthase *k*inase *3* (GSK3) is not affected by mTORC2 inactivation. This list is not exhaustive; more studies will be needed to get a clear vision of the subset of Akt targets that are directly dependent of mTORC2 activation. Indeed, western blotting with the phospho-Akt substrate antibody shows that phosphorylation of multiple proteins was decreased in the Rictor- and mSin1-null MEFs [13, 14]. The modification of the activation state of FOXO1a and FOXO3a transcription factors, which control the expression of genes involved in stress resistance, metabolism, cell-cycle arrest and apoptosis, of the proapoptotic molecule, BAD, and of AS160 which regulates GLUT4 trafficking could mediate part of mTORC2 response.

Knockdown of mSin1 in HeLa cells has been described to increase sensitivity to apoptosis inducer, etoposide, as judged by the increase of caspase-3 cleavage and *p*oly (*A*DP-*r*ibose) *p*olymerase (PARP) cleavage. This higher sensitivity to apoptosis has been correlated with a decrease of the phosphorylation of BAD observed after mSin1 knockdown, as it is well established that Akt phosphorylates BAD on Ser136 to inhibit apoptosis. So far, the effect of mLST8 or Rictor knockdown on apoptosis sensitivity has not been analyzed. On the other hand, Rictor-null MEFs present a slower growth rate than wild-type MEFs, associated with a slower metabolic activity when assessed by measure of redox potential, suggesting that cells lacking Rictor may be unable to increase their energy production, thereby resulting in an intracellular metabolic imbalance and eventual cell death [49]. In agreement with these results, downregulation of Rictor in colon cancer cells has been recently shown to significantly reduce cell proliferation [68]. Overexpression of Rictor, associated with an increase in mTORC2 activity, in low-grade glioma cell lines promotes clonogenic growth. The opposite effects are observed when Rictor expression is knockdown [69]. Finally, transformed human prostate epithelial cells lacking PTEN has been shown to require mTORC2 to form tumors when injected to nude mice [63]. All these studies reveal the key role played by mTORC2 in the control of cell survival, growth, and proliferation, even if the downstream signaling cascades responsible for these effects have not been entirely defined. Another important notion that has been recently raised is that mTORC2 seems selectively required for tumor development. Indeed, mTORC2 seems to be required for development of prostate cancer caused by PTEN deletion, harboring high levels of activated Akt, but not for normal prostate epithelial cells, suggesting that mTORC2 inhibitors may be of substantial clinical utility [63].

A role of mTORC2 in metabolism is starting to emerge. Recent studies indicate that Rictor/TORC2 regulates fat accumulation in *Caenorhabditis elegans*. Indeed, Rictor-null worms display a high body fat, associated with a mild development delay, a small size and are short-lived, suggesting that TORC2 functions as a negative regulator of lipid deposition [70, 71]. Interestingly, it was found that these effects are predominantly mediated by SGK1, and not Akt, as SGK1 deletion mimics the metabolic phenotype associated with Rictor loss, suggesting that SGK1 is the most significant TORC2 target in the worm. In agreement with a role of mTORC2 in lipid metabolism in yeast, cells deficient in mTORC2 activity show impairment in the biosynthesis of ceramide, which is a precursor to more complex sphingolipids [72]. So far, there is no clear evidence of a role of Rictor/mTORC2 in fat metabolism in mammals. Since whole-body knockout of mTORC2 members is embryonic lethal, conditional models need to be generated. Deletion of Rictor in the muscle leads to a very mild phenotype, with only a slight decrease in insulin stimulate glucose uptake associated with a decrease in AS160 phosphorylation level and an increase in basal glycogen synthase activity described by one group [51, 52]. In addition, mice where Rictor has been deleted in adipose tissue have been generated. In these models, an increase in lean organ size has been observed that seems to be the consequence of an increase in the circulating level of IGF-1, suggesting a negative control of whole-body growth by mTORC2 in adipose tissue [62, 73]. One group also showed that these mice present a defect in insulin-stimulated glucose transport due to a defect in GLUT4 translocation correlated with a decrease in AS160 phosphorylation, plus an increase in lipolysis in adipose tissue, leading to the development of glucose intolerance, hyperinsulinemia, insulin resistance in skeletal muscle and liver, and hepatic steatosis. This is the evidence of a role for Rictor/mTORC2 in fat cells in whole-body energy homeostasis [62]. However, there is a discrepancy between the two mice models of Rictor inactivation in adipose tissue, with a defect in glucose uptake and lipolysis only observed in one study. In addition, differences exist between the data obtained in worms and in mammals as Rictor/TORC2 inhibits fat storage in worm but inhibits lipolysis in adipose tissue in mammals. Additional studies are required to determine whether mTORC2 plays a significant role in fat metabolism in mammals.

IV. Regulation of mTORC2

In the recent years, much of progress has been made in understanding mTORC2 structure and the identification of its targets, giving fundamental clues on its function. Nevertheless, little is known about the signaling cascade leading to the activation of mTORC2 or any regulation that could directly target this complex, which represent today one of the main areas of uncertainty.

A. mTORC2 Expression Level and Cellular Localization

Even though each mTORC2 members has been described to be ubiquitously expressed, a tissue-specific regulation of mTORC2 activity cannot be excluded as conditional knockout of Rictor in the muscle reveals that mTORC2 seems to have a minor effect on the regulation of Akt phosphorylation level compared to what is observed after inactivation of Rictor in white adipose tissue [51, 52, 62, 73]. So far, the expression level of the different mTORC2 members has not been reported as highly regulated at transcriptional level. Nevertheless, Rictor protein and mRNA levels have been shown to be elevated and correlated with increased mTORC2 activity in some glioma cell lines, promoting anchorage-independent growth and tumor formation in xenografts [69], and total Rictor protein level is increased after PTEN deletion in prostate tissue [63]. In contrast, a recent study reports that, without requiring its DNA binding activity, FoxO1 elevates mRNA and protein level of Rictor, leading to an increase in mTORC2 activity that consequently activates Akt [74]. All these studies suggest that Rictor protein expression is directly correlated with mTORC2 activity, increasing the importance of defining mechanisms regulating the expression levels of the mTORC2 partner that could define regulation of mTORC2 activity. It is worth mentioning that mLST8 was originally described as regulated at the mRNA and protein levels by insulin in 3T3-L1 adipocytes. This is the only report mentioning a modification of mLST8 expression, and the potential effect on mTORC2 activity has not been studied [38].

There is no clear consensus on the exact localization of mTORC2 within the cell, but the subcellular localization of mTORC2 has been proposed to be important for its activity. Indeed, in syndecan-4 null endothelial cells, mTORC2 activity has been shown to be reduced, leading to a decrease of Akt activation, resulting from a decrease of Rictor, mLST8, and mSin1 levels in the lipid rafts, due to a defect in PKCα recruitment. However, the mechanisms by which PKCα regulates mTORC2 localization and by extension mTORC2 activity have not been described [75].

B. Posttranscriptional Modification of mTORC2

To try to get more knowledge about the regulation of mTORC2 activity, the identification of posttranscriptional modifications, such as phosphorylation, within mTORC2 has been a focus of recent activity. Phosphorylation sites have been identified in large-scale proteomic analyses within each of the mTORC2 members, but the roles or the kinases responsible for these modifications are still unknown in most cases.

Concerning mLST8, protor-1 and -2, only a few phosphorylation sites have been identified (human mLST8 Thr3, Ser4, Thr7; protor-1 Ser252, Ser254, Thr257; and protor-2 Ser28), with none of them having been related to any function or associated with any upstream kinases [76–79]. mSin1 is known to be modified by phosphorylation as phosphatase treatment increases its mobility on gel electrophoresis [13]. However, so far, even if some phosphorylation sites have been identified in large-scale proteomics analysis (Thr86, Ser185, Ser270, Ser510 in human), there is no link between these phosphorylations and any potential effect on mSin1 itself or on mTORC2 [76–81].

To date, 29 phosphorylation sites have been described within mTOR (full list displayed in Figure 5.4). Out of these 26 sites, only four have been validated *in vivo*: Ser1261, Thr2446, Ser2448, and Ser2481 [82–84]. Thr2446 within mTOR is phosphorylated in response to nutrients, but its role remains elusive and the intensity of this phosphorylation within mTORC1 versus mTORC2 has not been addressed [82]. Recently, the existence of mTORC-specific phosphorylation of mTOR has been highlighted. Indeed, mTOR seems to be predominantly phosphorylated at Ser2448 when associated to mTORC1, whereas within mTORC2 mTOR is predominantly phosphorylated on Ser2481 [85]. Ser2448 of mTOR has been shown to be phosphorylated in a rapamycin-sensitive manner by the ribosomal protein *S6 k*inase (S6K), in contrast to Ser2481, which has been shown to be a rapamycin-insensitive autophosphorylation site [84, 86, 87]. Both of these phosphorylation events, highly conserved in vertebrates, are dependent on intact mTORC1 and mTORC2, respectively. Ser2448 has recently been suggested to be a specific marker to detect intact mTORC2 within the cell and a biomarker for mTORC2 sensitivity to rapamycin [85]. These three phosphorylation sites, Ser2446, Ser2448, and Ser2481, all lie between the kinase domain and the FATC domain. To date, the functional importance of these phosphorylation events remains elusive, as do their relative roles in both mTOR complexes. The newly identified phosphorylation site within mTOR at Ser1261, which is promoted by insulin/PI3K signaling, is found in both complexes, but the potential role of this phosphorylation on mTORC2 function has not been addressed [88].

Rictor seems to be the mTORC2 subunit that is the most heavily phosphorylated, with 64 phosphorylation sites reported so far (full list displayed in Figure 5.5). Indeed, despite its high molecular weight, modification of Rictor's electrophoretic mobility has been detected, suggesting the existence of multiple phosphorylations [10, 13, 89]. These phosphorylation

mTOR phosphorylation sites identified by mass spectrometry		
Phosphorylation sites	**Surrounding sequences**	**Kinases**
S567	GLAHQLASPGLTTLP	
S916	SRDASAVSLSESKSS	
T1162	IIHPIVRTLDQSPEL	
S1261	PMKKLHVSTINLQKA	
T1262	MKKLHVSTINLQKAW	
T1829	GANITNATTAATTAA	
T1839	ATTAATATTTASTEG	
T1840	TTAATATTTASTEGS	
T1841	TAATATTTASTEGSN	
S1843	ATATTTASTEGSNSE	
S1847	TTASTEGSNSESEAE	
S1849	ASTEGSNSESEAEST	
S1851	TEGSNSESEAESTEN	
T1856	SESEAESTENSPTPS	
S1859	EAESTENSPTPSPLQ	
S1863	TENSPTPSPLQKKVT	
T1870	SPLQKKVTEDLSKTL	
S1874	KKVTEDLSKTLLMYT	
T1876	VTEDLSKTLLMYTVP	
Y1880	LSKTLLMYTVPAVQG	
T2444	KGNKRSRTRTDSYSA	
T2446	NKRSRTRTDSYSAGQ	
S2448	RSRTRTDSYSAGQSV	**S6K**
Y2449	SRTRTDSYSAGQSVE	
S2454	DSYSAGQSVEILDGV	
T2471	GEPAHKKTGTTVPES	
T2474	AHKKTGTTVPESIHS	
S2478	TGTTVPESIHSFIGD	
S2481	TVPESIHSFIGDGLV	**Autophosphorylation**

FIG. 5.4. List of mTOR phosphorylation sites identified by mass spectrometry. Only the one colored in gray have been validate in vivo.

events could modify Rictor interaction with mTOR, as the dephosphorylated form of Rictor has been described to have a weaker affinity for mTOR binding [13]. Most of these phosphorylation sites reported are concentrated in a region that is only conserved in vertebrate Rictor orthologs. Among this long list, only one phosphorylation site, the residue Thr1135, which is conserved among all vertebrate, has been validated *in vivo* [90–93]. Phosphorylation of Rictor at Thr1135 is stimulated by amino acids, growth factors, and oncogenic form of Ras or PI-3-kinase through a rapamycin-sensitive signaling cascade. Indeed, Rictor has been shown to be a direct target of ribosomal protein *S6 k*inase 1 (S6K1), revealing a cross talk between the mTORC1 pathway and mTORC2 [90–93]. In addition, these results reveal that Rictor, described as being the nutrient-insensitive companion of mTOR, is in fact phosphorylated in response to amino acids

Rictor phosphorylation sites identified by mass spectrometry		
Phosphorylation sites	**Surrounding sequences**	**Kinases**
S21	RVRGRND**S**GEENVPL	
S35	LDLTREP**S**DNLREIL	
Y262	APYTDFH**Y**RHSPDTA	
S265	TDFHYRH**S**PDTAEGQ	
Y863	KPVDGDN**Y**VRRSNQR	
S1026	PSTLSLN**S**ESTSSRH	
S1028	TLSLNSE**S**TSSRHNS	
S1030	SLNSEST**S**SRHNSES	
S1031	LNSESTS**S**RHNSESE	
S1035	STSSRHN**S**ESESVPS	
S1037	SSRHNSE**S**ESVPSSM	
T1103	NRILNSL**T**LPNKKHR	
T1135	NRRIRTL**T**EPSVDFN	**S6K1**
S1138	IRTLTEP**S**VDFNHSD	
T1156	PISTVQK**T**LQLETSF	
T1172	GNKHIED**T**GSTPSIG	
S1174	KHIEDTG**S**TPSIGEN	
T1175	HIEDTGS**T**PSIGEND	
S1177	EDTGSTP**S**IGENDLK	
S1217	SSHMKIR**S**QSFNTDT	
S1219	HMKIRSQ**S**FNTDTTT	
T1222	IRSQSFN**T**DTTTSGI	
T1224	SQSFNTD**T**TTSGISS	
T1225	QSFNTDT**T**TSGISSM	
T1226	SFNTDTT**T**SGISSMS	
S1227	FNTDTTT**S**GISSMSS	
S1230	DTTTSGI**S**SMSSSPS	
S1231	TTTSGIS**S**MSSSPSR	
S1235	GISSMSS**S**PSRETVG	
T1271	IKTSHYL**T**PQSNHLS	
S1274	SHYLTPQ**S**NHLSLSK	
S1282	NHLSLSK**S**NSVSLVP	
S1284	LSLSKSN**S**VSLVPPG	
S1286	LSKSNSV**S**LVPPGSS	
S1292	VSLVPPG**S**SHTLPRR	
S1293	SLVPPGS**S**HTLPRRA	
T1295	VPPGSSH**T**LPRRAQS	
S1302	TLPRRAQ**S**LKAPSIA	
S1313	PSIATIK**S**LADCNFS	
Y1321	LADCNFS**Y**TSSRDAF	
Y1330	SSRDAFG**Y**ATLKRLQ	
S1344	QQQRMHP**S**LSHSEAL	
S1346	QRMHPSL**S**HSEALAS	
S1348	MHPSLSH**S**EALASPA	
S1353	SHSEALA**S**PAKDVLF	
T1376	NSFESRL**T**PSRFMKA	
S1385	SRFMKAL**S**YASLDKE	
S1386	RFMKALS**Y**ASLDKED	
S1388	MKALSYA**S**LDKEDLL	
S1396	LDKEDLL**S**PINQNTL	
S1414	SSVRSMV**S**SATYGGS	
S1415	SVRSMVS**S**ATYGGSD	
T1417	RSMVSSA**T**YGGSDDY	
Y1418	SMVSSAT**Y**GGSDDYI	
S1421	SSATYGG**S**DDYIGLA	
S1470	HDAGGLP**S**GTGGLVK	
T1472	AGGLPSG**T**GGLVKNS	
S1479	TGGLVKN**S**FHLLRQQ	
S1571	EVVPSKF**S**GISGCSD	
S1574	PSKFSGI**S**GCSDGVS	
S1577	FSGISGC**S**DGVSQEG	
S1591	GSASSTK**S**TELLLGV	
T1592	SASSTKS**T**ELLLGVK	
T1605	VKTIPDD**T**PMCRILL	

FIG. 5.5. List of Rictor phosphorylation sites identified by mass spectrometry. Only the one colored in gray has been validate in vivo.

through an mTORC1-dependent signaling cascade, suggesting that Rictor could also play a role in the amino acid response. Expression of a recombinant Rictor T1135A mutant in Rictor-null MEFs restores Akt S473 phosphorylation to a higher degree than expressing recombinant wild-type Rictor, even if phosphorylation of Rictor at Thr1135 does not affect mTORC2 assembly, *in vitro* kinase activity, or cellular localization [90–93]. So far, the only other function for this phosphorylation event is the recruitment of 14-3-3 proteins to Rictor and mTORC2, although the functional significance of this is currently unknown [90, 91]. The richness in phosphorylation sites within Rictor suggests that it could integrate different inputs allowing fine control of mTORC2.

Currently, much effort is being put into the identification and characterization of phosphorylation events within mTORC2 members that could lead to the identification of new components of the mTORC2 signaling cascade.

C. Regulation of mTORC2 Activity

Regulation of mTORC2 activity is still the least well understood aspect of the mTORC2 signaling cascade (potential mechanisms involved in the control of mTORC2 activity are pictured in Figure 5.6).

1. Positive Regulators

Growth factors stimulate mTORC2 activity, as shown using *in vitro* kinase assays, and as mentioned above modulate phosphorylation level of some mTORC2 subunits, but the precise events leading to mTORC2 activation

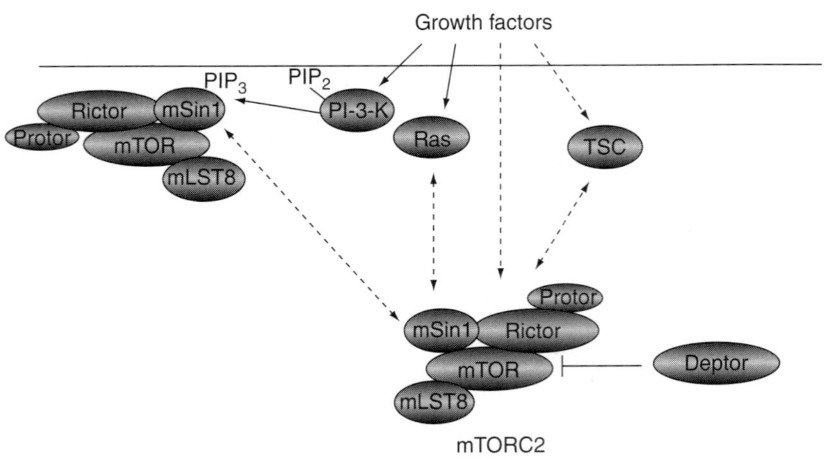

Fig. 5.6. Potential mechanisms involved in the control of mTORC2 activity.

are still unknown [12, 50]. mSin1, which contains a RBD and a PH domain in its C-terminal part, could be one of the key components of this activation cascade. Indeed, one isoform of mSin1 which lack part of its RBD and its PH domain has been described to phosphorylate Akt independently of growth factors stimulation, suggesting a role of mSin1 in the regulation of mTORC2 activity [12]. This evidence raises the possibility that Ras could perhaps directly influence mTORC2 activity by interacting with mSin1. In addition, growth factors stimulation of cells could potentially promote mTORC2 translocation to the membrane through the binding of phosphatidylinositol (3,4,5)-trisphosphate (PtdIns(3,4,5)P3) to the PH domain of mSin1 (reviewed in Ref. [94]). Indeed, in PTEN-null prostate epithelial cells, mTORC2 has been found strongly associated with the plasma membrane and, as mentioned above, mislocalization of mTORC2 in syndecan-4 null endothelial cells leads to a decrease of Akt activation [63, 75]. These studies suggest that mTORC2 localization could play an important role in the regulation of mTORC2 activity, although so far this is just speculation: additional work will be needed to support these hypotheses.

*T*uberous *S*clerosis *C*omplex 1/2 (TSC1/2), which function as a GAP toward the small GTPase *R*as *h*omolog *e*nriched in *b*rain (Rheb) that directly activates mTORC1, have been described as major regulators of mTORC1 activity [95–102]. The TSC1/2 complex has more recently been suggested to play as well a role in the control of mTORC2 activity [103–105]. The TSC1/2 complex has been shown to physically associate with mTORC2 and positively regulates mTORC2. Indeed, mTORC2 *in vitro* kinase activity was found to be severely blunted in a variety of cells lacking an intact TSC1/2 complex [104]. In addition, mTORC2 substrates are affected by loss of the TSC1/2 complex in cell culture models and kidney tumors from both $TSC2^{+/-}$ mice and TSC patients [105]. Rheb overexpression or downregulation is not able to modulate mTORC2 activity, suggesting that the action of TSC1/2 toward mTORC2 is independent of its GAP activity toward Rheb [103, 104].

Additional work is needed to gain a clear view of the mechanisms responsible for the activation of mTORC2 in response to growth factors.

2. *Negative Regulators*

A negative feedback loop whereby TSC/mTORC1/S6K1 activation inhibits PI-3-kinase/Akt signaling has been highlighted in many studies. The idea of a negative feedback loop was built on observations showing that chronic insulin treatment as well as *tsc*2-loss led to a downregulation of PI-3-kinase/Akt signaling [106–111]. Indeed, in a subset of cancer cell lines and patient tumors, pharmacological inhibition of mTORC1 induces Akt phosphorylation at the mTORC2 target site, the serine 473 [112, 113]. Additional studies provided some clues on the mechanisms responsible for this feedback. Notably, multiple phosphorylation and transcriptional repression of *i*nsulin *r*eceptor *s*ubstrate proteins (IRS) is induced by the mTORC1 target, S6K1 [114–117]. In agreement with these results, S6K1 knockout mice, due to loss of IRS inhibition, are hypersensitive to insulin [117]. But the existence of an IRS-independent feedback loop has emerged as other receptor tyrosine kinase pathways that do not depend on IRS are also likely subject to mTORC1-dependent negative regulation, such as *p*latelet-*d*erived *g*rowth *f*actor (PDGF) via the regulation of PDGF receptor expression [118]. The existence of other feedback loops is still a distinct possibility, and the S6K1-dependent phosphorylation of Rictor at threonine 1135 could be one of them [90–92].

More recently, a direct inhibitor of mTORC2 activity has been described. Deptor, in reference to the presence of two *D*ishevelled, *e*gl-10, *p*leckstrin (DEP) domains in its sequence (also known as DEPDC6), is a shared subunit with mTORC1 and mTORC2, found only in vertebrates, which binds mTOR near its kinase domain. Deptor, whose expression is negatively regulated by mTORC1 and mTORC2, has been shown to negatively regulate the kinase activity of both mTORC1 and mTORC2. It is the only regulator of mTORC2 activity identified so far [119]. Indeed, downregulation of DEPTOR expression activates mTORC1 and mTORC2 signaling cascades, promoting cell growth and survival. In contrast, overexpression of DEPTOR activates mTORC1, but by releasing the negative feedback from mTORC1 to PI3Kinase, activates Akt. Interestingly, DEPTOR expression levels, which have been shown to be very low in various type of cancer, are very high in a subset of myeloma. In this context, reduction of DEPTOR expression leads to the apoptosis of myeloma cells [119]. Modification of the expression level of DEPTOR appears to be a new way to regulate mTORC2 activity.

V. Potential of mTOR Inhibitors in Cancer Treatment

The power of mTOR in the control of cell growth, proliferation, and survival, and reports of its overactivation in a variety of cancer cells, make mTOR a valid target for cancer therapy. In addition, very recently, two different point mutations within mTOR have been identified in human cancer genomes that confer constitutive activation of mTORC1 signaling [120]. First generation mTOR inhibitors, called rapalogs as they are rapamycin derivative, are already approved for the treatment of patient with metastatic *r*enal-*c*ell *c*arcinoma (RCC); temsirolimus (also known as CCI-779) and everolimus (also known as RAD001). Temsirolimus is also approved for *m*antle-*c*ell *l*ymphomas (MCL) (reviewed in Ref. [121]). Rapalogs suppress mTOR activity through an allosteric mechanism that acts at a distance from the ATP-catalytic binding site. Rapamycin associates with an intracellular protein called *FK*506 *b*inding *p*rotein *12* (FKBP12), and together they bind the FRB domain of mTOR that is adjacent to its kinase domain, leading to mTORC1 inhibition. Recently, the effect of rapamycin on mTORC1 activity has been contested, as different studies reveal rapamycin-insensitive mTORC1 activity [122–124]. This rapamycin-resistance side of mTORC1 could in part explain the resistance of certain cancers to rapamycin treatment. In addition, as described above, inhibition of mTORC1 using rapamycin releases the inhibitory feedback loop from mTORC1/S6K on PI3K/Akt leading to an increase in Akt activity. Indeed, despite the fact that prolonged rapamycin treatment has been described to disrupt mTORC2 assembly in a subset of cells, rapamycin is not a general inhibitor of mTORC2, as the complex FKBP12/rapamycin cannot bind to mTOR when associates in mTORC2 [6, 10, 37]. For all these reasons, rapalogs show a limited clinical efficacy. Today, much effort is being put into the design of second generation mTOR inhibitors targeting the mTOR kinase domain, allowing full inhibition of mTORC1 and mTORC2 activity, such as Torin1, PP242, PP30, Ku-0063794, WAY-600, WYE-687, WYE-354, and AZD8055 [123–127]. As they target the catalytic site of both mTORC1 and mTORC2, all these ATP-competitive inhibitors of mTOR inhibit the kinase activity of mTOR within mTORC1 and mTORC2, as can be seen by inhibition of S6K1 phosphorylation at Thr389 and Akt phosphorylation at Ser473, respectively. In addition, these inhibitors have been shown to be more complete mTORC1 inhibitors. Indeed, PP242 has recently been shown to have marked therapeutic response in Akt-mediated tumors resistant to rapamycin. In this study, the therapeutic benefit of PP242 has been shown to be mediated through inhibition of mTORC1-dependent 4EBP-eIF4E hyperactivation, supporting a central role of the 4EBP-eIF4E axis as a vital therapeutic target and providing a mechanistic rationale for the therapeutic ineffectiveness of rapamycin in the treatment of many human cancers [128]. Interestingly, PP242, but not rapamycin, has also been reported to induce death of both mouse and human leukemia cells harboring the Philadelphia chromosome translocation [129]. These exciting studies reveal the potential of these ATP-competitive inhibitors of mTOR in cancer therapies. AZD8055 is currently tested in three clinical trials for cancer in phase I/II (D1600C00001, 2008-002606-19, NCT00731263; D1600C00014, NCT00999882; D1600C00003, NCT00973076).

It is certain that further generations of mTOR inhibitors will be developed in the coming years. A challenge could be to find an mTORC2-specific inhibitor, as mTORC2 has been described to play a key role in cancer establishment, for example, in a mouse model of prostate cancer driven by *pten* mutation [63]. The relatively minor consequences of ablation of mTORC2 activity in muscle, fat cells, and prostate give some hope that it will be possible to tolerate these kinds of inhibitors. To generate such specific inhibitors, more knowledge of mTORC2 complex structure and signaling pathways will be useful. So far, the 3D structure of mTORC2 has not been reported; this could help to design small molecules that could interfere with mTORC2 formation. Interference with substrate binding site or mTORC2 localization could be other strategies.

References

1. Heitman, J., Movva, N.R., and Hall, M.N. (1991). Targets for cell cycle arrest by the immunosuppressant rapamycin in yeast. *Science* 253:905–909.
2. Brown, E.J., Albers, M.W., Shin, T.B., Ichikawa, K., Keith, C.T., Lane, W.S., and Schreiber, S.L. (1994). A mammalian protein targeted by G1-arresting rapamycin-receptor complex. *Nature* 369:756–758.
3. Chiu, M.I., Katz, H., and Berlin, V. (1994). RAPT1, a mammalian homolog of yeast Tor, interacts with the FKBP12/rapamycin complex. *Proc Natl Acad Sci USA* 91:12574–12578.
4. Sabatini, D.M., Erdjument-Bromage, H., Lui, M., Tempst, P., and Snyder, S.H. (1994). RAFT1: a mammalian protein that binds to FKBP12 in a rapamycin-dependent fashion and is homologous to yeast TORs. *Cell* 78:35–43.
5. Sabers, C.J., Martin, M.M., Brunn, G.J., Williams, J.M., Dumont, F.J., Wiederrecht, G., and Abraham, R.T. (1995). Isolation of a protein target of the FKBP12-rapamycin complex in mammalian cells. *J Biol Chem* 270:815–822.
6. Jacinto, E., Loewith, R., Schmidt, A., Lin, S., Ruegg, M.A., Hall, A., and Hall, M.N. (2004). Mammalian TOR complex 2 controls the actin cytoskeleton and is rapamycin insensitive. *Nat Cell Biol* 6:1122–1128.
7. Kim, D.H., Sarbassov, D.D., Ali, S.M., King, J.E., Latek, R.R., Erdjument-Bromage, H., Tempst, P., and Sabatini, D.M. (2002). mTOR interacts with raptor to form a nutrient-sensitive complex that signals to the cell growth machinery. *Cell* 110:163–175.
8. Hara, K., Maruki, Y., Long, X., Yoshino, K., Oshiro, N., Hidayat, S., Tokunaga, C., Avruch, J., and Yonezawa, K. (2002). Raptor, a binding partner of target of rapamycin (TOR), mediates TOR action. *Cell* 110:177–189.
9. Loewith, R., Jacinto, E., Wullschleger, S., Lorberg, A., Crespo, J.L., Bonenfant, D., Oppliger, W., Jenoe, P., and Hall, M.N. (2002). Two TOR complexes, only one of which is rapamycin sensitive, have distinct roles in cell growth control. *Mol Cell* 10:457–468.
10. Sarbassov, D.D., Ali, S.M., Kim, D.H., Guertin, D.A., Latek, R.R., Erdjument-Bromage, H., Tempst, P., and Sabatini, D.M. (2004). Rictor, a novel binding partner of mTOR, defines a rapamycin-insensitive and raptor-independent pathway that regulates the cytoskeleton. *Curr Biol* 14:1296–1302.

11. Kim, D.H., Sarbassov, D.D., Ali, S.M., Latek, R.R., Guntur, K.V., Erdjument-Bromage, H., Tempst, P., and Sabatini, D.M. (2003). GbetaL, a positive regulator of the rapamycin-sensitive pathway required for the nutrient-sensitive interaction between raptor and mTOR. *Mol Cell* 11:895–904.
12. Frias, M.A., Thoreen, C.C., Jaffe, J.D., Schroder, W., Sculley, T., Carr, S.A., and Sabatini, D.M. (2006). mSin1 is necessary for Akt/PKB phosphorylation, and its isoforms define three distinct mTORC2s. *Curr Biol* 16:1865–1870.
13. Yang, Q., Inoki, K., Ikenoue, T., and Guan, K.L. (2006). Identification of Sin1 as an essential TORC2 component required for complex formation and kinase activity. *Genes Dev* 20:2820–2832.
14. Jacinto, E., Facchinetti, V., Liu, D., Soto, N., Wei, S., Jung, S.Y., Huang, Q., Qin, J., and Su, B. (2006). SIN1/MIP1 maintains rictor–mTOR complex integrity and regulates Akt phosphorylation and substrate specificity. *Cell* 127:125–137.
15. Pearce, L.R., Huang, X., Boudeau, J., Pawlowski, R., Wullschleger, S., Deak, M., Ibrahim, A.F., Gourlay, R., Magnuson, M.A., and Alessi, D.R. (2007). Identification of Protor as a novel Rictor-binding component of mTOR complex-2. *Biochem J* 405:513–522.
16. Woo, S.Y., Kim, D.H., Jun, C.B., Kim, Y.M., Haar, E.V., Lee, S.I., Hegg, J.W., Bandhakavi, S., Griffin, T.J., and Kim, D.H. (2007). PRR5, a novel component of mTOR complex 2, regulates platelet-derived growth factor receptor beta expression and signaling. *J Biol Chem* 282:25604–25612.
17. Thedieck, K., Polak, P., Kim, M.L., Molle, K.D., Cohen, A., Jeno, P., Arrieumerlou, C., and Hall, M.N. (2007). PRAS40 and PRR5-like protein are new mTOR interactors that regulate apoptosis. *PLoS One* 2:e1217.
18. Brown, E.J., Beal, P.A., Keith, C.T., Chen, J., Shin, T.B., and Schreiber, S.L. (1995). Control of p70 s6 kinase by kinase activity of FRAP in vivo. *Nature* 377:441–446.
19. Brunn, G.J., Williams, J., Sabers, C., Wiederrecht, G., Lawrence, J.C., Jr., and Abraham, R.T. (1996). Direct inhibition of the signaling functions of the mammalian target of rapamycin by the phosphoinositide 3-kinase inhibitors, wortmannin and LY294002. *EMBO J* 15:5256–5267.
20. Brunn, G.J., Fadden, P., Haystead, T.A., and Lawrence, J.C., Jr. (1997). The mammalian target of rapamycin phosphorylates sites having a (Ser/Thr)-Pro motif and is activated by antibodies to a region near its COOH terminus. *J Biol Chem* 272:32547–32550.
21. Brunn, G.J., Hudson, C.C., Sekulic, A., Williams, J.M., Hosoi, H., Houghton, P.J., Lawrence, J.C., Jr., and Abraham, R.T. (1997). Phosphorylation of the translational repressor PHAS-I by the mammalian target of rapamycin. *Science* 277:99–101.
22. Burnett, P.E., Barrow, R.K., Cohen, N.A., Snyder, S.H., and Sabatini, D.M. (1998). RAFT1 phosphorylation of the translational regulators p70 S6 kinase and 4E-BP1. *Proc Natl Acad Sci USA* 95:1432–1437.
23. Keith, C.T., and Schreiber, S.L. (1995). PIK-related kinases: DNA repair, recombination, and cell cycle checkpoints. *Science* 270:50–51.
24. Andrade, M.A., and Bork, P. (1995). HEAT repeats in the Huntington's disease protein. *Nat Genet* 11:115–116.
25. Kobe, B., Gleichmann, T., Horne, J., Jennings, I.G., Scotney, P.D., and Teh, T. (1999). Turn up the HEAT. *Structure* 7:R91–R97.
26. Perry, J., and Kleckner, N. (2003). The ATRs, ATMs, and TORs are giant HEAT repeat proteins. *Cell* 112:151–155.
27. Wullschleger, S., Loewith, R., Oppliger, W., and Hall, M.N. (2005). Molecular organization of target of rapamycin complex 2. *J Biol Chem* 280:30697–30704.

28. Takahara, T., Hara, K., Yonezawa, K., Sorimachi, H., and Maeda, T. (2006). Nutrient-dependent multimerization of the mammalian target of rapamycin through the N-terminal HEAT repeat region. *J Biol Chem* 281:28605–28614.
29. Kunz, J., Schneider, U., Howald, I., Schmidt, A., and Hall, M.N. (2000). HEAT repeats mediate plasma membrane localization of Tor2p in yeast. *J Biol Chem* 275:37011–37020.
30. Bosotti, R., Isacchi, A., and Sonnhammer, E.L. (2000). FAT: a novel domain in PIK-related kinases. *Trends Biochem Sci* 25:225–227.
31. Urano, J., Sato, T., Matsuo, T., Otsubo, Y., Yamamoto, M., and Tamanoi, F. (2007). Point mutations in TOR confer Rheb-independent growth in fission yeast and nutrient-independent mammalian TOR signaling in mammalian cells. *Proc Natl Acad Sci USA* 104:3514–3519.
32. Takahashi, T., Hara, K., Inoue, H., Kawa, Y., Tokunaga, C., Hidayat, S., Yoshino, K., Kuroda, Y., and Yonezawa, K. (2000). Carboxyl-terminal region conserved among phosphoinositide-kinase-related kinases is indispensable for mTOR function in vivo and in vitro. *Genes Cells* 5:765–775.
33. Chen, J., Zheng, X.F., Brown, E.J., and Schreiber, S.L. (1995). Identification of an 11-kDa FKBP12-rapamycin-binding domain within the 289-kDa FKBP12-rapamycin-associated protein and characterization of a critical serine residue. *Proc Natl Acad Sci USA* 92:4947–4951.
34. Choi, J., Chen, J., Schreiber, S.L., and Clardy, J. (1996). Structure of the FKBP12-rapamycin complex interacting with the binding domain of human FRAP. *Science* 273:239–242.
35. Sehgal, S.N., Baker, H., and Vezina, C. (1975). Rapamycin (AY-22, 989), a new antifungal antibiotic. II. Fermentation, isolation and characterization. *J Antibiot (Tokyo)* 28:727–732.
36. Vezina, C., Kudelski, A., and Sehgal, S.N. (1975). Rapamycin (AY-22, 989), a new antifungal antibiotic. I. Taxonomy of the producing streptomycete and isolation of the active principle. *J Antibiot (Tokyo)* 28:721–726.
37. Sarbassov, D.D., Ali, S.M., Sengupta, S., Sheen, J.H., Hsu, P.P., Bagley, A.F., Markhard, A.L., and Sabatini, D.M. (2006). Prolonged rapamycin treatment inhibits mTORC2 assembly and Akt/PKB. *Mol Cell* 22:159–168.
38. Rodgers, B.D., Levine, M.A., Bernier, M., and Montrose-Rafizadeh, C. (2001). Insulin regulation of a novel WD-40 repeat protein in adipocytes. *J Endocrinol* 168:325–332.
39. Guertin, D.A., Stevens, D.M., Thoreen, C.C., Burds, A.A., Kalaany, N.Y., Moffat, J., Brown, M., Fitzgerald, K.J., and Sabatini, D.M. (2006). Ablation in mice of the mTORC components raptor, rictor, or mLST8 reveals that mTORC2 is required for signaling to Akt-FOXO and PKCalpha, but not S6K1. *Dev Cell* 11:859–871.
40. Cheng, J., Zhang, D., Kim, K., Zhao, Y., Zhao, Y., and Su, B. (2005). Mip1, an MEKK2-interacting protein, controls MEKK2 dimerization and activation. *Mol Cell Biol* 25:5955–5964.
41. Schroder, W., Bushell, G., and Sculley, T. (2005). The human stress-activated protein kinase-interacting 1 gene encodes JNK-binding proteins. *Cell Signal* 17:761–767.
42. Makino, C., Sano, Y., Shinagawa, T., Millar, J.B., and Ishii, S. (2006). Sin1 binds to both ATF-2 and p38 and enhances ATF-2-dependent transcription in an SAPK signaling pathway. *Genes Cells* 11:1239–1251.
43. Schroder, W., Cloonan, N., Bushell, G., and Sculley, T. (2004). Alternative polyadenylation and splicing of mRNAs transcribed from the human Sin1 gene. *Gene* 339:17–23.
44. Schroder, W.A., Buck, M., Cloonan, N., Hancock, J.F., Suhrbier, A., Sculley, T., and Bushell, G. (2007). Human Sin1 contains Ras-binding and pleckstrin homology domains and suppresses Ras signalling. *Cell Signal* 19:1279–1289.

45. Johnstone, C.N., Castellvi-Bel, S., Chang, L.M., Sung, R.K., Bowser, M.J., Pique, J.M., Castells, A., and Rustgi, A.K. (2005). PRR5 encodes a conserved proline-rich protein predominant in kidney: analysis of genomic organization, expression, and mutation status in breast and colorectal carcinomas. *Genomics* 85:338–351.
46. Gatherar, I.M., Pollerman, S., Dunn-Coleman, N., and Turner, G. (2004). Identification of a novel gene hbrB required for polarised growth in *Aspergillus nidulans*. *Fungal Genet Biol* 41:463–471.
47. Gangloff, Y.G., Mueller, M., Dann, S.G., Svoboda, P., Sticker, M., Spetz, J.F., Um, S.H., Brown, E.J., Cereghini, S., Thomas, G., and Kozma, S.C. (2004). Disruption of the mouse mTOR gene leads to early postimplantation lethality and prohibits embryonic stem cell development. *Mol Cell Biol* 24:9508–9516.
48. Murakami, M., Ichisaka, T., Maeda, M., Oshiro, N., Hara, K., Edenhofer, F., Kiyama, H., Yonezawa, K., and Yamanaka, S. (2004). mTOR is essential for growth and proliferation in early mouse embryos and embryonic stem cells. *Mol Cell Biol* 24:6710–6718.
49. Shiota, C., Woo, J.T., Lindner, J., Shelton, K.D., and Magnuson, M.A. (2006). Multiallelic disruption of the rictor gene in mice reveals that mTOR complex 2 is essential for fetal growth and viability. *Dev Cell* 11:583–589.
50. Sarbassov, D.D., Guertin, D.A., Ali, S.M., and Sabatini, D.M. (2005). Phosphorylation and regulation of Akt/PKB by the rictor-mTOR complex. *Science* 307:1098–1101.
51. Kumar, A., Harris, T.E., Keller, S.R., Choi, K.M., Magnuson, M.A., and Lawrence, J.C., Jr. (2008). Muscle-specific deletion of rictor impairs insulin-stimulated glucose transport and enhances Basal glycogen synthase activity. *Mol Cell Biol* 28:61–70.
52. Bentzinger, C.F., Romanino, K., Cloetta, D., Lin, S., Mascarenhas, J.B., Oliveri, F., Xia, J., Casanova, E., Costa, C.F., Brink, M., Zorzato, F., Hall, M.N., *et al.* (2008). Skeletal muscle-specific ablation of raptor, but not of rictor, causes metabolic changes and results in muscle dystrophy. *Cell Metab* 8:411–424.
53. Viniegra, J.G., Martinez, N., Modirassari, P., Losa, J.H., Parada Cobo, C., Lobo, V.J., Luquero, C.I., Alvarez-Vallina, L., Ramon y Cajal, S., Rojas, J.M., and Sanchez-Prieto, R. (2005). Full activation of PKB/Akt in response to insulin or ionizing radiation is mediated through ATM. *J Biol Chem* 280:4029–4036.
54. Feng, J., Park, J., Cron, P., Hess, D., and Hemmings, B.A. (2004). Identification of a PKB/Akt hydrophobic motif Ser-473 kinase as DNA-dependent protein kinase. *J Biol Chem* 279:41189–41196.
55. Troussard, A.A., Mawji, N.M., Ong, C., Mui, A., St-Arnaud, R., and Dedhar, S. (2003). Conditional knock-out of integrin-linked kinase demonstrates an essential role in protein kinase B/Akt activation. *J Biol Chem* 278:22374–22378.
56. Bozulic, L., Surucu, B., Hynx, D., and Hemmings, B.A. (2008). PKBalpha/Akt1 acts downstream of DNA-PK in the DNA double-strand break response and promotes survival. *Mol Cell* 30:203–213.
57. McDonald, P.C., Oloumi, A., Mills, J., Dobreva, I., Maidan, M., Gray, V., Wederell, E.D., Bally, M.B., Foster, L.J., and Dedhar, S. (2008). Rictor and integrin-linked kinase interact and regulate Akt phosphorylation and cancer cell survival. *Cancer Res* 68:1618–1624.
58. Pearce, L.R., Komander, D., and Alessi, D.R. (2010). The nuts and bolts of AGC protein kinases. *Nat Rev Mol Cell Biol* 11:9–22.
59. Alessi, D.R., Andjelkovic, M., Caudwell, B., Cron, P., Morrice, N., Cohen, P., and Hemmings, B.A. (1996). Mechanism of activation of protein kinase B by insulin and IGF-1. *EMBO J* 15:6541–6551.

60. Alessi, D.R., James, S.R., Downes, C.P., Holmes, A.B., Gaffney, P.R., Reese, C.B., and Cohen, P. (1997). Characterization of a 3-phosphoinositide-dependent protein kinase which phosphorylates and activates protein kinase Balpha. *Curr Biol* 7:261–269.
61. Hresko, R.C., and Mueckler, M. (2005). mTOR.RICTOR is the Ser473 kinase for Akt/protein kinase B in 3T3-L1 adipocytes. *J Biol Chem* 280:40406–40416.
62. Kumar, A., Lawrence, J.C., Jr., Jung, D.Y., Ko, H.J., Keller, S.R., Kim, J.K., Magnuson, M.A., and Harris, T.E. (2010). Fat cell-specific ablation of rictor in mice impairs insulin-regulated fat cell and whole body glucose and lipid metabolism. *Diabetes* 59:1397–1406.
63. Guertin, D.A., Stevens, D.M., Saitoh, M., Kinkel, S., Crosby, K., Sheen, J.H., Mullholland, D.J., Magnuson, M.A., Wu, H., and Sabatini, D.M. (2009). mTOR complex 2 is required for the development of prostate cancer induced by Pten loss in mice. *Cancer Cell* 15:148–159.
64. Ikenoue, T., Inoki, K., Yang, Q., Zhou, X., and Guan, K.L. (2008). Essential function of TORC2 in PKC and Akt turn motif phosphorylation, maturation and signalling. *EMBO J* 27:1919–1931.
65. Facchinetti, V., Ouyang, W., Wei, H., Soto, N., Lazorchak, A., Gould, C., Lowry, C., Newton, A.C., Mao, Y., Miao, R.Q., Sessa, W.C., Qin, J., *et al.* (2008). The mammalian target of rapamycin complex 2 controls folding and stability of Akt and protein kinase C. *EMBO J* 27:1932–1943.
66. Garcia-Martinez, J.M., and Alessi, D.R. (2008). mTOR complex 2 (mTORC2) controls hydrophobic motif phosphorylation and activation of serum- and glucocorticoid-induced protein kinase 1 (SGK1). *Biochem J* 416:375–385.
67. Hauge, C., Antal, T.L., Hirschberg, D., Doehn, U., Thorup, K., Idrissova, L., Hansen, K., Jensen, O.N., Jorgensen, T.J., Biondi, R.M., and Frodin, M. (2007). Mechanism for activation of the growth factor-activated AGC kinases by turn motif phosphorylation. *EMBO J* 26:2251–2261.
68. Roulin, D., Cerantola, Y., Dormond-Meuwly, A., Demartines, N., and Dormond, O. (2010). Targeting mTORC2 inhibits colon cancer cell proliferation in vitro and tumor formation in vivo. *Mol Cancer* 9:57.
69. Masri, J., Bernath, A., Martin, J., Jo, O.D., Vartanian, R., Funk, A., and Gera, J. (2007). mTORC2 activity is elevated in gliomas and promotes growth and cell motility via over-expression of rictor. *Cancer Res* 67:11712–11720.
70. Soukas, A.A., Kane, E.A., Carr, C.E., Melo, J.A., and Ruvkun, G. (2009). Rictor/TORC2 regulates fat metabolism, feeding, growth, and life span in *Caenorhabditis elegans*. *Genes Dev* 23:496–511.
71. Jones, K.T., Greer, E.R., Pearce, D., and Ashrafi, K. (2009). Rictor/TORC2 regulates *Caenorhabditis elegans* fat storage, body size, and development through sgk-1. *PLoS Biol* 7:e60.
72. Aronova, S., Wedaman, K., Aronov, P.A., Fontes, K., Ramos, K., Hammock, B.D., and Powers, T. (2008). Regulation of ceramide biosynthesis by TOR complex 2. *Cell Metab* 7:148–158.
73. Cybulski, N., Polak, P., Auwerx, J., Ruegg, M.A., and Hall, M.N. (2009). mTOR complex 2 in adipose tissue negatively controls whole-body growth. *Proc Natl Acad Sci USA* 106:9902–9907.
74. Chen, C.C., Jeon, S.M., Bhaskar, P.T., Nogueira, V., Sundararajan, D., Tonic, I., Park, Y., and Hay, N. (2010). FoxOs inhibit mTORC1 and activate Akt by inducing the expression of Sestrin3 and Rictor. *Dev Cell* 18:592–604.
75. Partovian, C., Ju, R., Zhuang, Z.W., Martin, K.A., and Simons, M. (2008). Syndecan-4 regulates subcellular localization of mTOR Complex2 and Akt activation in a PKCalpha-dependent manner in endothelial cells. *Mol Cell* 32:140–149.

76. Daub, H., Olsen, J.V., Bairlein, M., Gnad, F., Oppermann, F.S., Korner, R., Greff, Z., Keri, G., Stemmann, O., and Mann, M. (2008). Kinase-selective enrichment enables quantitative phosphoproteomics of the kinome across the cell cycle. *Mol Cell* 31:438–448.
77. Oppermann, F.S., Gnad, F., Olsen, J.V., Hornberger, R., Greff, Z., Keri, G., Mann, M., and Daub, H. (2009). Large-scale proteomics analysis of the human kinome. *Mol Cell Proteomics* 8:1751–1764.
78. Zanivan, S., Gnad, F., Wickstrom, S.A., Geiger, T., Macek, B., Cox, J., Fassler, R., and Mann, M. (2008). Solid tumor proteome and phosphoproteome analysis by high resolution mass spectrometry. *J Proteome Res* 7:5314–5326.
79. Chen, R.Q., Yang, Q.K., Lu, B.W., Yi, W., Cantin, G., Chen, Y.L., Fearns, C., Yates, J.R., III, and Lee, J.D. (2009). CDC25B mediates rapamycin-induced oncogenic responses in cancer cells. *Cancer Res* 69:2663–2668.
80. Dephoure, N., Zhou, C., Villen, J., Beausoleil, S.A., Bakalarski, C.E., Elledge, S.J., and Gygi, S.P. (2008). A quantitative atlas of mitotic phosphorylation. *Proc Natl Acad Sci USA* 105:10762–10767.
81. Olsen, J.V., Blagoev, B., Gnad, F., Macek, B., Kumar, C., Mortensen, P., and Mann, M. (2006). Global, in vivo, and site-specific phosphorylation dynamics in signaling networks. *Cell* 127:635–648.
82. Cheng, S.W., Fryer, L.G., Carling, D., and Shepherd, P.R. (2004). Thr2446 is a novel mammalian target of rapamycin (mTOR) phosphorylation site regulated by nutrient status. *J Biol Chem* 279:15719–15722.
83. Nave, B.T., Ouwens, M., Withers, D.J., Alessi, D.R., and Shepherd, P.R. (1999). Mammalian target of rapamycin is a direct target for protein kinase B: identification of a convergence point for opposing effects of insulin and amino-acid deficiency on protein translation. *Biochem J* 344(Pt 2):427–431.
84. Peterson, R.T., Beal, P.A., Comb, M.J., and Schreiber, S.L. (2000). FKBP12-rapamycin-associated protein (FRAP) autophosphorylates at serine 2481 under translationally repressive conditions. *J Biol Chem* 275:7416–7423.
85. Copp, J., Manning, G., and Hunter, T. (2009). TORC-specific phosphorylation of mammalian target of rapamycin (mTOR): phospho-Ser2481 is a marker for intact mTOR signaling complex 2. *Cancer Res* 69:1821–1827.
86. Chiang, G.G., and Abraham, R.T. (2005). Phosphorylation of mammalian target of rapamycin (mTOR) at Ser-2448 is mediated by p70S6 kinase. *J Biol Chem* 280:25485–25490.
87. Holz, M.K., and Blenis, J. (2005). Identification of S6 kinase 1 as a novel mammalian target of rapamycin (mTOR)-phosphorylating kinase. *J Biol Chem* 280:26089–26093.
88. Acosta-Jaquez, H.A., Keller, J.A., Foster, K.G., Ekim, B., Soliman, G.A., Feener, E.P., Ballif, B.A., and Fingar, D.C. (2009). Site-specific mTOR phosphorylation promotes mTORC1-mediated signaling and cell growth. *Mol Cell Biol* 29:4308–4324.
89. Akcakanat, A., Singh, G., Hung, M.C., and Meric-Bernstam, F. (2007). Rapamycin regulates the phosphorylation of rictor. *Biochem Biophys Res Commun* 362:330–333.
90. Dibble, C.C., Asara, J.M., and Manning, B.D. (2009). Characterization of Rictor phosphorylation sites reveals direct regulation of mTOR complex 2 by S6K1. *Mol Cell Biol* 29:5657–5670.
91. Treins, C., Warne, P.H., Magnuson, M.A., Pende, M., and Downward, J. (2010). Rictor is a novel target of p70 S6 kinase-1. *Oncogene* 29:1003–1016.
92. Julien, L.A., Carriere, A., Moreau, J., and Roux, P.P. (2010). mTORC1-activated S6K1 phosphorylates Rictor on threonine 1135 and regulates mTORC2 signaling. *Mol Cell Biol* 30:908–921.

93. Boulbes, D., Chen, C.H., Shaikenov, T., Agarwal, N.K., Peterson, T.R., Addona, T.A., Keshishian, H., Carr, S.A., Magnuson, M.A., Sabatini, D.M., and Sarbassov, D.D. (2010). Rictor phosphorylation on the Thr-1135 site does not require mammalian target of rapamycin complex 2. *Mol Cancer Res.* 8(6):896–906.
94. Brazil, D.P., and Hemmings, B.A. (2001). Ten years of protein kinase B signalling: a hard Akt to follow. *Trends Biochem Sci* 26:657–664.
95. Ito, N., and Rubin, G.M. (1999). gigas, a *Drosophila* homolog of tuberous sclerosis gene product-2, regulates the cell cycle. *Cell* 96:529–539.
96. Gao, X., and Pan, D. (2001). TSC1 and TSC2 tumor suppressors antagonize insulin signaling in cell growth. *Genes Dev* 15:1383–1392.
97. Potter, C.J., Huang, H., and Xu, T. (2001). *Drosophila* Tsc1 functions with Tsc2 to antagonize insulin signaling in regulating cell growth, cell proliferation, and organ size. *Cell* 105:357–368.
98. Tapon, N., Ito, N., Dickson, B.J., Treisman, J.E., and Hariharan, I.K. (2001). The *Drosophila* tuberous sclerosis complex gene homologs restrict cell growth and cell proliferation. *Cell* 105:345–355.
99. Garami, A., Zwartkruis, F.J., Nobukuni, T., Joaquin, M., Roccio, M., Stocker, H., Kozma, S.C., Hafen, E., Bos, J.L., and Thomas, G. (2003). Insulin activation of Rheb, a mediator of mTOR/S6K/4E-BP signaling, is inhibited by TSC1 and 2. *Mol Cell* 11:1457–1466.
100. Saucedo, L.J., Gao, X., Chiarelli, D.A., Li, L., Pan, D., and Edgar, B.A. (2003). Rheb promotes cell growth as a component of the insulin/TOR signalling network. *Nat Cell Biol* 5:566–571.
101. Stocker, H., Radimerski, T., Schindelholz, B., Wittwer, F., Belawat, P., Daram, P., Breuer, S., Thomas, G., and Hafen, E. (2003). Rheb is an essential regulator of S6K in controlling cell growth in *Drosophila*. *Nat Cell Biol* 5:559–565.
102. Zhang, Y., Gao, X., Saucedo, L.J., Ru, B., Edgar, B.A., and Pan, D. (2003). Rheb is a direct target of the tuberous sclerosis tumour suppressor proteins. *Nat Cell Biol* 5:578–581.
103. Yang, Q., Inoki, K., Kim, E., and Guan, K.L. (2006). TSC1/TSC2 and Rheb have different effects on TORC1 and TORC2 activity. *Proc Natl Acad Sci USA* 103:6811–6816.
104. Huang, J., Dibble, C.C., Matsuzaki, M., and Manning, B.D. (2008). The TSC1–TSC2 complex is required for proper activation of mTOR complex 2. *Mol Cell Biol* 28:4104–4115.
105. Huang, J., Wu, S., Wu, C.L., and Manning, B.D. (2009). Signaling events downstream of mammalian target of rapamycin complex 2 are attenuated in cells and tumors deficient for the tuberous sclerosis complex tumor suppressors. *Cancer Res* 69:6107–6114.
106. Berg, C.E., Lavan, B.E., and Rondinone, C.M. (2002). Rapamycin partially prevents insulin resistance induced by chronic insulin treatment. *Biochem Biophys Res Commun* 293:1021–1027.
107. Greene, M.W., Sakaue, H., Wang, L., Alessi, D.R., and Roth, R.A. (2003). Modulation of insulin-stimulated degradation of human insulin receptor substrate-1 by Serine 312 phosphorylation. *J Biol Chem* 278:8199–8211.
108. Haruta, T., Uno, T., Kawahara, J., Takano, A., Egawa, K., Sharma, P.M., Olefsky, J.M., and Kobayashi, M. (2000). A rapamycin-sensitive pathway down-regulates insulin signaling via phosphorylation and proteasomal degradation of insulin receptor substrate-1. *Mol Endocrinol* 14:783–794.
109. Pirola, L., Bonnafous, S., Johnston, A.M., Chaussade, C., Portis, F., and Van Obberghen, E. (2003). Phosphoinositide 3-kinase-mediated reduction of insulin receptor substrate-1/

2 protein expression via different mechanisms contributes to the insulin-induced desensitization of its signaling pathways in L6 muscle cells. *J Biol Chem* 278:15641–15651.
110. Tremblay, F., and Marette, A. (2001). Amino acid and insulin signaling via the mTOR/p70 S6 kinase pathway. A negative feedback mechanism leading to insulin resistance in skeletal muscle cells. *J Biol Chem* 276:38052–38060.
111. Zhande, R., Mitchell, J.J., Wu, J., and Sun, X.J. (2002). Molecular mechanism of insulin-induced degradation of insulin receptor substrate 1. *Mol Cell Biol* 22:1016–1026.
112. O'Reilly, K.E., Rojo, F., She, Q.B., Solit, D., Mills, G.B., Smith, D., Lane, H., Hofmann, F., Hicklin, D.J., Ludwig, D.L., Baselga, J., and Rosen, N. (2006). mTOR inhibition induces upstream receptor tyrosine kinase signaling and activates Akt. *Cancer Res* 66:1500–1508.
113. Tabernero, J., Rojo, F., Calvo, E., Burris, H., Judson, I., Hazell, K., Martinelli, E., Ramon y Cajal, S., Jones, S., Vidal, L., Shand, N., Macarulla, T., *et al.* (2008). Dose- and schedule-dependent inhibition of the mammalian target of rapamycin pathway with everolimus: a phase I tumor pharmacodynamic study in patients with advanced solid tumors. *J Clin Oncol* 26:1603–1610.
114. Radimerski, T., Montagne, J., Hemmings-Mieszczak, M., and Thomas, G. (2002). Lethality of *Drosophila* lacking TSC tumor suppressor function rescued by reducing dS6K signaling. *Genes Dev* 16:2627–2632.
115. Harrington, L.S., Findlay, G.M., Gray, A., Tolkacheva, T., Wigfield, S., Rebholz, H., Barnett, J., Leslie, N.R., Cheng, S., Shepherd, P.R., Gout, I., Downes, C.P., *et al.* (2004). The TSC1-2 tumor suppressor controls insulin-PI3K signaling via regulation of IRS proteins. *J Cell Biol* 166:213–223.
116. Shah, O.J., Wang, Z., and Hunter, T. (2004). Inappropriate activation of the TSC/Rheb/mTOR/S6K cassette induces IRS1/2 depletion, insulin resistance, and cell survival deficiencies. *Curr Biol* 14:1650–1656.
117. Um, S.H., Frigerio, F., Watanabe, M., Picard, F., Joaquin, M., Sticker, M., Fumagalli, S., Allegrini, P.R., Kozma, S.C., Auwerx, J., and Thomas, G. (2004). Absence of S6K1 protects against age- and diet-induced obesity while enhancing insulin sensitivity. *Nature* 431:200–205.
118. Zhang, H., Cicchetti, G., Onda, H., Koon, H.B., Asrican, K., Bajraszewski, N., Vazquez, F., Carpenter, C.L., and Kwiatkowski, D.J. (2003). Loss of Tsc1/Tsc2 activates mTOR and disrupts PI3K-Akt signaling through downregulation of PDGFR. *J Clin Invest* 112:1223–1233.
119. Peterson, T.R., Laplante, M., Thoreen, C.C., Sancak, Y., Kang, S.A., Kuehl, W.M., Gray, N.S., and Sabatini, D.M. (2009). DEPTOR is an mTOR inhibitor frequently overexpressed in multiple myeloma cells and required for their survival. *Cell* 137:873–886.
120. Sato, T., Nakashima, A., Guo, L., Coffman, K., and Tamanoi, F. (2010). Single amino-acid changes that confer constitutive activation of mTOR are discovered in human cancer. *Oncogene* 29:2746–2752.
121. Dancey, J. (2010). mTOR signaling and drug development in cancer. *Nat Rev Clin Oncol* 7:209–219.
122. Choo, A.Y., Yoon, S.O., Kim, S.G., Roux, P.P., and Blenis, J. (2008). Rapamycin differentially inhibits S6Ks and 4E-BP1 to mediate cell-type-specific repression of mRNA translation. *Proc Natl Acad Sci USA* 105:17414–17419.
123. Thoreen, C.C., Kang, S.A., Chang, J.W., Liu, Q., Zhang, J., Gao, Y., Reichling, L.J., Sim, T., Sabatini, D.M., and Gray, N.S. (2009). An ATP-competitive mammalian target of rapamycin inhibitor reveals rapamycin-resistant functions of mTORC1. *J Biol Chem* 284:8023–8032.

124. Feldman, M.E., Apsel, B., Uotila, A., Loewith, R., Knight, Z.A., Ruggero, D., and Shokat, K.M. (2009). Active-site inhibitors of mTOR target rapamycin-resistant outputs of mTORC1 and mTORC2. *PLoS Biol* 7:e38.
125. Garcia-Martinez, J.M., Moran, J., Clarke, R.G., Gray, A., Cosulich, S.C., Chresta, C.M., and Alessi, D.R. (2009). Ku-0063794 is a specific inhibitor of the mammalian target of rapamycin (mTOR). *Biochem J* 421:29–42.
126. Yu, K., Toral-Barza, L., Shi, C., Zhang, W.G., Lucas, J., Shor, B., Kim, J., Verheijen, J., Curran, K., Malwitz, D.J., Cole, D.C., Ellingboe, J., *et al.* (2009). Biochemical, cellular, and in vivo activity of novel ATP-competitive and selective inhibitors of the mammalian target of rapamycin. *Cancer Res* 69:6232–6240.
127. Chresta, C.M., Davies, B.R., Hickson, I., Harding, T., Cosulich, S., Critchlow, S.E., Vincent, J.P., Ellston, R., Jones, D., Sini, P., James, D., Howard, Z., *et al.* (2010). AZD8055 is a potent, selective, and orally bioavailable ATP-competitive mammalian target of rapamycin kinase inhibitor with in vitro and in vivo antitumor activity. *Cancer Res* 70:288–298.
128. Hsieh, A.C., Costa, M., Zollo, O., Davis, C., Feldman, M.E., Testa, J.R., Meyuhas, O., Shokat, K.M., and Ruggero, D. (2010). Genetic dissection of the oncogenic mTOR pathway reveals druggable addiction to translational control via 4EBP-eIF4E. *Cancer Cell* 17:249–261.
129. Janes, M.R., Limon, J.J., So, L., Chen, J., Lim, R.J., Chavez, M.A., Vu, C., Lilly, M.B., Mallya, S., Ong, S.T., Konopleva, M., Martin, M.B., *et al.* (2010). Effective and selective targeting of leukemia cells using a TORC1/2 kinase inhibitor. *Nat Med* 16:205–213.

6

TORC2 and Chemotaxis in Dictyostelium discoideum

YOICHIRO KAMIMURA • HUAQING CAI •
PETER N. DEVREOTES

Department of Cell Biology
Johns Hopkins University
School of Medicine
Baltimore, Maryland, USA

I. Abstract

Chemotaxis involves the cellular reactions of motility and directional sensing, which enable cells to sense and move along extracellular chemical gradients. *Dictyostelium discoideum* cells display robust chemotactic responses to cAMP. Extensive characterization of these responses has provided insights into the mechanisms of chemotaxis. In this system, chemotaxis is composed of multiple signaling pathways including the TORC2–PDK–PKB module where target of rapamycin complex 2 (TORC2) and phosphoinositide-dependent kinase (PDK) function as upstream activators of protein kinase B (PKBs) through hydrophobic motif (HM) and activation loop (AL) phosphorylations, respectively. This module forms a unique signaling pathway where chemoattractant signals are separated and then converge on substrate phosphorylations mediated by two PKB homologues, phosphatidylinositol (3,4,5)-tris phosphate (PtdIns (3,4,5)P_3)-dependent PKBA and -independent PKBR1. PKBA and PKBR1 contribute minor, redundant and major activities, respectively. Consistently, *pkbR1*$^-$ cells are more severely impaired in chemotaxis than cells lacking PKBA or PtdIns (3,4,5)P_3. The TORC2–PDK–PKB module is selectively activated at the

ISSN NO: 1874-6047
DOI: 10.1016/S1874-6047(10)28006-X

front of morphologically polarized cells by an upstream regulator, RasC, during chemotaxis. Spatial dysregulation of this pathway, exemplified by *pten*$^{-}$ cells or cells expressing an active form of RasC where PtdIns (3,4,5) P_3-PKBA or TORC2-dependent PKBR1 is overactivated, respectively, leads to extension of pseudopods all around cells and to severe chemotaxis defects. These results strongly suggest that this pathway has critical roles in linking chemotactic stimuli to cytoskeletal rearrangements. These studies shed light on the function and regulation of the TORC2–PDK–PKB pathway from a new perspective.

II. Introduction

Chemotaxis, the directed motility of cells in chemical gradients, is a fundamental process displayed by cells throughout evolution. It is required for proper organ formation in embryogenesis, wound healing, and immune responses including inflammation and infection control. In microorganisms, it is used for forging and cell–cell interactions. Although human neutrophils and *D. discoideum* amoebae are widely separated evolutionally, they display very similar chemotactic behavior with high sensitivity to chemoattractants and rapid amoeboid movement and both systems have provided important insights to chemotaxis [1, 2]. *D. discoideum* is a powerful model for studies of chemotaxis. It has a sequenced 34Mbp haploid genome, and contains many signaling pathways important in higher eukaryotes, including PI3K–PTEN as well as TORC2–PDK–PKB systems [3]. Moreover, the chemotactic response is very robust and has been extensively studied.

3′,5′-Cyclic adenosine monophosphate (cAMP) acting as a chemoattractant triggers activation of a series of signaling pathways by binding to the cAMP receptor (cAR1). cAR1 is a G-protein coupled receptor (GPCR) that signals through the heterotrimeric G-proteins to mediate many events including activation of small GTPases, phosphatidylinositol (3,4,5)-tris phosphate (PtdIns (3,4,5)P_3) production, adenylyl cyclase activity (ACA) activation, and actin polymerization. Gradients of chemoattractants induce a polarized morphology and many of cAMP-stimulated events occur locally at the front or back. PtdIns (3,4,5)P_3 accumulates at the front of chemotaxing cells by the opposing actions of two enzymes, phosphatidylinositol-3-kinase (PI3K) and phosphatase and tensin homologue deleted on chromosome ten (PTEN) [4–7]. PI3K and PTEN, respectively, add or remove a phosphate at 3 position of the inositol ring of PtdIns (3,4,5)P_3 lipids. PtdIns (3,4,5)P_3 is synthesized at the pseudopod where actin polymerization occurs and drives the cell forward. *D. discoideum* cells lacking the PTEN form pseudopods all around the perimeter, resulting in a severe impairment of chemotaxis.

This argues for an important function of PtdIns (3,4,5)P_3 [7]. However, PtdIns (3,4,5)P_3 is not essential for chemotaxis because cells lacking PI3K activity can still carry out chemotaxis [8, 9]. This discrepancy appears to be explained by cAMP stimulation of PtdIns (3,4,5)P_3-dependent and -independent pathways which are mediated by two PKB homologues, PKBA and PKBR1 [10]. Activities of PKBA and PKBR1 are regulated in a PtdIns (3,4,5)P_3-dependent or -independent manner, respectively [5, 11]. Activation of both kinases requires phosphorylations by TORC2 and PDK [10, 12, 13]. While PKBA is also regulated by PtdIns (3,4,5)P_3 through its PH domain, PKBR1, a myristoylated molecule, is heavily dependent on TORC2 and PDK for activation. Recent studies have shown that TORC2 is regulated downstream of RasC activation. This review summarizes our recent understanding of chemotaxis, with a particular focus on the TORC2–PDK–PKB pathway.

III. The Life Cycle of *D. discoideum*

D. discoideum cells proliferate as a unicellular organism that feeds on bacteria and yeasts in the soil (Figure 6.1). Metabolites from microorganisms, for example folate from bacteria, act as chemoattractants. In nutrient poor environments, the cells begin a developmental program that leads to the formation of a multicellular structure. First, starvation triggers the expression of a series of genes required for early development. Cells begin to spontaneously secrete cAMP every 6 min, which functions as a chemoattractant. About a million cells aggregate to form a "mound." A tip forms on the mound and elongates into a slug-like structure. The "slug" moves around sensing temperature and light. Finally, it transforms into a fruiting body that is composed of a basal disc, a stalk, and spores that endure in harsh environments. These cells use chemotaxis repeatedly at various stages of their life cycles.

In the laboratory, we can grow cells in defined liquid nutrient media or together with bacteria on an agar surface where they form plaques, starve, and differentiate. Using this system, a wide variety of developmentally defective mutants have been isolated and characterized. The most severe ones are unable to aggregate on bacteria lawns and form "smooth" plaques. Other mutants arrest at various different stages of development, such as mound, slug, etc. The early developmental phase is easily recapitulated by transferring growing cells into a nonnutrient buffer and incubating them until they reach "aggregation competence." These cells are used to assess chemotactic ability and biochemical events during cAMP stimulation. Numerous assays for chemotaxis have been described. One key biochemical event is cAMP-mediated cAMP production. A method to measure

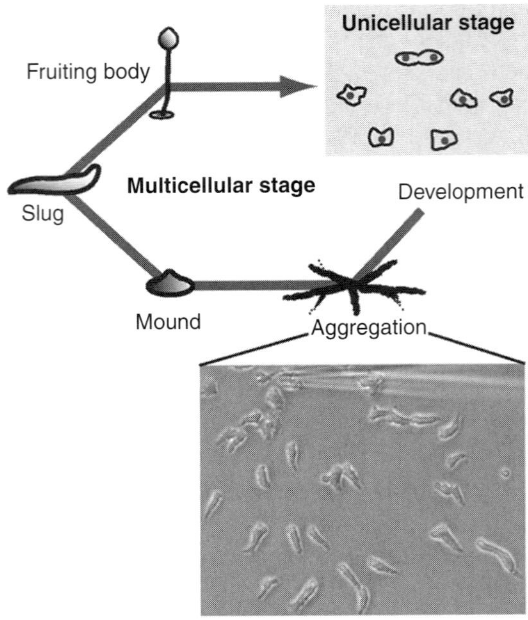

FIG. 6.1. The typical lifecycle of *D. discoideum* cells is depicted. In nutrient-rich environments, individual cells proliferate mitotically. However, once nutrients are deprived from their environment, cells begin a development program to form a multicellular aggregate by cAMP-mediated chemotaxis. An image of cells during chemotaxis toward a cAMP-filled micropipette is shown in the aggregation stage. The multicellular structure transforms to fruiting body through typical morphological steps, including mound and slug. When conditions are appropriate, spores germinate to grow vegetatively. More details are found in the text.

activation of ACA *in vitro* was established where nonhydrolyzable GTP, GTPγS, stimulates the enzymes acting through heterotrimeric G-proteins and small GTPases as described below. The cell lysate assay is also applied to a variety of other biochemical assays, such as TORC2–PDK-dependent PKB phosphorylations [10].

IV. The Components of TORC2–PDK–PKB Pathway in *D. discoideum*

A. PKB (PKBA and PKBR1)

AKT or PKB has important roles in diverse biological circumstances, including cell growth, development, and cell motility. The kinase domains of PKBs are highly conserved among different species. PKBA and PKBR1, two PKB homologues in *D. discoideum*, are structurally very similar except for the N-terminal region [5, 11] (Figure 6.2A). While PKBA is a canonical PKB and has a pleckstrin homology (PH) domain at the N-terminus, PKBR1 does not have PH domain but is myristoylated. Like mammalian PKB, PKBA localizes to PtdIns (3,4,5)P_3-enriched regions on the plasma membrane through its PH domain. In contrast, PKBR1 localizes to the plasma membrane constitutively and uniformly through its myristoylation.

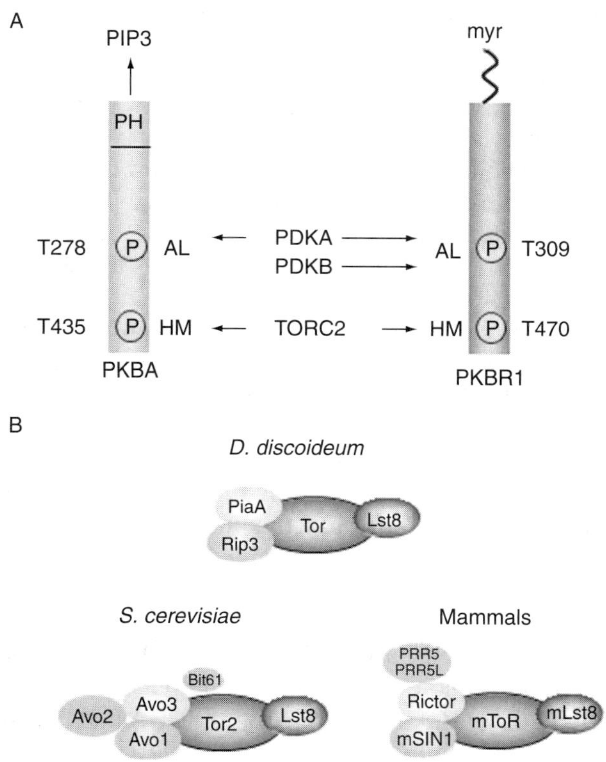

FIG. 6.2. The TORC2–PDK–PKB module in *D. discoideum* cells is shown. (A) The two PKB homologues, PKBA and PKBR1, are regulated by phosphorylations at the activation loop (AL) and the hydrophobic motif (HM). PKBA has a PtdIns (3,4,5)P_3 (PIP3)-dependent pleckstrin homology (PH) domain and PKBR1 is a myristoylated (myr) protein. Upon cAMP stimulation, PKBR1 and PKBA provide major and redundant, lesser activities, respectively. TORC2, composed of Tor, PiaA, Rip3, and Lst8, phosphorylates the HM of PKBA and PKBR1. The two PDK homologues, PDKA and PDKB, function as AL kinases. PDKA mainly phosphorylates PKBA and PKBR1 at the AL. PDKB phosphorylates only PKBR1. (B) The comparison of TORC2 structure in *D. discoideum*, *S. cerevisiae*, and mammals is shown.

Activation of PKB requires phosphorylations at the AL and at the HM. Extensive studies have shown that PDK and TORC2 are kinases responsible for the AL and HM phosphorylations [14–16]. PKBA and PKBR1 also have these consensus phosphorylation sites. The AL and the HM, respectively, at Thr278 and Thr435 in PKBA and at Thr309 and Thr470 in PKBR1 are essential for activity. Using PDK or TORC2 mutant cell lines, we found indeed PKBA and PKBR1 are phosphorylated at the AL and HM by PDK and TORC2, respectively [10, 13, 17]. However, the mechanisms of activation of PKBA and PKBR1 are somewhat complex. PKBA and PKBR1 with alanine mutations in the AL have impaired functions and, furthermore, reduced phosphorylations in the HM. The PKBs with alanine mutations in the HM are also not functional and display either no or greatly reduced AL phosphorylation for PKBR1 or PKBA, respectively [10, 17]. These results suggest that PKBA and PKBR1 activations require the initial phosphorylation at the HM by TORC2 and subsequent phosphorylation at the AL by PDKs. Moreover, the AL phosphorylation affects the stability or the efficiency of the HM phosphorylation. The reciprocal regulation between AL and HM is also supported by the experiments using *pdkA*$^-$*pdkB*$^-$ and TORC2 mutant cells [10, 13].

Mammalian AKT/PKB becomes constitutively active with phosphomimetic substitutions of both the AL and HM sites, suggesting that these phosphorylations are sufficient for its activation. However, activation of the PKBs in *D. discoideum* is not exactly the same as AKT/PKB in mammalian cells and requires further steps for activation [10]. First, PKBR1 with aspartate or glutamate in the AL is not active. Second, TORC2 appears to have additional functions beyond its role in HM phosphorylation because these substitutions at Thr470 of PKBR1 are sufficient for phosphorylation at the AL but not for function in cells lacking TORC2 subunit PiaA. We do not know whether PKBR1 with phosphomimetic substitutions at the HM in *piaA*$^-$ cells is impaired in kinase activity or another function, for example, substrate recognition. Although the exact mechanism remains unknown, another possible target of TORC2 is the turn motif (TM) of PKBs [18–21]. Several studies have shown the importance of TM phosphorylation by TORC2 in AGC kinases. Since both PKBA and PKBR1 conserve these sites, it is interesting to consider this possibility.

Cells lacking PKBA, PKBR1, or both PKBA and PKBR1 display different phenotypes, suggesting redundant and distinct functions *in vivo*. The *pkbA*$^-$ cells show weak chemotaxis defects and cells display slightly less polarity in early differentiation; however, they finally make normal fruiting bodies [5]. The *pkbR1*$^-$ cells are defective in chemotaxis and arrest at the "mound stage" [11]. While *pkbA*$^-$ or *pkbR1*$^-$ cells can grow in liquid medium, *pkbA*$^-$ *pkbR1*$^-$ cells cannot, suggesting overlapping functions for the two kinases in macropinocytosis [10, 11].

B. TORC2

D. discoideum express the counterparts of each of the subunits of TORC2, Tor, PiaA, Rip3, and Lst8 (Figure 6.2B). Tor encodes the catalytic subunit. The *tor* gene appears to be essential because targeted disruption of the gene by homologous recombination was not successful [12]. PianissimoA (PiaA) is an orthologue of Avo3 in budding yeast and RICTOR in mammals. PiaA was originally found in a genetic screen for cells defective in early development [22]. Independently, a mutant cell line that was temperature sensitive for early development was found to have an amino acid mutation of Asp at Gly917 in PiaA [23]. Ras interacting protein 3 (Rip3) is an orthologue of Avo1 in budding yeast and mSIN1 in mammals and was isolated as a factor that interacted with an activated form of mammalian Ha-Ras (Ha-Ras Gly12Val) in a yeast two-hybrid assay [24]. As the other homologues, Rip3 contains a Ras binding domain (RBD) and mutational analysis of the domain has demonstrated its functional importance [12]. The *lst8* gene was identified in the *D. discoideum* genome as an Lst8 orthologue of budding yeast or mammals [12].

Gene disruptions of all of the TORC2 subunits except for the *tor* catalytic gene were created and found to display similar phenotypes [10, 12, 22, 24]. First, they are unable to aggregate on bacteria lawns showing defects in early development upon nutrient depletion. Second, activation of PKBs, PKBA and PKBR1 upon cAMP stimulation of aggregation competent cells, are abolished. Furthermore, ACA activation and chemotaxis are abrogated in these mutants.

In addition to the similar phenotypes of the null mutant cells, the following observations also suggest that these factors function as a complex [12]. First, in coimmunoprecipitation experiments PiaA associated with Rip3 or Lst8, and both Tor and Lst8 were coimmunoprecipitated with PiaA in the presence of protein cross-linker. Second, in ACA activation experiments where extracts from the different mutant cell lines were mixed, no combinations yielded ACA activity, suggesting that a preformed complex is required. Recently, we found that PiaA proteins coimmunoprecipitate with the Tor catalytic subunit from wild-type cells but not from *rip3*$^{-}$ cells [25]. This suggests that TORC2 is unstable in *rip3*$^{-}$ cells. So far, purification of TORC2 has not been successful in *D. discoideum* and the exact composition of TORC2 awaits further examinations.

C. PDK (PDKA AND PDKB)

PDK1 belongs to the AGC family of kinases and was originally isolated as the kinase that catalyzes the AL phosphorylation of PKB in mammalian cells. *D. discoideum* cells have two PDK orthologues, PDKA and PDKB, that contain a kinase domain and a C-terminal PH domain. PDKA and PDKB share 67% and 58% similarities in the kinase and PH domains, respectively [13]. Although both PDKs are structurally similar, they are not completely redundant. PDKA is the major AL kinase of PKBA and PKBR1 and PDKB contributes to phosphorylation of only PKBR1. This preference could arise from the respective kinase domains, as the deletion of PH domain of PDKA does not change its specificity. Since PDKA and PDKB are the only structurally related homologues of PDK1 in *D. discoideum*, it was expected that *pdkA*$^-$ *pdkB*$^-$ cells would completely lack AL phosphorylations of PKBs. However, in *pdkA*$^-$ *pdkB*$^-$ cells, some residual AL phosphorylations of the PKBs remain, suggesting that another kinase is involved in the AL phosphorylation. Nonetheless, the *pdkA*$^-$ *pdkB*$^-$ cells display greatly reduced phosphorylations of the PKBs and show several phenotypes including the "mound arrest" and chemotaxis defects. These phenotypes are similar to those seen in *pkbR1*$^-$ or *pkbA*$^-$*pkbR1*$^-$ cells. These results confirm the fundamental role of the TORC2–PDK–PKB pathway in chemotaxis.

Although mammalian PDK1 requires its PH domain for binding to PtdIns (3,4,5)P_3 on the membrane and for its AL phosphorylation of AKT/PKB, the PH domains of the *D. discoideum* PDKs are not essential for their activities [13]. The localizations of PDKA and PDKB proteins are different. PDKB-GFP is found in the cytosol, while PDKA-GFP localizes in the cytosol and on the plasma membrane and translocates to the membrane during chemotactic stimulation. It is also found on macropinosomes which are membraneous vesicular structures enriched in PtdIns (3,4,5)P_3 and involved in endocytosis of liquid. To determine the dependency of the membrane localization of PDKA on PtdIns (3,4,5)P_3, cells lacking PI3K activity were examined. These cells indeed lack the PDKA-GFP signals on macropinosomes and the translocation does not occur. However, PDKA-GFP still localizes to the plasma membrane in unstimulated cells. In spite of these observations, PtdIns (3,4,5)P_3 binding of the PH domain of PDKA cannot be detected in biochemical assays using filters spotted with phospholipids [13, 17]. Together, these results suggest that the PH domain of PDKA can recognize PtdIns (3,4,5)P_3 and also other unidentified molecules on the membrane. Next, function of the PH domain of PDKA was examined by deleting it. Surprisingly, PDKAΔPH can rescue AL phosphorylation of the PKBs as well as the "mound arrest" defects of the *pdkA*$^-$ *pdkB*$^-$ cells. This result shows that the PH domain of PDKA is not essential for its function. However, when PDKAΔPH was combined with a point mutation in the kinase domain with reduced ATP binding activity, the double mutant showed reduced function compared with either single mutant, suggesting that the PH domain may enhance the function of the enzyme under certain conditions.

D. Other AGC Kinases in *D. discoideum*

In *D. discoideum*, 21 AGC kinases are assigned and some of them other than PKBA and PKBR1 have HM-like sequences. The genes carrying most closely related HMs to those found in the PKBs are uncharacterized DDB0220670 and DDB0220702. DDB0220670 also has a conserved TM and has a longer C-terminal tail following the HM than does PKBA and PKBR1. This structure is reminiscent of the budding yeast Sch9 and mammalian S6K1 [26]. Moreover, since DDB0220670 also has a motif in its C-terminus similar to the autoinhibitory pseudosubstrate domain of S6K1, the same molecular regulation may also be conserved in DDB0220670. Sch9 and S6K1 have been shown to be phosphorylated by TORC1 and involved in cellular growth. The *D. discoideum* genome encodes a homologue of the budding yeast Kog1/mammalian RAPTOR suggesting that TORC1 is present. Together this evidence suggests that DDB0220670 might be an orthologue of Sch9 and S6K1 and regulated by TORC1. It is also possible that DDB0220670 is a substrate of TORC2.

As described above, since upon nutrient starvation TORC2 mutant cell lines do not initiate the development process resulting in a "smooth plaque" while *pkbA*$^-$ *pkbR1*$^-$ cells arrest at the "mound stage," TORC2 may have an additional target in early development other than the PKBs. DDB0220702 could be one of the candidates.

V. The Signal Transduction Pathway for Chemotaxis

Chemoattractant-triggered signaling networks integrate processes of motility, directional sensing, and polarity to achieve efficient chemotaxis [1]. Studies of cAMP-mediated chemotaxis of *D. discoideum* cells in early development have provided the general insights into the process. These amoeboid cells move around randomly even in the absence of a chemoattractant gradient. Motility is driven largely by remodeling of the actin cytoskeleton to generate pseudopodial extensions and contraction force. Directional sensing is the process that decodes the extracellular gradient of chemoattractants. Cells detect gradient steepness even with differences of only 2% across the cell, over a wide range of midpoint concentrations. Cells amplify the gradient to produce localized changes in intracellular signaling molecules. Chemotaxing cells display morphologically distinct fronts and backs which are referred to as polarity. Pseudopod extensions occur at the front of cells and myosin II-mediated contraction occurs at the back. At the molecular level, many events also are spatially localized in a coordinated fashion that facilitates efficient chemotaxis. A working

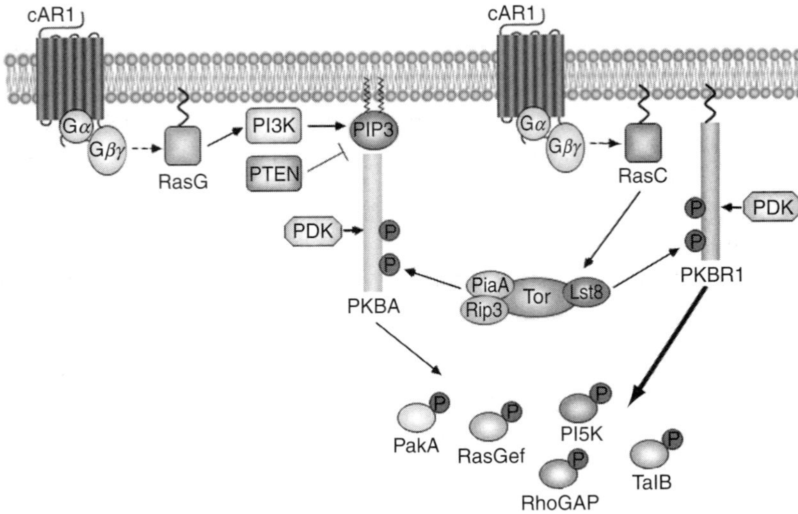

Fig. 6.3. The signaling network of cAMP-stimulated chemotaxis is summarized. cAMP binding to cAR1 stimulates dissociation between gα2 and gβγ subunit of the heterotrimeric G protein. RasG and RasC, small GTPases, are locally activated downstream of the heterotrimeric G protein at the front of cells. PI3 kinase (PI3K) is also recruited to the front and produces PtdIns (3,4,5)P_3 (PIP3) with RasG. PIP3 is detected by PIP3-specific PH domain containing proteins including PKBA. RasC functions as an activator of the PIP3-independent TORC2–PDK–PKBR1 pathway. PKBR1 and PKBA provide major and redundant, lesser activities, respectively. Both PKBs contribute to the phosphorylation of common substrates including TalB, RasGEF, PakA, PI5K, and RhoGAP linking chemotactic stimuli to rearrangement of cytoskeleton.

model of the signaling networks involved in chemotaxis is summarized in Figure 6.3 and is discussed below in more detail.

A. GPCR and Heterotrimeric G-Protein

cAMP released from cells is recognized by a series of cAMP receptors (cAR1–4) that comprise a class of GPCR [27]. Guanine nucleotide exchange within the α2 subunit, one of 12 encoded α subunits, leads to dissociation of the βγ subunit, which are unique in the genome. In *cAR1*$^-$, *gα2*$^-$, *gβ*$^-$, and *gγ*$^-$ cells, all early development events including chemotaxis are impaired [27]. Using fluorescently labeled cAMP and single molecule imaging, it was shown that cAR1 binds with affinities of 2–200 nM and an average half-life of about 1.5 s [28]. Fluorescent resonance energy transfer (FRET) experiments designed to monitor the dissociation between the gα2- and the gβ-subunit showed that the extent of subunit dissociation parallels the external gradient, suggesting that the G-protein activation is not amplified [29]. This implies that amplification occurs downstream of this module. Although the direct target of the heterotrimeric G-proteins is unknown, they mediate events including small GTPases activation, PtdIns (3,4,5)P_3 production, ACA activation, and actin polymerization. Most of these downstream events are polarized to the front or back of cells.

B. PtdIns (3,4,5)P_3-Dependent Pathway

PtdIns (3,4,5)P_3 accumulation at the front of chemotaxing cells is controlled by reciprocal regulation of PI3K and PTEN. The *D. discoideum* genome encodes five typical type-I PI3Ks, which have a RBD and a catalytic domain at the C-terminus [9]. PI3K1 and PI3K2, the major expressed PI3Ks, localize at the front of cells during chemotaxis through the N-terminal regions, which lack known motifs [6]. The RBDs of these PI3Ks have been shown to bind to an activated form of RasG, one of 15 Ras homologues in *D. discoideum*, in a yeast two-hybrid assay. RasG as well as other Ras proteins are activated at the front of cells. RasG activation depends on GefR as RasGEF and DdNF1 as RasGAP. Cells lacking GefR show only partially reduced RasG activation suggesting there are additional RasGEFs for RasG [30]. DdNF1, a homologue of human RasGAP neurofibromin 1 (NF1), found by the screening for impaired chemotaxis among RasGAP deletions was shown to be involved in RasG deactivation [31]. Importantly, cells lacking DdNF1 display abnormal activation of not only RasG but also PI3Ks resulting in chemotaxis defects with spatially unregulated pseudopod formation. Together, both recruitment to the membrane and activation by RasG contribute to the spatially localized activation of the PI3Ks. It is important to determine the binding sites for PI3Ks at the plasma membrane and determine the exact role of RasG in activation. As described above, PTEN localizes at the back of wild-type cells and helps confine accumulation of PtdIns (3,4,5)P_3 to the front. Cells lacking PTEN lose polarity and normal chemotactic activity since they accumulate PtdIns (3,4,5)P_3 along the entire cell.

The localized signal provided by PtdIns (3,4,5)P_3 is conveyed by PtdIns (3,4,5)P_3-specific PH domain containing proteins including PKBA, cytosolic regulator of adenylate cyclase (CRAC), and PH domain protein A (PHDA) in *D. discoideum* [4, 5, 32, 33]. These proteins translocate to the front of cells and when disrupted cause slight defects in chemotaxis. CRAC has an additional essential function for ACA activation [34]. Since CRAC and PHDA have no obvious domains, other than a PH domain, identification of interacting proteins will be important for determining the functions of these proteins. Among these PH domain containing proteins, PKBA appears to have an important role in chemotaxis, since the chemotaxis defects of *pten*$^-$ cells are largely suppressed by concomitantly deleting the *pkbA* gene (M. Tang, M. Iijima, Y. Kamimura, and P. Devreotes, manuscript in preparation). Although they still display highly elevated PtdIns (3,4,5)P_3 levels on the plasma membrane, *pten*$^-$ *pkbA*$^-$ cells display much stronger polarity than the *pten*$^-$ cells, with spatially restricted pseudopod extensions occurring primarily at the front of cells.

C. TORC2–PDK–PKB Pathway

Studies of the phosphorylation of PKB substrates have shown that most of these phosphorylation events are unaffected in the cells lacking PtdIns (3,4,5)P_3 production or PKBA but significantly reduced in *pkbR1*$^-$ cells [10]. This shows that PKBR1 and PKBA contribute major and redundant, lesser activities, respectively. Furthermore, in spite of the uniform localization of PKBR1 on the plasma membrane, its activation, as that of PKBA, occurs selectively at the front of cells. Consistently, chemotaxis is abrogated much more severely in *pkbR1*$^-$ cells than in *pkbA*$^-$ cells. Furthermore, cell lines lacking the other key activators of PKBs, TORC2 and PDKs, also display reduced PKB activities as well as chemotaxis defects. These results show that the TORC2- and PDK-mediated activation of PKBR1 is an important PtdIns (3,4,5)P_3-independent pathway in chemotaxis. Most of the phenotypes of mutants related to PKBA and PKBR1 can be explained by the following model (see Figure 6.3). In wild-type cells, both PKBR1 and PKBA are activated. Cells lacking PtdIns (3,4,5)P_3 or PKBA still retain enough PKBR1 activity for almost normal chemotaxis. This explanation implies that TORC2 as well as PDKs can be activated independently of PtdIns (3,4,5)P_3. However, since *pten*$^-$ cells overaccumulate PtdIns (3,4,5) P_3, PKBA misregulation leads to severe chemotaxis defects.

Implication of Ras proteins in TORC2 activation was suggested by several early findings. First, Rip3, a subunit of TORC2, has a RBD which binds to an activated form of RasG [24]. Second, GTPγS addition to cell extracts from aggregation competent cells can trigger TORC2-dependent PKBR1 activation [10]. Third, in *rasC*$^-$ *rasG*$^-$ cells, phosphorylation of the HM of PKBR1 is decreased by 70% [10]. Recently, we have shown that in *rasC*$^-$ cells PKB phosphorylation is equally reduced [25]. Moreover, *rasC*$^-$ cells share some phenotypes with TORC2 mutant cells, including a failure to aggregate and a reduction of ACA activation, although chemotaxis defects of *rasC*$^-$ cells are milder. Moreover, expression of an activated form of RasC(Gln62Leu) prolongs PKB activation and causes a severe loss of polarity and chemotaxis defects. These results assign RasC, but not RasG, as an activator of TORC2. Early screens for chemotaxis mutants led to identification of Aimless (AleA) which carried a RasGEF domain and was later demonstrated to be an exchange factor specific for RasC [30, 35]. The GAP for RasC has not been identified. Consistently, *aleA*$^-$ and *rasC*$^-$ cells show similar decreased levels of PKB phosphorylation [10, 25]. These results indicate the importance of AleA- and RasC-mediated activation of TORC2 in PKB regulation. Although the exact mechanism of this activation is unknown, we can speculate as to whether activated RasC directly regulates TORC2 or renders PKBs as more suitable substrates for TORC2.

D. Other PtdIns (3,4,5)P_3-Independent Pathway

In addition to the TORC–PDK–PKBR1 pathway, additional PtdIns (3,4,5)P_3-independent pathways appear to be needed for proper chemotaxis. Phospholipase A_2 (PLA_2) was found by screening for mutants that were supersensitive to PI3K inhibitors in early development [8]. In addition, depletion of PLA2 activity enhances the chemotaxis defects of PI3K mutant cells and of cells treated with PI3K inhibitors. These observations strongly suggest that PLA_2 also has a role in chemotaxis in the PtdIns (3,4,5)P_3-independent manner.

Yet, another PtdIns (3,4,5)P_3-independent pathway is cGMP-mediated Myosin II filament formation [36]. cGMP production upon chemoattractant addition is regulated by membrane-bound and soluble guanylyl cyclases, GCA and sGC, and a cGMP-specific phosphodiesterase, PdeD. Myosin II molecules, composed of Myosin heavy chain (MHC) and light chain (MLC), mostly form filaments at the back of cells and aid in producing contraction force during chemotaxis. Accordingly, *mhcA*$^-$ cells show reduced motility speed compared to wild-type cells and cells lacking guanylyl cyclase activity or the cGMP binding protein (GbpC) have a similar phenotype to that of *mhcA*$^-$ cells. Thus, these factors are thought to regulate Myosin II filament formation.

E. Downstream Events

One of most important downstream events is pseudopod formation through actin polymerization. The TORC2–PDK–PKB pathway has functions in actin polymerization, since dysregulated actin polymerizations occur in *pten*$^-$ cells or Ras(Gln62Leu) expressing cells [7, 25]. Some PKB substrates have already been identified, for example, Talin B (TalB), RasGEFs (GefS and GefN), p21 activated kinase (PakA), RhoGAP

(GacG and GacQ), and PI5K (Table 6.1) [10, 37] (M. Tang, M. Iijima, Y. Kamimura, and P. Devreotes, manuscript in preparation). Since many of these homologues are reported to have roles in actin polymerization dynamics in other organisms, further characterization is likely to link these factors to regulation of pseudopod formation.

cAMP-dependent ACA activation is critical for the signal relay responses between cells needed to efficiently form multicellular aggregates. Curiously, this pathway shares components with chemotactic responses, including the TORC2–PDK–PKB pathway. ACA is regulated by TORC2 and CRAC, independently, since ACA activation can be restored by mixing cell lysates from *piaA*$^-$ and *crac*$^-$ cells while only wild-type lysates will restore activation of *piaA*$^-$ *crac*$^-$ cells [22]. In *pten*$^-$ cells or in cells expressing RasC(Gln62Leu) ACA activation is prolonged [7, 25]. The effects of active RasC(Gln62Leu) are suppressed in *piaA*$^-$ cells, again confirming that TORC2 functions downstream of RasC. Furthermore, we have recently shown that *pkbA*$^-$ *pkbR1*$^-$ cells also impair ACA activation. Together, these results suggest that a PKB substrate, regulated by phosphorylation of TORC2-mediated PKB, could be an essential component for ACA activation.

TABLE 6.1

IDENTIFIED cAMP-STIMULATED PKB SUBSTRATES IN *D. DISCOIDEUM*

Gene DDB number	MW (kDa)	Putative PKB phosphorylation sites (RxRxxS/T)	Localization during chemotaxis	Motifs et al.	Ref.
talB DDB0191526	277.7	RIRGQT396	ND	Talin homologue	[10]
gefN DDB0167277	185.4	RLRSFT749 RSRAQT800	ND	RasGEF	[10]
pakA DDB0191313	134.6	RSRSHT585	Back	p21 activated kinase homologue	[37]
gacG DDB0233879	147.0	RRRTSS365 RERSSS480	ND	RhoGAP	[a]
gefS DDB0191324	116.9	RVRHST387 RIRSPS623	Cytosol	RasGEF	[10]
N.D. DDB0234212	80.5	RVRLNT262	Plasma membrane	Phosphatidylinositol-4-phosphate 5 kinase	[10]
gacQ DDB0233774	59.0	RQRSNT528	Cytosol	RhoGAP	[10]

[a]M. Tang, M. Iijima, Y. Kamimura, and P. Devreotes, manuscript in preparation.
ND is not determined.

TABLE 6.2

Identified Folate-Stimulated PKB Substrates in *D. discoideum*

Gene DDB number	MW (kDa)	Putative PKB phosphorylation sites (K/RxK/RxxS/T)	Localization during chemotaxis	Motifs et al.	Ref.
PHAPS DDB0307127	85.9	KKRTTT102	ND	SAM, PH	[17]
SHAPS DDB0306661	52.0	RPRASTT281	ND	BAR, SH3	[17]

ND is not determined. SAM is sterile alpha motif. PH is pleckstrin homology. BAR is Bin/Amphiphysin/Rvs.

F. Folate Chemotaxis

D. discoideum cells use folate as a chemoattractant for seeking bacteria. Although folate chemotaxis is less efficient than cAMP chemotaxis, the two phenomena appear to share similar but not identical signaling pathways. While folate receptors have not been identified, they are certainly not cAR1. They are likely GPCRs since Gα4 (instead of Gα2) and Gβγ are required for folate chemotaxis [27]. Whereas PKBR1 is the major PKB activity stimulated by cAMP, folate primarily activates PKBA [17]. Moreover, the phosphorylation profiles of PKB substrates differ during folate versus cAMP stimulation. In folate-stimulated cells, PKBA preferentially phosphorylates PH/AKT-preferential substrate (PHAPS) and SH3/AKT-preferential substrate (SHAPS) (Table 6.2). Interestingly, although the transient stimulation of PKBA is prevented, basal phosphorylations of PKBA at both the AL and HM are retained in TORC2 mutant cell lines, *piaA*$^-$, *rip3*$^-$, and *lst8*$^-$ cells. In contrast to PKBA, PKBR1 phosphorylations are completely lost in TORC2 mutant cells. This result suggests that under certain circumstances PKBA can be phosphorylated by an unidentified kinase. It is important to determine how folate chemotaxis is affected in cells lacking components of the TORC2–PDK–PKB pathway.

VI. Conclusion

The TORC2–PDK–PKB module undergoes spatially confined activation and has critical roles in chemotaxis. The spatial activation of this module is dependent on RasC, a subtype of Ras small GTPases. Since little is known about TORC2 regulation in other organisms, further characterization is needed to understand the molecular basis of activation mechanism. Especially, it is important to discriminate whether the target of RasC is TORC2,

PKBs, or both. As described here, the activation of PKBs by TORC2 is complex. Our evidence suggests that PKBs require regulation in addition to phosphorylation in the HM. One possible mechanism might be phosphorylation of the TM. Alternatively, TORC2 also might have uncharacterized mechanism for the PKB activation. Related to this, AL phosphorylation of PKBs in *D. discoideum* cells is interesting. The PDK homologues can phosphorylate PKBs independently of both PtdIns (3,4,5)P_3 and TORC2, as long as the HM site is substituted with a phosphomimetic residue. Complete understanding of PKB activation by TORC2 and PDK could provide general insights for the spatial and temporal activation of this module.

The TORC2 complex appears to be fragile in cell lysates. Although interactions between TORC2 subunits can be seen, it has not been possible to purify the TORC2 complex to homogeneity. This may suggest that complex formation itself is regulatable. Like mammalian cells, *D. discoideum* cells share a common catalytic subunit between TORC1 and TORC2. The transition between these two different complexes could be important for certain cellular phenomenon. Furthermore, the TORC2 complex is more unstable in *rip3*$^-$ cells than in wild-type cells, suggesting, perhaps, that some subunits may regulate the amount of functional TORC2. To address these questions, it is essential to establish a biochemical method for purification of TORC2.

In addition to its function in chemotaxis, TORC2 is also required for the early developmental program itself. When starved, TORC2 mutant cell lines cannot initiate the developmental process, while cells lacking PKB activity progress further. This result suggests that TORC2 has substrates other than the PKBs. These may be other AGC family kinases or other substrates such as those found in yeast. Once additional substrates are identified, it will be interesting to explore RasC dependency as well as the cellular sites of activation.

Acknowledgments

Annotations of genes were provided by dictyBase. This work was supported by a Helen Hay Whitney Fellowship (to H. C.) and by National Institute of Health Grants GM 28007 and GM 34933 (to P. N. D.).

References

1. Franca-Koh, J., Kamimura, Y., and Devreotes, P. (2006). Navigating signaling networks: chemotaxis in *Dictyostelium discoideum*. *Curr Opin Genet Dev* 16:333–338.
2. Stephens, L., Milne, L., and Hawkins, P. (2008). Moving towards a better understanding of chemotaxis. *Curr Biol* 18:R485–R494.

3. Eichinger, L., Pachebat, J.A., Glockner, G., Rajandream, M.A., Sucgang, R., Berriman, M., Song, J., Olsen, R., Szafranski, K., Xu, Q., Tunggal, B., Kummerfeld, S., *et al.* (2005). The genome of the social amoeba *Dictyostelium discoideum*. *Nature* 435:43–57.
4. Parent, C.A., Blacklock, B.J., Froehlich, W.M., Murphy, D.B., and Devreotes, P.N. (1998). G protein signaling events are activated at the leading edge of chemotactic cells. *Cell* 95:81–91.
5. Meili, R., Ellsworth, C., Lee, S., Reddy, T.B., Ma, H., and Firtel, R.A. (1999). Chemoattractant-mediated transient activation and membrane localization of Akt/PKB is required for efficient chemotaxis to cAMP in *Dictyostelium*. *EMBO J* 18:2092–2105.
6. Funamoto, S., Meili, R., Lee, S., Parry, L., and Firtel, R.A. (2002). Spatial and temporal regulation of 3-phosphoinositides by PI 3-kinase and PTEN mediates chemotaxis. *Cell* 109:611–623.
7. Iijima, M., and Devreotes, P. (2002). Tumor suppressor PTEN mediates sensing of chemoattractant gradients. *Cell* 109:599–610.
8. Chen, L., Iijima, M., Tang, M., Landree, M.A., Huang, Y.E., Xiong, Y., Iglesias, P.A., and Devreotes, P.N. (2007). PLA2 and PI3K/PTEN pathways act in parallel to mediate chemotaxis. *Dev Cell* 12:603–614.
9. Hoeller, O., and Kay, R.R. (2007). Chemotaxis in the absence of PIP3 gradients. *Curr Biol* 17:813–817.
10. Kamimura, Y., Xiong, Y., Iglesias, P.A., Hoeller, O., Bolourani, P., and Devreotes, P.N. (2008). PIP3-independent activation of TorC2 and PKB at the cell's leading edge mediates chemotaxis. *Curr Biol* 18:1034–1043.
11. Meili, R., Ellsworth, C., and Firtel, R.A. (2000). A novel Akt/PKB-related kinase is essential for morphogenesis in *Dictyostelium*. *Curr Biol* 10:708–717.
12. Lee, S., Comer, F.I., Sasaki, A., McLeod, I.X., Duong, Y., Okumura, K., Yates, J.R., III, Parent, C.A., and Firtel, R.A. (2005). TOR complex 2 integrates cell movement during chemotaxis and signal relay in *Dictyostelium*. *Mol Biol Cell* 16:4572–4583.
13. Kamimura, Y., and Devreotes, P.N. (2010). Phosphoinositide-dependent protein kinase (PDK) activity regulates phosphatidylinositol 3, 4, 5-trisphosphate-dependent and -independent protein kinase B activation and chemotaxis. *J Biol Chem* 285:7938–7946.
14. Alessi, D.R., Andjelkovic, M., Caudwell, B., Cron, P., Morrice, N., Cohen, P., and Hemmings, B.A. (1996). Mechanism of activation of protein kinase B by insulin and IGF-1. *EMBO J* 15:6541–6551.
15. Alessi, D.R., James, S.R., Downes, C.P., Holmes, A.B., Gaffney, P.R., Reese, C.B., and Cohen, P. (1997). Characterization of a 3-phosphoinositide-dependent protein kinase which phosphorylates and activates protein kinase Balpha. *Curr Biol* 7:261–269.
16. Sarbassov, D.D., Guertin, D.A., Ali, S.M., and Sabatini, D.M. (2005). Phosphorylation and regulation of Akt/PKB by the rictor–mTOR complex. *Science* 307:1098–1101.
17. Liao, X.H., Buggey, J., and Kimmel, A.R. (2010). Chemotactic activation of *Dictyostelium* AGC-family kinases AKT and PKBR1 requires separate but coordinated functions of PDK1 and TORC2. *J Cell Sci* 123:983–992.
18. Matsuo, T., Kubo, Y., Watanabe, Y., and Yamamoto, M. (2003). *Schizosaccharomyces pombe* AGC family kinase Gad8p forms a conserved signaling module with TOR and PDK1-like kinases. *EMBO J* 22:3073–3083.
19. Kamada, Y., Fujioka, Y., Suzuki, N.N., Inagaki, F., Wullschleger, S., Loewith, R., Hall, M.N., and Ohsumi, Y. (2005). Tor2 directly phosphorylates the AGC kinase Ypk2 to regulate actin polarization. *Mol Cell Biol* 25:7239–7248.
20. Hauge, C., Antal, T.L., Hirschberg, D., Doehn, U., Thorup, K., Idrissova, L., Hansen, K., Jensen, O.N., Jorgensen, T.J., Biondi, R.M., and Frodin, M. (2007). Mechanism for

activation of the growth factor-activated AGC kinases by turn motif phosphorylation. *EMBO J* 26:2251–2261.
21. Facchinetti, V., Ouyang, W., Wei, H., Soto, N., Lazorchak, A., Gould, C., Lowry, C., Newton, A.C., Mao, Y., Miao, R.Q., Sessa, W.C., Qin, J., *et al.* (2008). The mammalian target of rapamycin complex 2 controls folding and stability of Akt and protein kinase C. *EMBO J* 27:1932–1943.
22. Chen, M.Y., Long, Y., and Devreotes, P.N. (1997). A novel cytosolic regulator, Pianissimo, is required for chemoattractant receptor and G protein-mediated activation of the 12 transmembrane domain adenylyl cyclase in *Dictyostelium*. *Genes Dev* 11:3218–3231.
23. Pergolizzi, B., Peracino, B., Silverman, J., Ceccarelli, A., Noegel, A., Devreotes, P., and Bozzaro, S. (2002). Temperature-sensitive inhibition of development in *Dictyostelium* due to a point mutation in the piaA gene. *Dev Biol* 251:18–26.
24. Lee, S., Parent, C.A., Insall, R., and Firtel, R.A. (1999). A novel Ras-interacting protein required for chemotaxis and cyclic adenosine monophosphate signal relay in *Dictyostelium*. *Mol Biol Cell* 10:2829–2845.
25. Cai, H., Das, S., Kamimura, Y., Long, Y., Parent, C., and Devreotes, P.N. (2010). Ras-mediated activation of the TORC2-PKB pathway is critical for chemotaxis. *J Cell Biol* 190:233–245.
26. Jacinto, E., and Lorberg, A. (2008). TOR regulation of AGC kinases in yeast and mammals. *Biochem J* 410:19–37.
27. Manahan, C.L., Iglesias, P.A., Long, Y., and Devreotes, P.N. (2004). Chemoattractant signaling in *Dictyostelium discoideum*. *Annu Rev Cell Dev Biol* 20:223–253.
28. Ueda, M., Sako, Y., Tanaka, T., Devreotes, P., and Yanagida, T. (2001). Single-molecule analysis of chemotactic signaling in *Dictyostelium* cells. *Science* 294:864–867.
29. Janetopoulos, C., Jin, T., and Devreotes, P. (2001). Receptor-mediated activation of heterotrimeric G-proteins in living cells. *Science* 291:2408–2411.
30. Kae, H., Kortholt, A., Rehmann, H., Insall, R.H., Van Haastert, P.J., Spiegelman, G.B., and Weeks, G. (2007). Cyclic AMP signalling in *Dictyostelium*: G-proteins activate separate Ras pathways using specific RasGEFs. *EMBO Rep* 8:477–482.
31. Zhang, S., Charest, P.G., and Firtel, R.A. (2008). Spatiotemporal regulation of Ras activity provides directional sensing. *Curr Biol* 18:1587–1593.
32. Comer, F.I., Lippincott, C.K., Masbad, J.J., and Parent, C.A. (2005). The PI3K-mediated activation of CRAC independently regulates adenylyl cyclase activation and chemotaxis. *Curr Biol* 15:134–139.
33. Funamoto, S., Milan, K., Meili, R., and Firtel, R.A. (2001). Role of phosphatidylinositol 3′ kinase and a downstream pleckstrin homology domain-containing protein in controlling chemotaxis in *Dictyostelium*. *J Cell Biol* 153:795–810.
34. Insall, R., Kuspa, A., Lilly, P.J., Shaulsky, G., Levin, L.R., Loomis, W.F., and Devreotes, P. (1994). CRAC, a cytosolic protein containing a pleckstrin homology domain, is required for receptor and G protein-mediated activation of adenylyl cyclase in *Dictyostelium*. *J Cell Biol* 126:1537–1545.
35. Insall, R.H., Borleis, J., and Devreotes, P.N. (1996). The aimless RasGEF is required for processing of chemotactic signals through G-protein-coupled receptors in *Dictyostelium*. *Curr Biol* 6:719–729.
36. Bosgraaf, L., and van Haastert, P.J. (2006). The regulation of myosin II in *Dictyostelium*. *Eur J Cell Biol* 85:969–979.
37. Chung, C.Y., Potikyan, G., and Firtel, R.A. (2001). Control of cell polarity and chemotaxis by Akt/PKB and PI3 kinase through the regulation of PAKa. *Mol Cell* 7:937–947.

7

The TOR-Mediated Regulation of Autophagy in the Yeast Saccharomyces cerevisiae☆

YOSHIAKI KAMADA[a] • YOSHINORI OHSUMI[b]

[a]*Division of Cell Biology, National Institute for Basic Biology, Okazaki, Japan*

[b]*Integrated Research Institute, Tokyo Institute of Technology, Yokohama, Japan*

I. Abstract

Autophagy is a bulk protein degradation process that is well conserved from yeast to human and plant. Atg (autophagy-related) proteins also conserved among eukaryotes cooperatively play essential roles in autophagy. Autophagy is induced by nutrient limitation or rapamycin treatment due to inactivation of Tor complex1 (TORC1), a nutrient-sensor protein kinase complex that controls various aspects of cell growth. Thus, TORC1 is involved in regulating Atg proteins to prevent autophagy under nutrient replete condition. In this review chapter, we focus on yeast Atg proteins during autophagy-inducing conditions, and discuss the molecular mechanisms of which TORC1 controls autophagy induction in response to cellular nutrient environments.

☆ This chapter is dedicated to the memory of Kazuyoshi Yonezawa, our sincere collaborator and a great scientist who devoted his life to TOR research.

THE ENZYMES, Vol. XXVIII

ISSN NO: 1874-6047

DOI: 10.1016/S1874-6047(10)28007-1

II. Autophagy and *ATG* Genes in Yeast

A. Autophagy (Self-Eating System) in Yeast Cell

Tell me what you eat, and I will tell you what you are.
Brillat-Savarin

A eukaryotic cell has the lysosome/vacuole as a lytic compartment. In the lysosome/vacuole there exist various lytic enzymes, such as proteases, peptidases, lipases, and phosphatases [1]. Since these enzyme activities are harmful in the cytoplasm, their localization is strictly restricted in the lysosome/vacuole, and substrates for vacuolar degradation should be delivered from the cytoplasm into the vacuole via closed membrane-mediated transport, that is, autophagy (Figure 7.1) [2]. One of the most crucial events in autophagy is the sequestration of cytosolic materials by forming a new compartment, called autophagosome (Figure 7.1A and B) [2–7]. In the budding yeast *Saccharomyces cerevisiae*, the autophagosome is observed as a double-membrane organelle (400–900 nm diameter) (Figure 7.1B), and when it meets the vacuole, the outer membrane of the autophagosome is fused with the vacuolar membrane to transport its contents (cytoplasmic materials enclosed by the inner membrane (autophagic body, see Figure 7.1C and D)) into the vacuolar lumen to be degraded by the vacuolar enzymes (Figure 7.1A) [8–10]. When vacuolar proteases (e.g., proteinase A, Pep4) are inactivated, the degradation of the autophagic body is inhibited, and its accumulation can be observed by microscope (Figure 7.1C and D) [8].

B. Start from *ATG* Genes to Understand the Molecular Mechanism of Autophagy

Study of yeast autophagy was started with observing autophagic bodies (Figure 7.1B–D) [8, 9]. Next followed genetic screens identifying group of genes essential for autophagy [11]. Independent screening of autophagy-deficient mutants produced different gene names (*apg*, *aut*, *cvt*, *gsa*, *pag*, *paz*, *pdd*) that have been unified to *ATG* (autophagy-related) [12]. So far, 34 *ATG* genes have been identified to be responsible for various subtypes of autophagy [6, 13–15].

III. Induction of Autophagy by Nutrient Limitation

A. Fundamental Machinery of Autophagy

Autophagy can be classified roughly into two subtypes, induced autophagy and constitutive autophagy [2]. The former is induced by several cellular stresses, such as nutrient starvation, change of carbon source, and

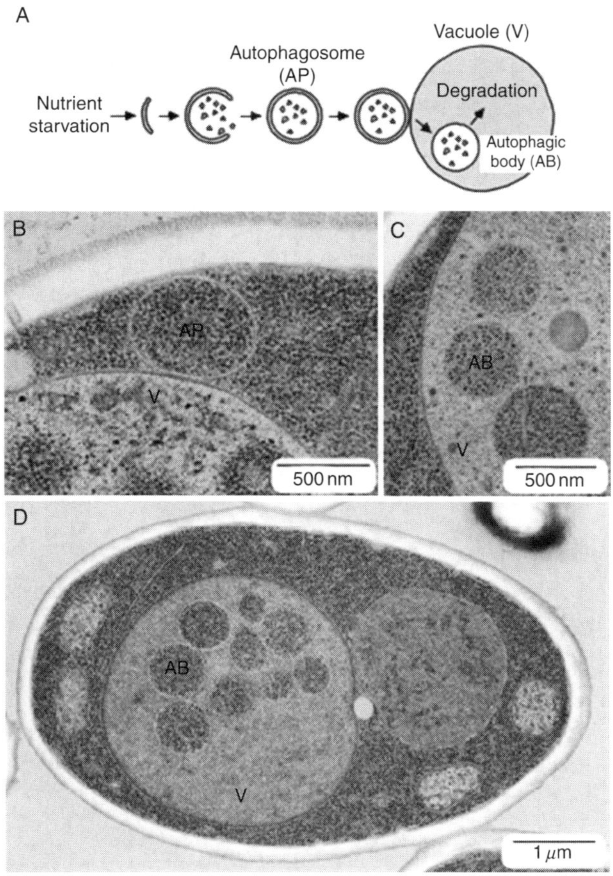

FIG. 7.1. Autophagy in yeast. (A) Starvation-induced autophagy. In response to nutrient starvation, Atg proteins assemble to the preautophagosomal structure (PAS), a putative site for the autophagosome formation. At the PAS, isolation membrane enlarges and sequesters cytosolic materials to become the autophagosome (AP). The outer membrane of the autophagosome fuses to the vacuole, and the autophagic body (the inner membrane-bound component, AB) is degraded by vacuolar hydrolases. (B–D) Electron microscope images of yeast autophagy. (B) Image of the double-membrane-bound autophagosome (AP). Vacuolar protease-deficient cell is treated with nitrogen depleted medium. V, vacuole. (C) Autophagic body is filled with cytosolic materials, such as ribosomes (seen as electron-dense particles). (D) Accumulation of autophagic body (AB) in the vacuole. These photographs are kindly supplied by M. Baba (Kogakuin University).

ER stress [7, 8, 16]. In the budding yeast, the latter is exemplified by the Cvt (cytoplasm-to-vacuole targeting) pathway that mediates the biosynthetic transport of the vacuolar aminopeptidase 1 (Ape1) from the cytoplasm into the vacuole [2]. In the Cvt pathway, a small (150 nm diameter) autophagosome-like double-membrane compartment (Cvt vesicle) is formed instead of autophagosome [17]. Among 34 Atg proteins, 15 Atg proteins (called as core Atg proteins) are required for both the pathways, suggesting that these proteins act as the fundamental machinery for the biogenesis of autophagy-related membranes [5, 6]. Characterization of these 15 Atg proteins revealed that they consist of five subgroups: the Atg1 kinase and its regulator Atg13 [18–20]; phosphatidylinositol 3-kinase complex (Vps15, Vps30/Atg6, Vps34, Atg14) [21, 22]; the Atg12 conjugation system (Atg5, Atg7, Atg10, Atg12, Atg16) [23–27]; the Atg8 conjugation system (Atg3, Atg4, Atg7, Atg8) [28–33]; and a subgroup of functionally unknown proteins (Atg2, Atg9, Atg18) [34–40]. Accumulation of evidences has proposed that the core Atg proteins localize together to a perivacuolar site to form autophagosome and Cvt vesicle, and the proposed site is designated as PAS (preautophagosomal structure) (see Figure 7.2) [41, 42].

B. Physiological Role of Starvation-Induced Autophagy

He hath filled the hungry with good things
Luke 1:53

When a cell is deprived of nutrients, it must procure nutrients by degrading its own contents to survive such a harsh condition. Autophagy is induced by nutrient starvation [8]. In the budding yeast, autophagy is induced by the deprivation of various nutrients, for example, nitrogen,

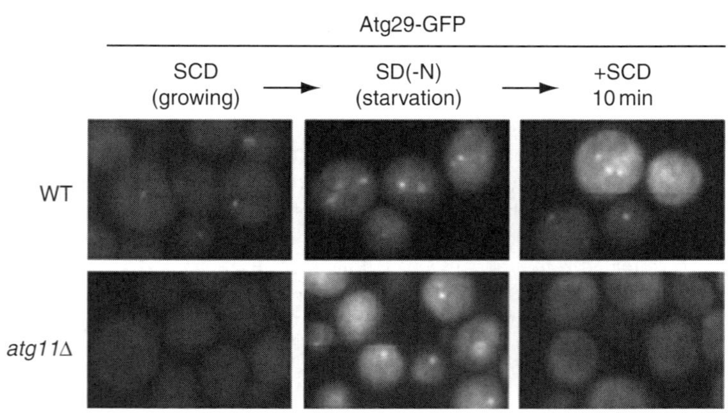

Fig. 7.2. Nutrient-sensitive localization of Atg protein to the PAS. The recruitment of GFP-fused Atg29 to the PAS in wild-type and *atg11Δ* cells is observed by fluorescent microscope. In *atg11Δ* cells, Atg29 localizes to the PAS in response to starvation.

amino acid (in case of amino acid auxotroph strain), carbon, sulfur, and phosphate. On the other hand, autophagy is not induced under uracil-limiting condition in uracil auxotroph strain. Starvation-induced autophagy is thought to play two important roles in remodeling cell to adapt starvation. One is degrading energy-consuming cellular materials (such as ribosomes), which are synthesized in great amounts during vegetative growing phase, but which are better to be removed under nutrient-limiting condition (see Figure 7.1B–D). The other is recycling nutrients (e.g., amino acids) supplied by the degradation products to use for adaptation. Starvation-induced autophagy employs three autophagy-specific Atg proteins (Atg17, Atg29, and Cis1/Atg31) that are required for starvation-induced autophagy but dispensable for the Cvt pathway [20, 43, 44] in addition to the core 15 Atg proteins to form autophagosome [6]. Defect of any of these 18 *ATG* genes causes decreased viability under nitrogen starvation, demonstrating that autophagy is essential for survival against starvation [11, 18, 19, 21, 23, 26, 45].

C. Recruitment of Atg Proteins to the PAS, the Site for Autophagosome Formation

Are we all then assembled? Then stay no longer:
Wagner, Die Walküre

Atg proteins are localized to the PAS irrespectively of nutrient condition, because the Cvt vesicle is formed at the PAS in nutrient-rich condition (Figure 7.2) [41]. The starvation-induced-autophagy-specific Atg proteins (Atg17, Atg29, and Atg31) are also localized to the PAS in nutrient-rich condition, though their physiological function at the PAS under nutrient-rich condition is unknown [43, 44]. Thus, the PAS was previously thought as a static structure; however, recent studies have revealed dynamic aspects of the PAS, which is controlled by cellular nutrient conditions. For example, in *atg11Δ/cvt9Δ* cell in which the Cvt pathway is abrogated, starvation-responsive assembly of Atg proteins to the PAS can be observed (Figure 7.2) [46–48]. Atg proteins localized throughout the cytoplasm in growing cell are recruited to the PAS in response to nitrogen depletion. Their recruitment is quickly (within 10 min) canceled, when the starved cell is replenished with nitrogen source [48]. Replenishment of nutrient to starved cell also immediately terminates autophagic activity, well corresponding to the PAS disassembly [48]. These results suggest that organization of the PAS is regulated by nutrient conditions.

IV. Induction of Autophagy by TOR Inactivation

TOR pathway is involved in regulation of autophagy, because rapamycin, a specific inhibitor of TOR, mimics starvation condition and induces autophagy (Figure 7.3A) [49]. The budding yeast has two homologous TOR proteins, Tor1 and Tor2 [50], which form distinct TOR complexes, TORC (Tor complex) 1 and TORC2 [51, 52]. TORC1 contains Kog1, Tco89, Lst8, and either Tor1 or Tor2, regulating rapamycin-sensitive branch of TOR pathway [53, 54]. On the other hand, TORC2 contains Avo1, Avo2, Avo3, Bit61, Lst8, and Tor2 (but not Tor1) [55]. TORC2 function is insensitive to rapamycin. A couple of evidence demonstrate that TORC1 branch but not TORC2 regulates autophagy: autophagy is induced by rapamycin; autophagy is induced in *tor1*Δ *tor2*^{ts} mutant (dysfunction of both TORC1 and TORC2) but not in TOR1 *tor2*^{ts} (dysfunction of TORC2) at nonpermissive

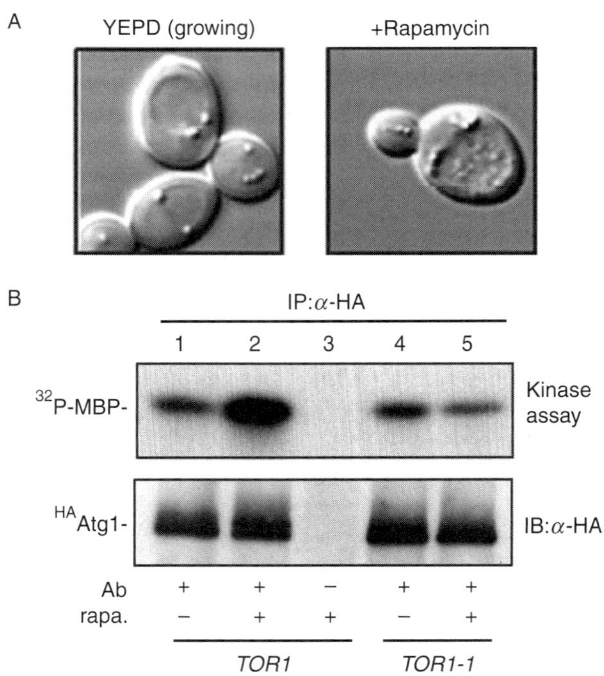

FIG. 7.3. TOR-mediated regulation of autophagy. (A) Induction of autophagy by rapamycin (accumulation of autophagic bodies in *pep4*Δ cell) is observed by light microscope. (B) Activation of Atg1 kinase by rapamycin. HA-tagged Atg1 (^{HA}Atg1) is immunoprecipitated and subjected to *in vitro* kinase assay with myelin basic protein (MBP) and [γ-^{32}P]ATP as substrates. Noted that activation of Atg1 is canceled in rapamycin-insensitive *TOR1* mutant (*TOR1-1*).

temperature [49] (Kamada and Ohsumi, unpublished results). Temperature-sensitive mutant of *KOG1* does not induce autophagy at nonpermissive temperature, but it is still likely that TORC1 is responsible for autophagy regulation [56]. This mutant (*kog-105*) displays very distinct characteristics from that of rapamycin-treated cells, thus only some part of TORC1 functions is disturbed in *kog1-105* cells. Neither Tap42 nor Sch9, two major TORC1 substrates [57, 58], is involved in autophagy, suggesting that a novel branch of TORC1 signaling is responsible for autophagy induction [20, 59]. Addition of cAMP abrogates the effect of rapamycin on autophagy, suggesting that PKA (cAMP-dependent protein kinase) has an inhibitory role in regulating autophagy [49, 60, 61]. However, studies using ATP-analog-sensitive PKA mutant are controversial [59, 62]. Since it is reported that PKA is under the control of TORC1 pathway [63], role of PKA and TORC1 in cell growth and autophagy should be further investigated, for example, whether or not complete inactivation of PKA affects TORC1 function.

V. Regulation of Atg1 Kinase Complex by TOR Complex1

A. Atg1, a Protein Kinase

ATG1 encodes a Ser/Thr protein kinase, whose catalytic activity is essential for autophagy [18, 64]. Out of 21 *atg1* mutant collection, we have 11 alleles expressing the full-length Atg1 protein, and all of these mutants have point mutations in the conserved kinase region, confirming the indispensable role of Atg1 activity in autophagy [11, 45]. *In vitro* protein kinase assay of Atg1 showed that the catalytic activity of Atg1 is highly elevated in the cells shifted to starvation condition [20]. Atg1 activation is also observed in rapamycin-treated cells, suggesting that Atg1 activation is required for the induction of autophagy and that it is negatively regulated by TORC1 (Figure 7.3B). Once there was a report that cast doubt on the requirement of Atg1 kinase activity for autophagy, because Atg1 kinase-deficient mutant, *atg1*^{K54A} can still induce autophagy under starvation condition [45, 47, 65]. We precisely reexamined this question. As shown in Figure 7.4, Atg1^{K54A} does not display any detectable kinase activity *in vitro*, though the autophagic activity is not completely diminished. Autophagy is completely abolished in the other *atg1* kinase-deficient mutants, such as D211A and A235V (a.k.a. *atg1-1*) [11, 18], suggesting that K54A may maintain undetectable but certain kinase activity to slightly induce autophagy (Figure 7.4B). Additional mutation at M102 (to A) to K54A also abolishes autophagy induction [47].

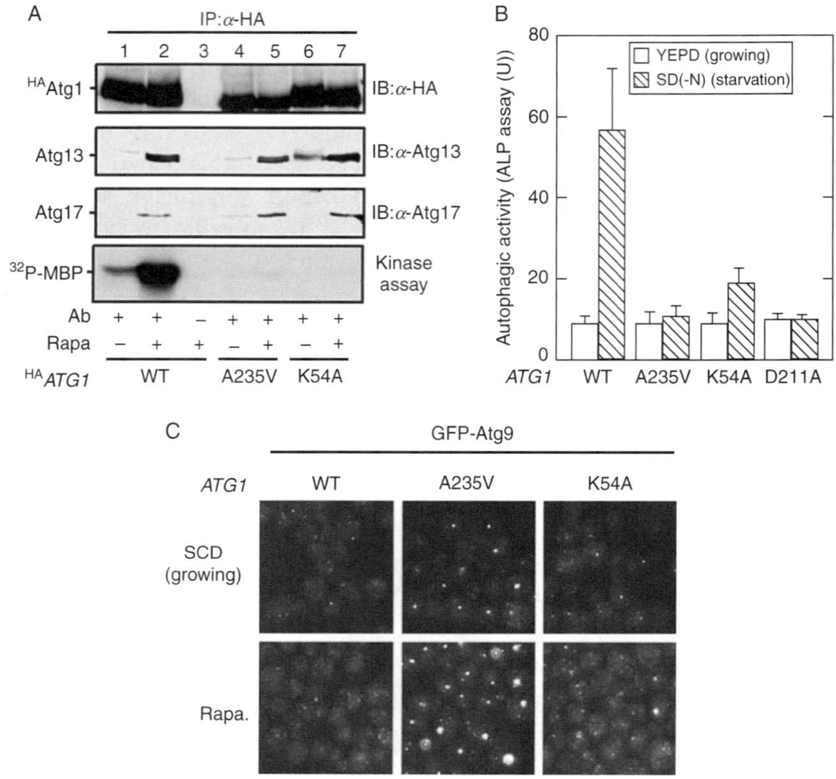

FIG. 7.4. Requirement of the Atg1 activity for autophagy. (A) Atg1 kinase activity is not required for Atg1 complex formation. HAAtg1 is immunoprecipitated for kinase assay, and coprecipitated components of Atg1 complex are detected. (B) Autophagic activity of *atg1* kinase-deficient mutants. Autophagic activity is estimated by ALP assay [83]. (C) Localization of Atg9 to the PAS is enhanced in *atg1* mutants. This photograph is kindly supplied by T. Sekito (Ehime University).

B. Cvt-Specific Proteins

Several studies including two-hybrid screen revealed that Atg1 interacts with Atg13, Atg11, Atg17, Atg20, Atg24, Atg29, Atg31, and Vac8 (Figure 7.5A) [20, 43, 44, 66–70]. Atg11, Atg20, Atg24, and Vac8 are exclusively involved in the Cvt pathway [5, 6, 67, 68, 71]. Biochemical or cell biological characteristics of these proteins under autophagy-inducing conditions have not yet been determined. Atg11, an adapter protein for precursor of Ape1, a major cargo of the Cvt pathway, binds to Atg1 under growing and starvation conditions (Figure 7.5B) [48, 67]. Deletion of *ATG11* gene does not affect Atg1 kinase activity, supporting that Atg11

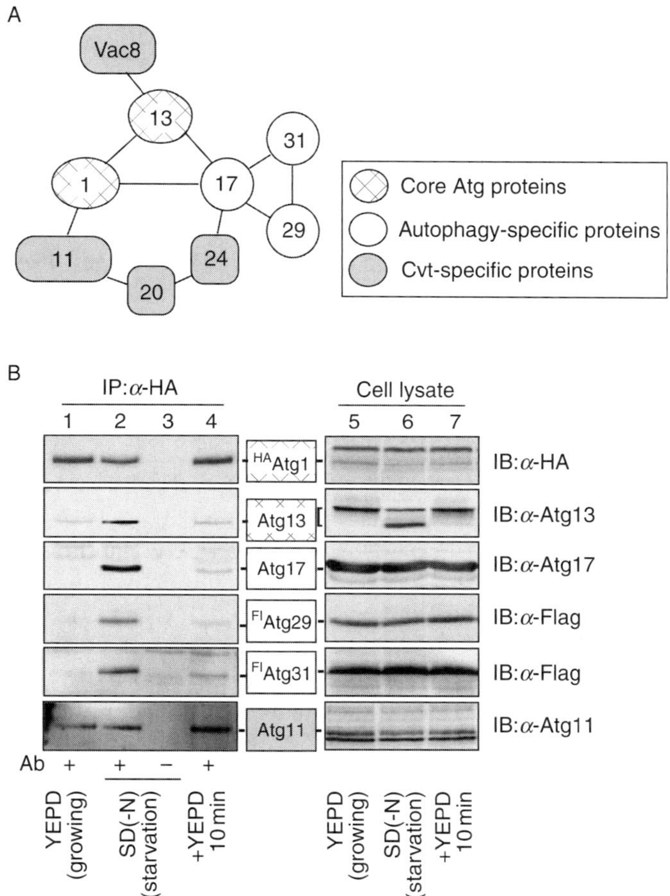

FIG. 7.5. The Atg1 complex. (A) Detailed interaction among components of the Atg1 complex. (B) Nutrient-sensitive formation of the Atg1 complex. ^{HA}Atg1 is immunoprecipitated to examine association of components of the Atg1 complex. Dephosphorylation of Atg13 in response to starvation (lane 6) triggers Atg1 complex formation (lane 2). Noted that Atg11 dispensable for starvation-induced autophagy binds to Atg1 irrespective to nutrient conditions.

is dispensable for starvation-induced autophagy [20]. Vac8, a vacuolar membrane protein, is reported that it is not required for autophagy [66]. But it may have some role in autophagy, because autophagic activity is partially but significantly affected in *vac8Δ* cell [66, 71]. Palmitoylation is required for the localization of Vac8 to the vacuole, but not for the Cvt pathway [71], suggesting that the unmodified form might associate with Atg1 complex at the PAS.

C. Atg17, Atg29, and Atg31: The Autophagy-Specific Proteins

Three heads are better than one.

As mentioned above, Atg17, Atg29, and Atg31 are specifically required for starvation-induced autophagy, and they are dispensable for the Cvt pathway [20, 43, 44]. This fact strongly suggests that their characterization leads a clue to distinguish between the starvation-induced autophagy and the Cvt pathway. In our previous study, we described that "abnormally small autophagosome" (diameter of ~200 nm) is formed in *atg17*Δ cells [45], but now this structure should be noted as "large Cvt vesicle" rather than autophagosome. The autophagy-specific proteins form a stable ternary complex (estimated molecular weight is ~600 kDa determined by gel filtration) *in vivo*, and complex formation is not affected by cellular nutrient conditions [48, 72]. *In vitro* study using recombinant proteins to reconstitute the complex indicates that the protein complex comprises equal number of Atg17, Atg29, and Atg31, possibly dimeric complex of Atg17–Atg29–Atg31 unit [72].

Among the components of the ternary complex, Atg17 is the most characterized. Atg17 was first identified as an Atg1-binding protein by two-hybrid screen [20]. Atg17 is required for the activation of Atg1 [20, 45]. Atg17 has five putative coiled-coil motifs (CC1–CC5), and CC2 domain is responsible for binding to Atg29 [48, 73]. CC3 is essential, but not enough for binding to Atg1 and Atg13 [73]. These coiled-coil motifs, but CC4 is inevitable for autophagy. To our knowledge, CC5 has an important function for self-dimerization (Kamada and Ohsumi, unpublished data). In two-hybrid analysis, Atg17 binds to Atg1 in *atg13*Δ cell and it binds to Atg13 in *atg1*Δ cell, suggesting that Atg17 has distinct Atg1-binding and Atg13-binding site [45]. In contrast, coimmunoprecipitation experiments show that Atg13 is required for Atg17 association to Atg1 (but Atg17 is not required for Atg13 association to Atg1), suggesting that *in vivo* Atg1–Atg13 binding is important for the assembly of Atg17 (presumably the ternary complex) to form whole Atg1 complex [45, 74].

ATG29 was isolated as an *ATG* gene by genome-wide screen using a collection of yeast knockout strains [43]. Atg29 has a coiled-coil motif with which it binds to Atg17 [43]. Point mutations at this region abolish autophagic activity due to the inability to bind Atg17. And recruitment of Atg29 to the PAS requires Atg17 and Atg31 in *atg11*Δ cell. These results indicate that the Atg17–Atg29–Atg31 complex formation is essential for autophagy [48]. In two-hybrid assay, Atg17 but not Atg29 binds to Atg1 and Atg13, suggesting that the ternary complex binds to Atg1 and Atg13 via Atg17 (Figure 7.5A). This idea is supported by the fact that Atg17 but not Atg29 is required for Atg1 activation [43].

Atg31/Cis1 is found as an Atg17-binding protein in the *Saccharomyces* genome database[1] [44]. It is originally isolated as a multicopy suppressor of *cik1* mutant; however, the suppressive activity of *ATG31/CIS1* is not obvious [44]. Atg31 is phosphorylated in response to nutrient starvation [72].

D. Atg13, the Regulator of Atg1 in a TORC1-Dependent Manner

With me, you don't feel any night too long.
R. Strauss, Der Rosenkavalier

ATG13 was first described as an *ATG* gene, whose mutation is suppressed by an overexpression of Atg1 [19]. Next, it was isolated as an Atg1-binding protein by two-hybrid screen [20]. Atg13 does not have any known conserved motif, but Atg1-binding region. An Atg13 mutant whose Atg1-binding site is partially truncated displays complete defect in autophagy but only partial defect in the Cvt pathway, suggesting that Atg1–Atg13 binding is more important for autophagy than the Cvt pathway. What is most important about Atg13 is that it is highly phosphorylated under nutrient-rich condition and is quickly dephosphorylated in response to starvation (within 30 min after nitrogen depletion) [20]. The phosphorylation state of Atg13 is extremely sensitive to nutrient condition, because the dephosphorylated Atg13 in starved cells is rapidly phosphorylated after the resupply of nutrient (see Figure 7.5B lane 5–7). Atg13 is dephosphorylated by rapamycin treatment (within 5 min after rapamycin addition), indicating that phosphorylation of Atg13 is controlled by TORC1. As for phosphatase (s) responsible for Atg13 dephosphorylation, 30 deletion mutants lacking nonessential phosphatase were examined, but phosphorylation state of Atg13 was not affected in any phosphatase mutant (Funakoshi and Ohsumi, unpublished results) [75]. Therefore, Atg13 might not be dephosphorylated by specific phosphatase whose activity is stimulated by starvation, and TORC1 activity mainly regulates phosphorylation status of Atg13. Immunoprecipitation analyses revealed that the TORC1-dependent phosphorylation of Atg13 plays a pivotal role in autophagy induction (Figure 7.5B) [20]. The phosphorylated Atg13 in growing cell has nearly undetectable affinity to Atg1. Since Atg13 is needed for Atg17 association with Atg1, the autophagy-specific ternary complex (Atg17–Atg29–Atg31) does not bind to Atg1, either [45, 48]. In contrast, immediate dephosphorylation in response to starvation confers Atg13 the ability to bind to Atg1. Subsequently, the ternary complex associates with Atg1–Atg13 to form the Atg1 complex. When cell is replenished with nutrient, Atg13 is quickly

[1] http://www.yeastgenome.org/

rephosphorylated to abolish the complex formation [48]. The Atg1 complex formation triggers autophagy at least in two ways.

First, the recruitment of Atg proteins to the PAS is stimulated by the Atg1 complex [47, 48]. As mentioned above, in *atg11*Δ cells core Atg proteins dispersed in the cytoplasm under nutrient replete condition assemble at the PAS in response to starvation. The lack of any components of the Atg1 complex or their mutants (e.g., *atg13*$^{1-448}$, *atg29*^{F67L}) unable to form the complex abrogates the recruitment of core Atg proteins (including Atg1 and Atg13) to the PAS as well as autophagy-specific ternary complex (Atg17–Atg29–Atg31) [48]. Therefore, the Atg1 complex is likely to function as a conductor in organizing the PAS for autophagosome formation [6, 47, 48]. Actually, in wild-type (*ATG11*) cells core Atg proteins already localize to the PAS before autophagy induction, thus it might be more appropriate to describe that the Atg1 complex formation "re"-organizes the PAS to autophagy from the Cvt pathway in response to nutrient limitation [6]. This reorganization is dependent on nutrient condition, and rephosphorylation of Atg13 in nutrient-replenished cells cancels Atg proteins' assembly at the PAS via disruption of the Atg1 complex [48].

Second, Atg1 kinase activity is enhanced by the Atg1 complex formation [20, 45]. The lack of Atg13 and Atg17 and their mutations (e.g., *atg13*$^{1-448}$, *atg17*^{C24R}) affect Atg1 kinase activity. Especially, Atg1 activity is largely attenuated in *atg13*Δ cells [3, 20]. On the other hand, Atg1 activation in response to rapamycin is normally observed in *atg29*Δ cell [43], supporting the result of two-hybrid assay that the ternary complex binds to Atg1 via Atg17. Studies using kinase-deficient (e.g., *atg1*^{K54A} and *atg1*^{D211A}) and temperature-sensitive (*atg1*^{ts}, three point mutations within kinase domain) mutants of Atg1 demonstrate that the protein kinase activity is not required for formation of the Atg1 complex or the PAS assembly of the most Atg proteins (Figure 7.4A and C) [37, 39, 41, 47, 48]. So far, only Atg2 exceptionally requires Atg1 activity for its localization to the PAS [39]. In the above *atg1* mutants, Atg proteins appear to accumulate at the PAS under autophagy-inducing condition, suggesting that Atg1 kinase activity has an essential role in autophagosome formation after the PAS organization (Figure 7.4C). Once Atg1 is activated by the complex formation, the activated state is maintained even after the complex is broken apart upon resupply of nutrients (Kamada and Ohsumi, unpublished results). Since nutrient supply terminates organization of the PAS (for autophagosome formation) and autophagy induction, it is natural to assume that Atg1 phosphorylates its substrate(s) at the PAS to trigger a downstream event in autophagy. Although several Atg proteins are phosphorylated in an Atg1-dependent manner both *in vivo* and *in vitro*, authentic substrate(s) of Atg1 still remains elusive (Kamada and Ohsumi, unpublished observation).

VI. Phosphorylation of Atg13 by TORC1 to Regulate Autophagy

A. Identification of Phosphorylation Sites of Atg13 by TORC1

While it is proposed that dephosphorylation of Atg13 is one of the initial steps of autophagy [76], it remains as an unanswered question whether Atg13 is a direct target of TORC1 or dephosphorylation of Atg13 is sufficient for induction of autophagy. Because TORC1 controls various aspects of cell growth [53, 54], it is also to be answered whether or not more than two downstream pathways may cooperatively regulate autophagy. To address these questions, we tried to determine the phosphorylation sites of Atg13 using mass spectrometry/mass spectrometry [77]. As the result, eight Ser sites (S348 S437 S438 S496 S535 S541 S646 S649) were identified to be phosphorylated. *In vitro* TORC1 assay confirms that TORC1 directly phosphorylates Atg13 and that these Ser residues are major target sites. In mammal, mTORC1 substrates, such as S6K, 4E-BP1 [78, 79], and PRAS40 [80], have conserved TOS (TOR signaling) motif (F–D/E–Φ–D/E–Φ, Φ; hydrophobic residue) which is required for substrate recognition and binding to TORC1 [81]. However, Atg13 does not have obvious TOS motif, and neither Atg1 nor Atg13 protein is coimmunoprecipitated with TORC1. Sch9, another yeast TORC1 substrate does not have obvious TOS motif, either [58], while Ypk2, a TORC2 substrate has a "TOS-like" region (F-D-V-T-R) [82]. Thus, recognition of TOS motif by TORC1 may not be conserved in yeast cells.

B. Induction of Autophagy by Atg13-8SA Bypassing TOR Pathway

Your way is shorter; my purposes do draw me much about:
Shakespeare, Antony, and Cleopatra

Ser to Ala substituted mutant of Atg13 (S348A S437A S438A S496A S535A S541A S646A S649A, Atg13-8SA) is barely phosphorylated by TORC1 both *in vivo* and *in vitro*, suggesting that Atg13-8SA acts as a dephosphorylated form of Atg13 [77]. Expression of Atg13-8SA in vegetatively growing cell brings about the formation of Atg1 complex, activation of Atg1 kinase, and organization of the PAS without inhibiting TORC1 function (Figure 7.6A). Strikingly, induction of autophagy is further observed by the expression of Atg13-8SA (Figure 7.6B and C) [77]. Autophagic activity induced by Atg13-8SA reaches 50–70% of that of nitrogen starved and rapamycin-treated cells (Figure 7.6B). Autophagic activity (ALP assay using *pho8*Δ60 strain) is thought to be correlated to the volume

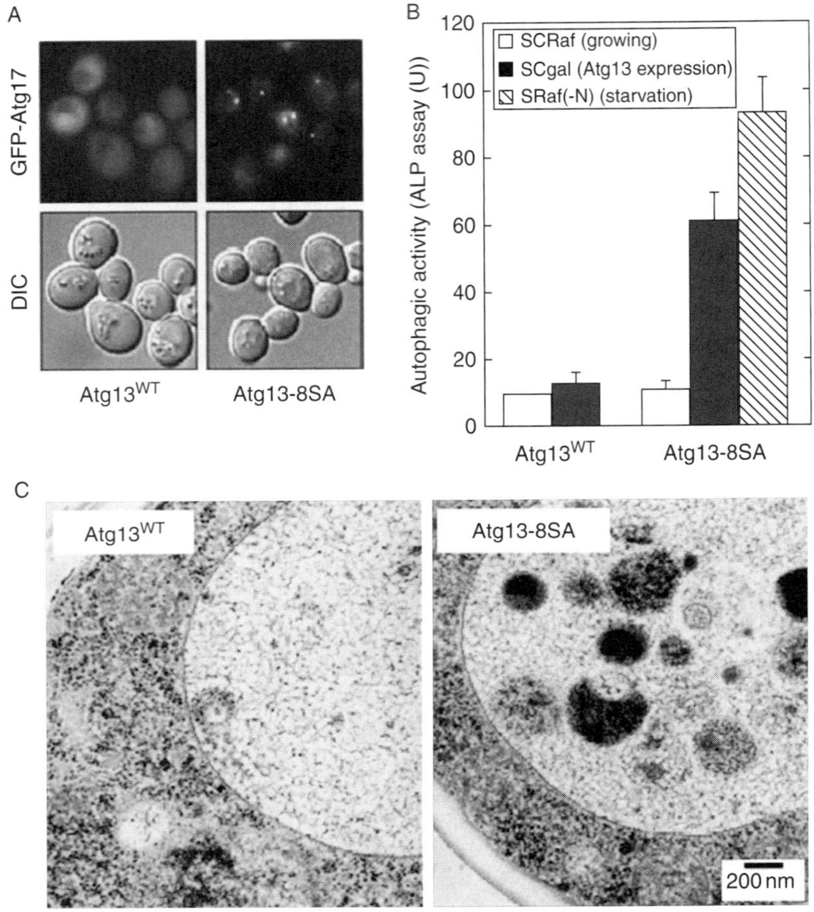

FIG. 7.6. Induction of autophagy by expression of unphosphorylatable form of Atg13 (Atg13-8SA). (A) The organization of the PAS (marked by GFP-Atg17) is triggered by Atg13-8SA expression. (B) Expression of Atg13-8SA induces autophagy. Autophagic activity is estimated by ALP assay. (C) Images of electron microscopy of Atg13^{WT} (left) and Atg13-8SA (right) expressing cell.

of cytoplasm which is sequestered by autophagosomes [83], thus basically in proportion to cube of diameter of autophagosome. Then it is estimated that 80–90% of the autophagosomal membrane (i.e., surface area of autophagosome) can be generated during Atg13-8SA-induced autophagy compared to starvation-induced autophagy. Expression of Atg13-8SA had only a slight synergistic effect on rapamycin-induced autophagy in Atg13-8SA-expressing cells, excluding the possibility that any other TORC1 signaling branch might be involved in autophagy induction [77]. These results clearly demonstrate that dephosphorylation of Atg13 acts as a molecular switch of autophagy induction. It should be noted that Atg13-8SA expression did not disturb cell growth at all, indicating that Tor is not inhibited by Atg13-8SA expression or Atg1 activation, and that the subsequent induction of autophagy does not affect cell growth [77].

It is known that Atg8 expression is stimulated at a transcriptional level by Tor inactivation, such as nitrogen starvation [30]. It is not dramatically changed by Atg13-8SA expression, suggesting that upregulation of the Atg8 protein does not play a key role at least in the induction of autophagy. Because the amount of Atg8 has large influence on the size of autophagosome [84], this might be a reason of difference in autophagic activity between Atg13-8SA and nitrogen starvation. Therefore, TORC1 regulates autophagy at least in two independent procedures; induction of autophagy by phosphorylation of Atg13 and enlargement of phagophore membrane by expression of Atg8.

VII. ULK Complex: Mammalian Counterpart of Yeast Atg1 Complex

Mammalian counterparts of the components of the Atg1 complex were recently reported, thus the TORC1–Atg1 complex signaling module may be conserved among most eukaryotes to regulate autophagy [85]. Atg1 counterpart in *Caenorhabditis elegans* and mammals, unc-51 and ULK1/2 (unc-51-like protein kinase), respectively, has distinct dual functions, for autophagy (as Atg1) and axonal elongation (as unc-51) [86–92]. ULK is phosphorylated by mTORC1 [93, 94]. Components of ULK complex regulating autophagy have been identified, FIP200, mAtg13, and Atg101 [93–99]. FIP200, containing multiple coiled-coil domains, is thought as a counterpart of Atg17, because both of FIP200 and Atg17 are required for ULK/Atg1 activity [95]. FIP200 also associates with various (non-Atg) proteins, suggesting that it has multiple functions besides autophagy. The mammalian authologue of Atg13 (mAtg13) is identified by PSI-BLAST and high-throughput LC–MS/MS project [93, 94, 96, 98]. Like the yeast counterpart, mAtg13 positively regulates ULK activity, and it is phosphorylated by mTORC1 *in vivo* and *in vitro* [93, 94]. However, dephosphorylation of mAtg13 in response to mTORC1 inactivation is not obvious in immunoblot, presumably because mAtg13 is also phosphorylated by ULK *in vitro* [93, 94, 96]. As for the yeast Atg13, *in vivo* ^{32}P-labeling experiment shows that Atg13 is still phosphorylated to some level under starvation condition [20], presumably by activated Atg1. When Atg1 is overexpressed,

Atg13-8SA is detected as upshift band in immunoblot, suggesting that Atg1 phosphorylates different Ser/Thr residues from TORC1-target sites (Kamada and Ohsumi, unpublished results). However, it is not yet known whether Atg1-dependent phosphorylation of (m)Atg13 is important for autophagy. Atg101 is identified as a mAtg13-binding protein and it does not show any homology with yeast components of Atg1 complex [98, 99].

On the other hand, UNC-14, SynGAP, and Syntenin were identified as ULK-binding proteins to function together for axonal elongation [92, 100]. It is not yet known whether this complex is also regulated by TORC1 by direct phosphorylation.

Unlike yeast, formation of mammalian ULK complex is not regulated by nutrient condition; ULK–mAtg13–FIP200–Atg101 association is constitutively detected as a ~3 MDa complex [93, 94, 99]. We assume that this difference comes from the dual functions of ULK. Constitutive binding of mAtg13–FIP200–Atg101 to ULK may restrict one function (for autophagy as Atg1 complex) from the other (for neural development as UNC-51 complex). Since UNC-51 complex may not act at downstream of TORC1, constitutive formation of ULK–mAtg13–FIP200–Atg101 complex allows ULK to receive TORC1 signal under nutrient replete condition. As well as in yeast, mTORC1 regulates the ULK complex through direct phosphorylation of its component(s).

VIII. Concluding Remarks

In this review, we introduce studies showing the TORC1-mediated regulatory mechanism of autophagy. According to the recent findings, we propose a current model of autophagy induction (Figure 7.7A). Under nutrient replete condition, active TORC1 phosphorylates Atg13 at multiple sites to inhibit Atg1 complex formation. In contrast, upon TORC1 inactivation due to nutrient starvation and rapamycin treatment, Atg13 is quickly dephosphorylated. Dephosphorylation of Atg13 allows Atg1–Atg13 binding and further complex formation with autophagy-specific Atg proteins. The Atg1 complex formation recruits another core Atg proteins to the PAS, where Atg1 phosphorylates its physiological substrate(s) to induce autophagy.

As described above, two yeast TORC1 downstreams, Tap42 and Sch9 are unlikely to be involved in autophagy. Thus, above findings also indicate that TORC1–Atg1 complex comprises a novel branch of TORC1 signaling, presumably specified for autophagy induction.

Expression of unphosphorylatable Atg13 (Atg13-8SA) can bypass TORC1 dysfunction to induce autophagy. This suggests that any other

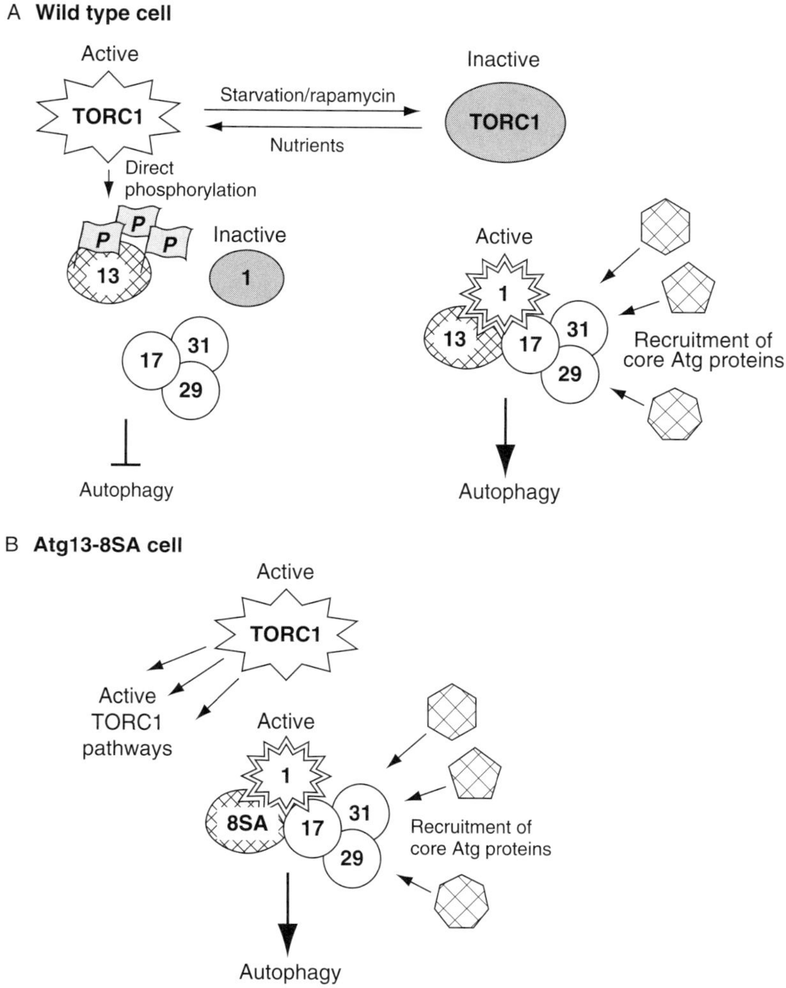

FIG. 7.7. Schematic model of TORC1-mediated regulation of autophagy via direct phosphorylation of Atg13. See text for details.

branches of TORC1 pathway is not involved in autophagy regulation but Atg13 (Figure 7.7B). Most molecular events of Atg proteins during autophagosome formation take place in Atg13-8SA-expressing cell, demonstrating that Atg13 itself acts as a molecular switch turning on/off autophagy.

Accumulation of evidences has unveiled that autophagy is involved in various physiological events, such as aging, cancer, neurodegeneration, and cell death [101, 102]. However, most of these studies have examined nutrient-starved or rapamycin-treated cells in which Tor function and its diverse downstream targets are affected, for example, protein synthesis and turnover. Thus, it is important to distinguish the role of autophagy *per se* from (a mixture of) many responses against nutrient starvation. Based on this study, Atg13-8SA may be a useful tool for these purposes.

Acknowledgments

We thank M. Baba, T. Funakoshi, M. N. Hall, Y. Imamura, F. Inagaki, Y. Kabeya, T. Kawamata, D. J. Klionsky, C. Kondo, R. Loewith, Y. Maruki, K. Nagano, A. Nakashima, N. N. Noda, M. Ohsumi, T. Sekito, T. Shintani, H. Takikawa, L. Weisman, K. Yonezawa, and K. Yoshino for their fruitful collaborations. We also thank members of Ohsumi laboratory.

References

1. Klionsky, D.J., Herman, P.K., and Emr, S.D. (1990). The fungal vacuole: composition, function, and biogenesis. *Microbiol Rev* 54:266–292.
2. Klionsky, D.J., and Ohsumi, Y. (1999). Vacuolar import of proteins and organelles from the cytoplasm. *Annu Rev Cell Dev Biol* 15:1–32.
3. Kamada, Y., Sekito, T., and Ohsumi, Y. (2004). Autophagy in yeast: a TOR-mediated response to nutrient starvation. *Curr Top Microbiol Immunol* 279:73–84.
4. Cebollero, E., and Reggiori, F. (2009). Regulation of autophagy in yeast *Saccharomyces cerevisiae*. *Biochim Biophys Acta* 1793:1413–1421.
5. Xie, Z., and Klionsky, D.J. (2007). Autophagosome formation: core machinery and adaptations. *Nat Cell Biol* 9:1102–1109.
6. Nakatogawa, H., Suzuki, K., Kamada, Y., and Ohsumi, Y. (2009). Dynamics and diversity in autophagy mechanisms: lessons from yeast. *Nat Rev Mol Cell Biol* 10:458–467.
7. He, C., and Klionsky, D.J. (2009). Regulation mechanisms and signaling pathways of autophagy. *Annu Rev Genet* 43:67–93.
8. Takeshige, K., Baba, M., Tsuboi, S., Noda, T., and Ohsumi, Y. (1992). Autophagy in yeast demonstrated with proteinase-deficient mutants and conditions for its induction. *J Cell Biol* 119:301–311.
9. Baba, M., Takeshige, K., Baba, N., and Ohsumi, Y. (1994). Ultrastructural analysis of the autophagic process in yeast: detection of autophagosomes and their characterization. *J Cell Biol* 124:903–913.
10. Baba, M., Osumi, M., and Ohsumi, Y. (1995). Analysis of the membrane structures involved in autophagy in yeast by freeze-replica method. *Cell Struct Funct* 20:465–471.
11. Tsukada, M., and Ohsumi, Y. (1993). Isolation and characterization of autophagy-defective mutants of *Saccharomyces cerevisiae*. *FEBS Lett* 333:169–174.
12. Klionsky, D.J., *et al.* (2003). A unified nomenclature for yeast autophagy-related genes. *Dev Cell* 5:539–545.
13. Okamoto, K., Kondo-Okamoto, N., and Ohsumi, Y. (2009). Mitochondria-anchored receptor Atg32 mediates degradation of mitochondria via selective autophagy. *Dev Cell* 17:87–97.
14. Kanki, T., Wang, K., Cao, Y., Baba, M., and Klionsky, D.J. (2009). Atg32 is a mitochondrial protein that confers selectivity during mitophagy. *Dev Cell* 17:98–109.

15. Kanki, T., *et al.* (2009). A genomic screen for yeast mutants defective in selective mitochondria autophagy. *Mol Biol Cell* 20:4730–4738.
16. Yorimitsu, T., Nair, U., Yang, Z., and Klionsky, D.J. (2006). Endoplasmic reticulum stress triggers autophagy. *J Biol Chem* 281:30299–30304.
17. Baba, M., Osumi, M., Scott, S.V., Klionsky, D.J., and Ohsumi, Y. (1997). Two distinct pathways for targeting proteins from the cytoplasm to the vacuole/lysosome. *J Cell Biol* 139:1687–1695.
18. Matsuura, A., Tsukada, M., Wada, Y., and Ohsumi, Y. (1997). Apg1p, a novel protein kinase required for the autophagic process in *Saccharomyces cerevisiae*. *Gene* 192:245–250.
19. Funakoshi, T., Matsuura, A., Noda, T., and Ohsumi, Y. (1997). Analyses of *APG13* gene involved in autophagy in yeast, *Saccharomyces cerevisiae*. *Gene* 192:207–213.
20. Kamada, Y., Funakoshi, T., Shintani, T., Nagano, K., Ohsumi, M., and Ohsumi, Y. (2000). Tor-mediated induction of autophagy via an Apg1 protein kinase complex. *J Cell Biol* 150:1507–1513.
21. Kametaka, S., Okano, T., Ohsumi, M., and Ohsumi, Y. (1998). Apg14p and Apg6/Vps30p form a protein complex essential for autophagy in the yeast, *Saccharomyces cerevisiae*. *J Biol Chem* 273:22284–22291.
22. Kihara, A., Noda, T., Ishihara, N., and Ohsumi, Y. (2001). Two distinct Vps34 phosphatidylinositol 3-kinase complexes function in autophagy and carboxypeptidase Y sorting in *Saccharomyces cerevisiae*. *J Cell Biol* 152:519–530.
23. Kametaka, S., Matsuura, A., Wada, Y., and Ohsumi, Y. (1996). Structural and functional analyses of *APG5*, a gene involved in autophagy in yeast. *Gene* 178:139–143.
24. Mizushima, N., *et al.* (1998). A protein conjugation system essential for autophagy. *Nature* 395:395–398.
25. Shintani, T., Mizushima, N., Ogawa, Y., Matsuura, A., Noda, T., and Ohsumi, Y. (1999). Apg10p, a novel protein-conjugating enzyme essential for autophagy in yeast. *EMBO J* 18:5234–5241.
26. Mizushima, N., Noda, T., and Ohsumi, Y. (1999). Apg16p is required for the function of the Apg12p-Apg5p conjugate in the yeast autophagy pathway. *EMBO J* 18:3888–3896.
27. Hanada, T., *et al.* (2007). The Atg12-Atg5 conjugate has a novel E3-like activity for protein lipidation in autophagy. *J Biol Chem* 282:37298–37302.
28. Schlumpberger, M., Schaeffeler, E., Straub, M., Bredschneider, M., Wolf, D.H., and Thumm, M. (1997). AUT1, a gene essential for autophagocytosis in the yeast *Saccharomyces cerevisiae*. *J Bacteriol* 179:1068–1076.
29. Lang, T., Schaeffeler, E., Bernreuther, D., Bredschneider, M., Wolf, D.H., and Thumm, M. (1998). Aut2p and Aut7p, two novel microtubule-associated proteins are essential for delivery of autophagic vesicles to the vacuole. *EMBO J* 17:3597–3607.
30. Kirisako, T., *et al.* (1999). Formation process of autophagosome is traced with Apg8/Aut7p in yeast. *J Cell Biol* 147:435–446.
31. Kirisako, T., *et al.* (2000). The reversible modification regulates the membrane-binding state of Apg8/Aut7 essential for autophagy and the cytoplasm to vacuole targeting pathway. *J Cell Biol* 151:263–276.
32. Ichimura, Y., *et al.* (2000). A ubiquitin-like system mediates protein lipidation. *Nature* 408:488–492.
33. Nakatogawa, H., Ichimura, Y., and Ohsumi, Y. (2007). Atg8, a ubiquitin-like protein required for autophagosome formation, mediates membrane tethering and hemifusion. *Cell* 130:165–178.

34. Shintani, T., Suzuki, K., Kamada, Y., Noda, T., and Ohsumi, Y. (2001). Apg2p functions in autophagosome formation on the perivacuolar structure. *J Biol Chem* 276:30452–30460.
35. Wang, C.W., *et al.* (2001). Apg2 is a novel protein required for the cytoplasm to vacuole targeting, autophagy, and pexophagy pathways. *J Biol Chem* 276:30442–30451.
36. Noda, T., *et al.* (2000). Apg9p/Cvt7p is an integral membrane protein required for transport vesicle formation in the Cvt and autophagy pathways. *J Cell Biol* 148:465–480.
37. Reggiori, F., Tucker, K.A., Stromhaug, P.E., and Klionsky, D.J. (2004). The Atg1-Atg13 complex regulates Atg9 and Atg23 retrieval transport from the pre-autophagosomal structure. *Dev Cell* 6:79–90.
38. He, C., Baba, M., Cao, Y., and Klionsky, D.J. (2008). Self-interaction is critical for Atg9 transport and function at the phagophore assembly site during autophagy. *Mol Biol Cell* 19:5506–5516.
39. Sekito, T., Kawamata, T., Ichikawa, R., Suzuki, K., and Ohsumi, Y. (2009). Atg17 recruits Atg9 to organize the pre-autophagosomal structure. *Genes Cells* 14:525–538.
40. Efe, J.A., Botelho, R.J., and Emr, S.D. (2007). Atg18 regulates organelle morphology and Fab1 kinase activity independent of its membrane recruitment by phosphatidylinositol 3, 5-bisphosphate. *Mol Biol Cell* 18:4232–4244.
41. Suzuki, K., Kirisako, T., Kamada, Y., Mizushima, N., Noda, T., and Ohsumi, Y. (2001). The pre-autophagosomal structure organized by concerted functions of APG genes is essential for autophagosome formation. *EMBO J* 20:5971–5981.
42. Suzuki, K., Kubota, Y., Sekito, T., and Ohsumi, Y. (2007). Hierarchy of Atg proteins in pre-autophagosomal structure organization. *Genes Cells* 12:209–218.
43. Kawamata, T., *et al.* (2005). Characterization of a novel autophagy-specific gene, ATG29. *Biochem Biophys Res Commun* 338:1884–1889.
44. Kabeya, Y., Kawamata, T., Suzuki, K., and Ohsumi, Y. (2007). Cis1/Atg31 is required for autophagosome formation in *Saccharomyces cerevisiae*. *Biochem Biophys Res Commun* 356:405–410.
45. Kabeya, Y., Kamada, Y., Baba, M., Takikawa, H., Sasaki, M., and Ohsumi, Y. (2005). Atg17 functions in cooperation with Atg1 and Atg13 in yeast autophagy. *Mol Biol Cell* 16:2544–2553.
46. Shintani, T., and Klionsky, D.J. (2004). Cargo proteins facilitate the formation of transport vesicles in the cytoplasm to vacuole targeting pathway. *J Biol Chem* 279:29889–29894.
47. Cheong, H., Nair, U., Geng, J., and Klionsky, D.J. (2008). The Atg1 kinase complex is involved in the regulation of protein recruitment to initiate sequestering vesicle formation for nonspecific autophagy in *Saccharomyces cerevisiae*. *Mol Biol Cell* 19:668–681.
48. Kawamata, T., Kamada, Y., Kabeya, Y., Sekito, T., and Ohsumi, Y. (2008). Organization of the pre-autophagosomal structure responsible for autophagosome formation. *Mol Biol Cell* 19:2039–2050.
49. Noda, T., and Ohsumi, Y. (1998). Tor, a phosphatidylinositol kinase homologue, controls autophagy in yeast. *J Biol Chem* 273:3963–3966.
50. Kunz, J., Henriquez, R., Schneider, U., Deuter-Reinhard, M., Movva, N.R., and Hall, M.N. (1993). Target of rapamycin in yeast, TOR2, is an essential phosphatidylinositol kinase homolog required for G1 progression. *Cell* 73:585–596.
51. Loewith, R., *et al.* (2002). Two TOR complexes, only one of which is rapamycin sensitive, have distinct roles in cell growth control. *Mol Cell* 10:457–468.
52. Wedaman, K.P., Reinke, A., Anderson, S., Yates, J., III, McCaffery, J.M., and Powers, T. (2003). Tor kinases are in distinct membrane-associated protein complexes in *Saccharomyces cerevisiae*. *Mol Biol Cell* 14:1204–1220.

53. Jacinto, E., and Hall, M.N. (2003). Tor signalling in bugs, brain and brawn. *Nat Rev Mol Cell Biol* 4:117–126.
54. Rohde, J.R., Bastidas, R., Puria, R., and Cardenas, M.E. (2008). Nutritional control via Tor signaling in *Saccharomyces cerevisiae. Curr Opin Microbiol* 11:153–160.
55. Cybulski, N., and Hall, M.N. (2009). TOR complex 2: a signaling pathway of its own. *Trends Biochem Sci* 34:620–627.
56. Nakashima, A., *et al.* (2008). The yeast Tor signaling pathway is involved in G2/M transition via polo-kinase. *PLoS ONE* 3:e2223.
57. Jiang, Y., and Broach, J.R. (1999). Tor proteins and protein phosphatase 2A reciprocally regulate Tap42 in controlling cell growth in yeast. *EMBO J* 18:2782–2792.
58. Urban, J., *et al.* (2007). Sch9 is a major target of TORC1 in *Saccharomyces cerevisiae. Mol Cell* 26:663–674.
59. Yorimitsu, T., Zaman, S., Broach, J.R., and Klionsky, D.J. (2007). Protein kinase A and Sch9 cooperatively regulate induction of autophagy in *Saccharomyces cerevisiae. Mol Biol Cell* 18:4180–4189.
60. Budovskaya, Y.V., Stephan, J.S., Reggiori, F., Klionsky, D.J., and Herman, P.K. (2004). The Ras/cAMP-dependent protein kinase signaling pathway regulates an early step of the autophagy process in *Saccharomyces cerevisiae. J Biol Chem* 279:20663–20671.
61. Budovskaya, Y.V., Stephan, J.S., Deminoff, S.J., and Herman, P.K. (2005). An evolutionary proteomics approach identifies substrates of the cAMP-dependent protein kinase. *Proc Natl Acad Sci USA* 102:13933–13938.
62. Stephan, J.S., Yeh, Y.Y., Ramachandran, V., Deminoff, S.J., and Herman, P.K. (2009). The Tor and PKA signaling pathways independently target the Atg1/Atg13 protein kinase complex to control autophagy. *Proc Natl Acad Sci USA* 106:17049–17054.
63. Martin, D.E., Soulard, A., and Hall, M.N. (2004). TOR regulates ribosomal protein gene expression via PKA and the Forkhead transcription factor FHL1. *Cell* 119:969–979.
64. Straub, M., Bredschneider, M., and Thumm, M. (1997). AUT3, a serine/threonine kinase gene, is essential for autophagocytosis in *Saccharomyces cerevisiae. J Bacteriol* 179:3875–3883.
65. Abeliovich, H., Zhang, C., Dunn, W.A., Jr., Shokat, K.M., and Klionsky, D.J. (2003). Chemical genetic analysis of Apg1 reveals a non-kinase role in the induction of autophagy. *Mol Biol Cell* 14:477–490.
66. Scott, S.V., *et al.* (2000). Apg13p and Vac8p are part of a complex of phosphoproteins that are required for cytoplasm to vacuole targeting. *J Biol Chem* 275:25840–25849.
67. Kim, J., *et al.* (2001). Cvt9/Gsa9 functions in sequestering selective cytosolic cargo destined for the vacuole. *J Cell Biol* 153:381–396.
68. Nice, D.C., Sato, T.K., Stromhaug, P.E., Emr, S.D., and Klionsky, D.J. (2002). Cooperative binding of the cytoplasm to vacuole targeting pathway proteins, Cvt13 and Cvt20, to phosphatidylinositol 3-phosphate at the pre-autophagosomal structure is required for selective autophagy. *J Biol Chem* 277:30198–30207.
69. Cao, Y., Nair, U., Yasumura-Yorimitsu, K., and Klionsky, D.J. (2009). A multiple ATG gene knockout strain for yeast two-hybrid analysis. *Autophagy* 5:699–705.
70. Yorimitsu, T., and Klionsky, D.J. (2005). Atg11 links cargo to the vesicle-forming machinery in the cytoplasm to vacuole targeting pathway. *Mol Biol Cell* 16:1593–1605.
71. Wang, Y.X., Catlett, N.L., and Weisman, L.S. (1998). Vac8p, a vacuolar protein with armadillo repeats, functions in both vacuole inheritance and protein targeting from the cytoplasm to vacuole. *J Cell Biol* 140:1063–1074.
72. Kabeya, Y., Noda, N.N., Fujioka, Y., Suzuki, K., Inagaki, F., and Ohsumi, Y. (2009). Characterization of the Atg17-Atg29-Atg31 complex specifically required for starvation-

induced autophagy in *Saccharomyces cerevisiae. Biochem Biophys Res Commun* 389:612–615.
73. Cheong, H., Yorimitsu, T., Reggiori, F., Legakis, J.E., Wang, C.W., and Klionsky, D.J. (2005). Atg17 regulates the magnitude of the autophagic response. *Mol Biol Cell* 16:3438–3453.
74. Cao, Y., Cheong, H., Song, H., and Klionsky, D.J. (2008). In vivo reconstitution of autophagy in *Saccharomyces cerevisiae. J Cell Biol* 182:703–713.
75. Sakumoto, N., *et al.* (1999). A series of protein phosphatase gene disruptants in *Saccharomyces cerevisiae. Yeast (Chichester, England)* 15:1669–1679.
76. Klionsky, D.J. (2005). The molecular machinery of autophagy: unanswered questions. *J Cell Sci* 118:7–18.
77. Kamada, Y., *et al.* (2010). Tor directly controls the Atg1 kinase complex to regulate autophagy. *Mol Cell Biol* 30:1049–1058.
78. Brunn, G.J., *et al.* (1997). Phosphorylation of the translational repressor PHAS-I by the mammalian target of rapamycin. *Science (New York, N.Y.)* 277:99–101.
79. Harris, T.E., and Lawrence, J.C., Jr. (2003). TOR signaling. *Sci STKE* 2003:re15.
80. Oshiro, N., *et al.* (2007). The proline-rich Akt substrate of 40 kDa (PRAS40) is a physiological substrate of mammalian target of rapamycin complex 1. *J Biol Chem* 282:20329–20339.
81. Schalm, S.S., and Blenis, J. (2002). Identification of a conserved motif required for mTOR signaling. *Curr Biol* 12:632–639.
82. Kamada, Y., *et al.* (2005). Tor2 directly phosphorylates the AGC kinase Ypk2 to regulate actin polarization. *Mol Cell Biol* 25:7239–7248.
83. Noda, T., Matsuura, A., Wada, Y., and Ohsumi, Y. (1995). Novel system for monitoring autophagy in the yeast *Saccharomyces cerevisiae. Biochem Biophys Res Commun* 210:126–132.
84. Xie, Z., Nair, U., and Klionsky, D.J. (2008). Atg8 controls phagophore expansion during autophagosome formation. *Mol Biol Cell* 19:3290–3298.
85. Wullschleger, S., Loewith, R., and Hall, M.N. (2006). TOR signaling in growth and metabolism. *Cell* 124:471–484.
86. Ogura, K., *et al.* (1994). *Caenorhabditis elegans* unc-51 gene required for axonal elongation encodes a novel serine/threonine kinase. *Genes Dev* 8:2389–2400.
87. Kuroyanagi, H., *et al.* (1998). Human ULK1, a novel serine/threonine kinase related to UNC-51 kinase of *Caenorhabditis elegans*: cDNA cloning, expression, and chromosomal assignment. *Genomics* 51:76–85.
88. Okazaki, N., *et al.* (2000). Interaction of the Unc-51-like kinase and microtubule-associated protein light chain 3 related proteins in the brain: possible role of vesicular transport in axonal elongation. *Brain Res* 85:1–12.
89. Chan, E.Y., Kir, S., and Tooze, S.A. (2007). siRNA screening of the kinome identifies ULK1 as a multidomain modulator of autophagy. *J Biol Chem* 282:25464–25474.
90. Scott, R.C., Juhasz, G., and Neufeld, T.P. (2007). Direct induction of autophagy by Atg1 inhibits cell growth and induces apoptotic cell death. *Curr Biol* 17:1–11.
91. Melendez, A., Talloczy, Z., Seaman, M., Eskelinen, E.L., Hall, D.H., and Levine, B. (2003). Autophagy genes are essential for dauer development and life-span extension in *C. elegans. Science (New York, N.Y.)* 301:1387–1391.
92. Tomoda, T., Kim, J.H., Zhan, C., and Hatten, M.E. (2004). Role of Unc51.1 and its binding partners in CNS axon outgrowth. *Genes Dev* 18:541–558.
93. Jung, C.H., *et al.* (2009). ULK-Atg13-FIP200 complexes mediate mTOR signaling to the autophagy machinery. *Mol Biol Cell* 20:1992–2003.

94. Hosokawa, N., *et al.* (2009). Nutrient-dependent mTORC1 association with the ULK1-Atg13-FIP200 complex required for autophagy. *Mol Biol Cell* 20:1981–1991.
95. Hara, T., *et al.* (2008). FIP200, a ULK-interacting protein, is required for autophagosome formation in mammalian cells. *J Cell Biol* 181:497–510.
96. Chan, E.Y., Longatti, A., McKnight, N.C., and Tooze, S.A. (2009). Kinase-inactivated ULK proteins inhibit autophagy via their conserved C-terminal domains using an Atg13-independent mechanism. *Mol Cell Biol* 29:157–171.
97. Chang, Y.Y., and Neufeld, T.P. (2009). An Atg1/Atg13 complex with multiple roles in TOR-mediated autophagy regulation. *Mol Biol Cell* 20:2004–2014.
98. Mercer, C.A., Kaliappan, A., and Dennis, P.B. (2009). A novel, human Atg13 binding protein, Atg101, interacts with ULK1 and is essential for macroautophagy. *Autophagy* 5:649–662.
99. Hosokawa, N., Sasaki, T., Iemura, S., Natsume, T., Hara, T., and Mizushima, N. (2009). Atg101, a novel mammalian autophagy protein interacting with Atg13. *Autophagy* 5:973–979.
100. Ogura, K., and Goshima, Y. (2006). The autophagy-related kinase UNC-51 and its binding partner UNC-14 regulate the subcellular localization of the Netrin receptor UNC-5 in *Caenorhabditis elegans*. *Development (Cambridge, England)* 133:3441–3450.
101. Mizushima, N., Levine, B., Cuervo, A.M., and Klionsky, D.J. (2008). Autophagy fights disease through cellular self-digestion. *Nature* 451:1069–1075.
102. Alvers, A.L., *et al.* (2009). Autophagy and amino acid homeostasis are required for chronological longevity in *Saccharomyces cerevisiae*. *Aging Cell* 8:353–369.

8

Conservation of the Tsc/Rheb/TORC1/S6K/S6 Signaling in Fission Yeast

AKIO NAKASHIMA[1] • FUYUHIKO TAMANOI

Department of Microbiology, Immunology and Molecular Genetics
Molecular Biology Institute, Jonsson Comprehensive Cancer Center
University of California
Los Angeles, California, USA

I. Abstract

The TSC/Rheb/TORC1/S6K/S6 signaling pathway plays critical roles in regulating protein synthesis and growth in eukaryotes. Our recent work using fission yeast *Schizosaccharomyces pombe* revealed that this signaling pathway is conserved from humans to fission yeast. In addition to target of rapamycin (TOR) homologs and tuberous sclerosis complex (TSC) homologs, fission yeast but not budding yeast, has a functional homolog of Rheb, a small G-protein acting as an activator of TOR complex 1 (TORC1). Several lines of genetic evidence suggest that the Tsc1–Tsc2 complex and Rheb act as upstream players of TORC1 in fission yeast. We have recently demonstrated that TORC1, but not TORC2, regulates phosphorylation of ribosomal protein S6 in response to nutrient availability. Candidate S6 kinase (S6K) protein has been identified. In addition, we find that rapamycin prevents a subset of TORC1 activity to regulate S6 phosphorylation in fission yeast.

[1] Current address: Biosignal Research Center, Kobe University, Kobe, Japan

THE ENZYMES, Vol. XXVIII

ISSN NO: 1874-6047
DOI: 10.1016/S1874-6047(10)28008-3

II. Introduction

Evolutionarily conserved mechanisms by which protein phosphorylation mediates numerous cellular activities, such as cell cycle progression, differentiation, and metabolism, in response to alteration of environmental conditions operate in eukaryotic cells. TOR is a highly conserved serine/threonine kinase of the phosphatidylinositol kinase-related kinase family in eukaryotes. TOR has pivotal roles in cell proliferation, cell growth, and metabolism in response to environmental stimuli, such as nutrients, growth factors, and stresses. TOR exists as the catalytic subunit in two distinct multiprotein complexes, TORC1 and TORC2. The TORC1 signaling promotes anabolic processes, such as protein synthesis and ribosome biogenesis, and inhibits catabolic processes such as autophagy. The immunosuppressant and anticancer drug rapamycin preferentially inhibits TORC1 activity but not TORC2 through the binding to immunophilin FKBP12 [1, 2].

In mammalian cells, mammalian TORC1 (mTORC1) contributes to cell growth by directly phosphorylating translation regulators, S6K and eIF4E-binding protein (4E-BP) [3]. A member of the Ras superfamily G-protein, Rheb, positively regulates mTORC1 but not mTORC2 when bound to GTP. On the other hand, the tumor suppressor proteins, tuberous sclerosis complex 1 (TSC1, also known as hamartin) and TSC2 (also known as tuberin), associate with each other and negatively control mTORC1 functions through inactivation of Rheb. In particular, TSC2 contains a domain that shares homology with the catalytic domain of GTPase-activating proteins (GAPs). The TSC–Rheb pathway mediates inputs from numerous signaling, such as growth factor, hypoxia, or energy status, to mTORC1 to control cell growth and other cellular processes [4, 5].

The fission yeast *S. pombe* has a similar TOR signaling system as described below. First, fission yeast possesses two TOR genes, *tor1*$^{+}$ and *tor2*$^{+}$. Second, the microbe also has TORC1, which consists of Tor2, a raptor homolog Mip1, an mLST8 homolog Wat1 (also known as Pop3), Toc1, and Tco89, as well as TORC2, which consists of Tor1, a rictor homolog Ste20, Sin1, Wat1, and Bit61. TORC1 controls the switch between cell cycle and cell differentiation by sensing nitrogen availability, whereas TORC2 participates in cell proliferation under stress conditions and in sexual differentiation under nitrogen starvation. Third, unlike budding yeast, the structural and functional homologs of Rheb and the TSC1–TSC2 complex are conserved as Rhb1 and the Tsc1–Tsc2 complex, respectively. It is believed that Rhb1 positively and the TSC complex negatively modulate TORC1 function. Therefore, fission yeast has emerged as an ideal system to study functions of the TOR signaling [6, 7].

Our study in fission yeast led first to the idea that the function of the TSC/Rheb/TORC1 in fission yeast is very similar to that in mammalian cells. Our further characterization of this model system led to the identification of S6 homolog in fission yeast. These results reveal that the entire TSC/Rheb/TORC1/S6K/S6 signaling pathway is conserved from human to fission yeast. We also found that this pathway in fission yeast is inhibited by rapamycin [8]. This is interesting, as, unlike some mammalian cell lines and budding yeast, rapamycin does not prevent the proliferation of the prototrophic cells in fission yeast [9, 10]. In this review, we will focus on fission yeast and describe the results obtained concerning the TSC/Rheb/TORC1/S6K/S6 signaling pathway.

III. Overview of the TSC/Rheb/TORC1 Signaling in Fission Yeast

A. TSC1–TSC2

The *TSC1* and *TSC2* genes were first identified as the genetic loci mutated in the autosomal dominant disorder TSC in human genome [11, 12]. In fission yeast, the *tsc* genes, $tsc1^+$ and $tsc2^+$, were identified as structural homologs of human, *TSC1* and *TSC2*, respectively [13]. The gene products Tsc1 and Tsc2 form a heterodimer and promote nutrient uptake, such as amino acids and nucleotides through regulation of localization and expression of transporters, such as amino acid transporter [13–17]. Disruption of either the *tsc1* or the *tsc2* gene markedly decreases uptake of leucine and basic amino acids, such as arginine, lysine, and histidine. Therefore, null mutant of either the *tsc1* or *tsc2* gene exhibits poor growth in media containing low concentration of leucine when the cells possess a leucine auxotroph. Conversely, decrease in uptake of the basic amino acids in the *tsc* null mutants confers resistance to canavanine and thialysine, which are toxic analogs of arginine and lysine, respectively [14, 15, 17]. It is noteworthy that the $tsc2^{N1191K}$ mutant (the mutation corresponds to human TSC2 mutation derived from patients of TSC) loses function in amino acid uptake, suggesting the use of *S. pombe* TSC as a model system for human TSC [14].

We have demonstrated that mislocalization of Cat1, a cationic amino acid transporter, is responsible for resistance to canavanine in *tsc2*Δ cells. Cat1 mostly localizes to the golgi-like punctates in the cytoplasm in *tsc2*Δ cells, whereas in the wild-type cells it is predominantly found on the periphery of cell tips with minor internal staining. Although it is not known how the TSC complex regulates localization of amino acid transporters, loss of Pub1, an E3 ubiquitin ligase, restores proper localization of Cat1 and sensitivity to canavanine in *tsc2Δ*, suggesting that the Tsc complex might affect localization of Cat1 through regulation of Pub1 function [17]. In budding yeast, RSP5, a homolog of Pub1, is also involved in endocytotic downregulation of several amino acid transporters via RSP5-mediated ubiquitination [18–20].

It is reported that aberrant amino acid uptake in *tsc2Δ* is suppressed by downregulation of Rhb1 (Rheb) function, as described below. Expression of a dominant negative *rhb*1 mutant such as *rhb1D60K* but not the wild-type gene suppresses mislocalization of Cat1 and canavanine resistance in *tsc2Δ* [14, 17, 21, 22]. Cpp1 encodes the β-subunit of protein farnesyltransferase that catalyzes farnesylation of Rhb1. This lipid modification is important for the function of the small G-protein [15, 23, 24]. Mutation of the *cpp1* gene, leading to downregulation of Rhb1 function, causes hypersensitivity to canavanine and suppressed decreased amino acid uptake in *tsc2Δ*. By contrast, these phenotypes of the *cpp1* mutation are suppressed by the expression of a geranylgeranylated mutant, $rhb1^{CVIL}$, which can bypass farnesylation [15, 23]. These genetic interactions suggest that Rhb1 is a downstream player of the Tsc complex in amino acid uptake (other relationships between Tsc and Rhb1 are discussed below).

Tsc1 and Tsc2 are also involved in cell response to nitrogen starvation. Fission yeast cells arrest in G1 phase under nitrogen starvation and subsequently proceed to mating, meiosis, and sporulation if different mating type cells exist. These events are known as sexual differentiation. Disruption of either the *tsc1* or the *tsc2* gene results in a delay in G1 arrest and decrease in mating efficiency. Microarray analysis has shown that disruption of either of the *tsc* genes causes reduction of gene expression, such as $fnx1^+$, $mei2^+$, and $sxa2^+$, that are involved in nitrogen starvation response and sexual differentiation under nitrogen starvation [13, 15, 25]. van Slegtenhorst et al. have shown that Pas1, a G1 cyclin, contributes to a delay in G1 phase arrest in the *tsc* mutants under nitrogen starvation [22]. However, it has not been determined how the activity of the Tsc complex is regulated when cells perceive alteration of nutrient conditions such as nitrogen source and amino acids.

B. Rheb

Fission yeast $rhb1^+$ was first identified as a homolog of human Rheb by genome sequencing [26, 27]. Cells disrupted in $rhb1^+$ do not grow, indicating that $rhb1^+$ is essential for cell proliferation [24, 27]. Interestingly, expression of the human *Rheb* gene, but not the budding yeast *RHB1*, restores cell proliferation in *rhb1* null mutant, suggesting a functional relationship between the human Rheb and the fission yeast Rhb1 [24]. Furthermore, decrease in *rhb1*$^{+}$ expression or overexpression of dominant negative *rhb1* mutations causes growth defect, cell cycle arrest in G1 phase and small round cell [24, 27, 28]. These terminal phenotypes are similar to those of nitrogen-starved cells. Indeed, expression of *fnx1*$^{+}$ and *mei2*$^{+}$, whose transcription is induced by nitrogen starvation, is elevated by repressing *rhb1* expression [27]. In addition, downregulation of Rhb1 function by the failure of its farnesylation in the *cpp1* mutant results in the accumulation of G0/G1 phase cells under stationary phase and suppressing the defect of gene expression in *tsc2Δ* cells in response to nitrogen starvation [15, 24]. These lines of evidence suggest that Rhb1 mediates nutrient signaling to regulation of cell cycle progression and gene expression.

We and others have identified hyperactivating mutations in the *rhb1* gene. Most *rhb1* hyperactivating mutations except for *rhb1-DA8* are caused by a single amino acid change that occurs at conserved amino acids in G1 and G4 boxes or near to G5, whose boxes are implicated in the recognition of guanine nucleotides [21, 25]. Mutations of Rhb1^{V17G} and Rhb1^{S21G} in G1 box are located close to the phosphates of GTP, whereas those of Rhb1^{K120R} and Rhb1^{N153T} that are in or around G4 and G5 boxes, respectively, are located close to the guanine ring. Biochemical characterization of these *Rhb1* mutants has revealed that the four mutants exhibit decrease in GDP binding, and GTPase activity is decreased with Rhb1^{V17G} and Rhb1^{S21G} mutants, but not in Rhb1^{K120R} and Rhb1^{N153T} [21]. On the other hand, the dominant negative mutation, *rhb1-D60V*, occurs at a conserved aspartic acid in G3 box. The Rhb1-D60V mutant loses GTP-binding ability and exhibits preferential GDP binding [28].

The *rhb1* hyperactivating mutations confer resistance to canavanine and thialysine and cause decrease in leucine uptake, similar to the disruption of the *tsc* genes [21, 25]. Murai et al. have shown that two distinct *rhb1* hyperactivating mutants, *rhb1-DA4* and *rhb1-DA8*, exhibit distinct phenotypes under nitrogen starvation. Although both these mutants exhibit canavanine resistance, only *rhb1-DA4* shows reduced cell response to nitrogen starvation, namely, the phenotypes of *rhb1-DA4* are largely consistent with those of *tsc2Δ* [25]. Taken together, these molecular genetic studies indicate that Rhb1 acts as a downstream factor of the TSC complex and that Rhb1 and the TSC complex have opposite effects on amino acid uptake and nitrogen starvation response. These opposite functions are also seen in the TSC–Rheb pathway in higher eukaryotes. Interestingly, the *rhb1-DA8* cells are able to respond to nitrogen starvation similar to the wild-type cells. The *rhb1-DA8* mutation further suppresses defect of cell response of *tsc2Δ* in the *rhb1-DA8 tsc2Δ* double mutant [25].

C. TORC1 Signaling

Several groups have independently demonstrated that Tor2, a catalytic subunit of TORC1, monitors nitrogen conditions to determine cell fate between cell proliferation (and growth) and sexual differentiation (or quiescence). The conditional knockdown of *tor2* function by temperature-sensitive (ts) *tor2* mutations or by inhibition of $tor2^+$ expression results in mimicking nitrogen starvation, such as cell cycle arrest in G1 phase and entry to sexual differentiation even in nitrogen-rich conditions [16, 29–32]. Microarray analysis of a *tor2-ts* mutant has revealed that downregulation of *tor2* function leads to alteration of expression of a large number but not all of nitrogen-starvation responsive genes [31]. These phenotypes are similar to those caused by the downregulation of *rhb1* function and are opposite to those of the *rhb1* active mutants or the *tsc*Δ mutants as described above. On the other hand, the *tor2-ts* mutations suppress defect of adenine uptake in the *tsc2*Δ background [31]; however, it is reported that a *tor2-ts* mutation does not suppress canavanine resistance in the *rhb1* active mutant or *tsc2*Δ [25].

In screening for downstream factors of $rhb1^+$, we have identified a point mutation in *tor2* that is able to suppress growth defect of the *rhb1* null mutant. Further study identified a total of 20 single amino acid mutations at 18 positions located in the *tor2* gene. It is noteworthy that those mutations are clustered mainly in the FAT and the kinase domains [33]. This may suggest that the FAT and the kinase domains are probably involved in the regulation of Tor2 activity by Rhb1. Cells expressing these mutations also exhibited a delay of cell cycle arrest in G1 phase, maintained cell size, and exhibited lower sporulation efficiency under nitrogen depletion conditions [33]. Similar phenomena upon nitrogen starvation are observed in cells carrying the *rhb1* activating mutations and the *tsc* disruptions [13, 25, 33]. In addition to these genetic relationships, we and others have shown that Rhb1 binds to Tor2 [21, 30].

Taken together, these studies on $tor2^+$, $tsc1^+$–$tsc2^+$, and $rhb1^+$ show that Tor2 in TORC1 functions as a downstream player of the TSC–Rhb1 pathway, and this signaling is conserved between fission yeast and higher eukaryotes (Figure 8.1). In contrast to fission yeast, the counterparts of the TSC complex in budding yeast have not been identified so far. Furthermore, although a Rheb homolog has been found as RHB1 in budding yeast, this small G-protein appears to have no effect on the TOR signaling [34]. Therefore, fission yeast has provided an ideal system to study the TORC1 signaling. We have recently demonstrated that TORC1 activity assessed by the phosphorylation of ribosomal protein S6 is indeed influenced by the TSC–Rhb1 pathway as described in detail below [8].

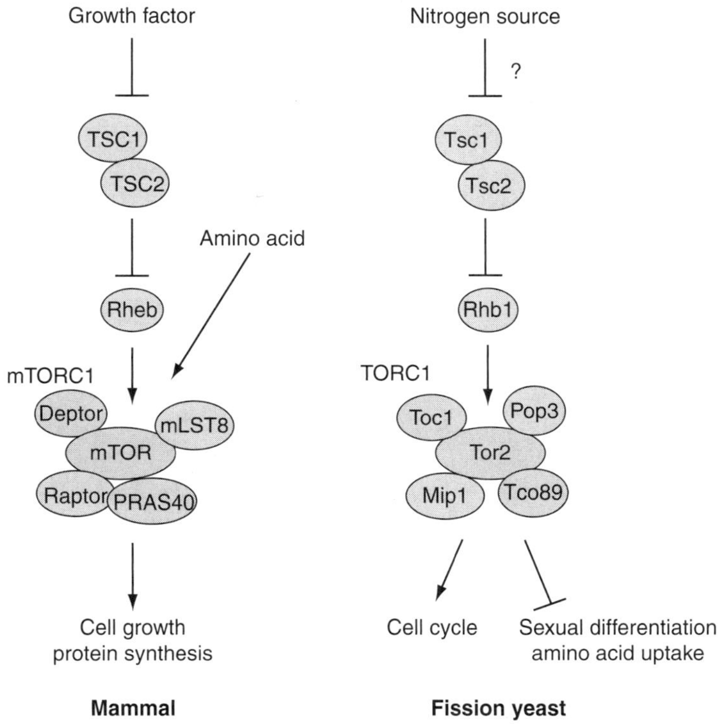

FIG. 8.1. Conservation of the TORC1 signaling in mammal and fission yeast. Note that the TSC/Rheb/TORC1 signaling functions similarly in both systems and that the components of TORC1 are similar.

IV. PAS Assay and Detection of S6 in Fission Yeast

A. Ribosomal Protein S6

In contrast to upstream factors of TORC1, its downstream factor(s) had not been identified in fission yeast. We have identified downstream molecules of TORC1 by using the anti-phospho-Akt-substrate (PAS) antibody that can recognize the consensus phosphorylation motif, R/K–X–R/K–X–X–pS/pT, where X represents any amino acid, and pS and pT is phosphorylated serine and threonine, respectively [8]. The consensus recognition motif of the PAS antibody is roughly consistent with a site phosphorylated by Akt and p70S6K [35–37]. Akt and S6K are well-known substrates of mTORC2 and mTORC1, respectively [38]; thus, substrates of Akt and S6K are predicted to be located in downstream of the TOR complexes.

In mammalian cells, the PAS antibody indeed recognizes phosphorylation of ribosomal protein S6 that is directly phosphorylated by S6K regulated by mTORC1 in a nutrient-dependent manner [35, 39]. In the experiments using the PAS antibody, we find several nitrogen source (ammonium)-dependent phospho-proteins in fission yeast (Figure 8.2). Among these, phosphorylation of p27 is quickly and completely abolished under nitrogen starvation. Our analysis has further revealed that p27 corresponds to ribosomal protein (Rp) S6s, which are encoded by two distinct genes, *rps601*$^{+}$ and *rps602*$^{+}$. We also showed that phosphorylation of Rps6 is regulated by nitrogen availability. Downregulation of Rps6 phosphorylation by nitrogen depletion is suppressed by the *tor2* activating mutations, which confer an *rhb1*-independent growth as described above (Figure 8.3A). Conversely, the *tor2-ts* mutations at nonpermissive temperature diminish the increase of Rps6 phosphorylation in response to the addition of ammonium (Figure 8.3B). By contrast, TORC2 and other nutritional-stress-related pathways, such as the cyclic AMP–PKA and the stress-activated protein kinases signaling cascades, are not mainly involved in the regulation of Rps6 phosphorylation [8]. Accordingly, phosphorylation of Rps6 regulated by nitrogen status is mediated by TORC1 that contains Tor2.

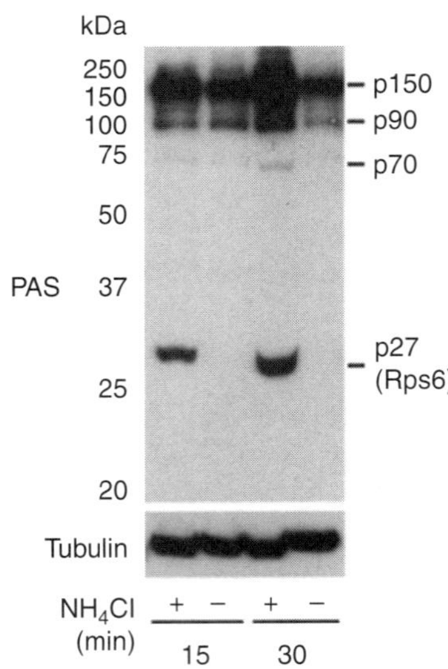

FIG. 8.2. Nitrogen source-dependent phospho-proteins recognized with the PAS antibody. JUp1204 cells were grown to exponential phase and then incubated in medium with or without ammonium for the indicated times. Cell extracts were run on a SDS polyacrylamide gel and probed with the PAS antibody. Tubulin is shown as a loading control. (From Ref. [8], Fig. 1A, p. 778.)

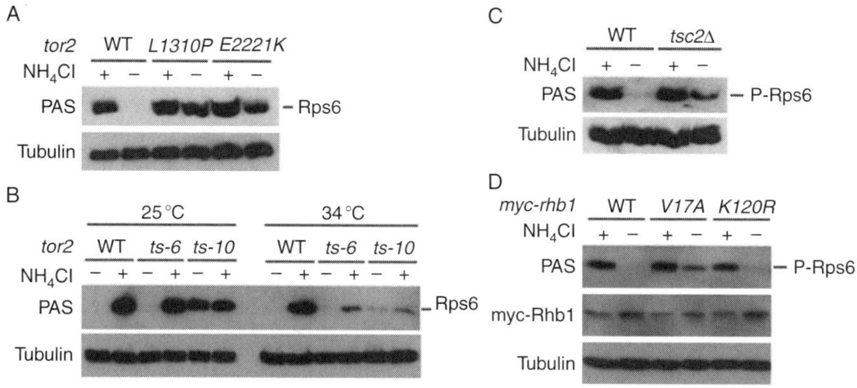

FIG. 8.3. The TSC–Rhb1–TORC1 signaling is involved in the regulation of Rps6 phosphorylation. (A) Cells expressing the mutant Tor2 (L1310P or E2221K), as well as the wild-type Tor2 were grown and incubated with or without ammonium. Extracts were immunoblotted with the PAS antibody. (B) Tor2 ts mutants, ts-6 and ts-10, as well as the wild-type cells were grown to exponential phase, incubated at the permissive or nonpermissive temperature and then incubated with or without ammonium. Extracts were immunoblotted with the PAS antibody. (C) The wild-type and the *tsc2* mutant were grown and incubated with or without ammonium. Extracts were immunoblotted with the PAS antibody. (D) Cells expressing mutant Rhb1V17A, and K120R, as well as the wild-type Rhb1 were grown and incubated with or without ammonium. Extracts were immunoblotted with the PAS antibody, as well as anti-myc antibody. Tubulin is shown as a loading control. (From Ref. [8], Figs. 2A and C and 3A and C, p. 779 and p. 780.)

The TSC–Rhb1 pathway is also involved in the regulation of the Tor2-mediated Rps6 phosphorylation, namely, a *tsc2* deletion mutation or an *rhb1* constitutively active mutation leads to attenuation of Rps6 dephosphorylation after the shift to nitrogen starvation (Figure 8.3C and D). These observations provide a model in which the TSC complex negatively and Rhb1 positively regulate TORC1 activity in fission yeast (Figure 8.1). It is of interest that there is a difference in the strength of effects on Rps6 phosphorylation between different *rhb1* active mutants. To confer a partial resistance of Rps6 phosphorylation to nitrogen depletion, $rhb1^{V17A}$ but not $rhb1^{K120R}$ is sufficient when expressed from its own promoter (lower expression), whereas $rhb1^{K120R}$ and $rhb1^{N153T}$ require overexpression using multicopy plasmid (higher expression). It is noteworthy that Tor2 activity in the *tor2* activating mutants, for example in *rhb1*-independent *tor2* mutants, a *tsc2* deletion mutant, or a *rhb1* constitutive activation mutant, is eventually abolished under nitrogen starvation [8]. Similar to this finding, it appears that in the mammalian system, mTORC1 activity in fibroblast cells lacking *TSC2* is also inhibited by amino acid depletion even though a high proportion of GTP-bound Rheb in the mutant fibroblasts is not affected [40, 41]. These imply the existence of other pathway(s) that mediates nutrient signaling to TORC1 other than the TSC–Rheb pathway. It has been reported that Rag/Gtr, heterodimeric small GTPases, mediate nutrient signaling to TORC1 in mammalian cells and budding yeast [42–45]. Fission yeast also possesses the predicted homologs of Rag/Gtr, suggesting that the fission yeast Rag/Gtr proteins may participate in the nutrient-TORC1 signaling.

B. Phosphorylation Sites in Rps6

Ribosomal protein S6 is highly conserved from yeast to higher eukaryotes including plants (Figure 8.4). It has been known that the mammalian S6 has five phosphorylatable serine residues (Ser235, Ser236, Ser240, Ser244, and Ser247) in the region close to the carboxy terminus. Among the serine residues, Ser235 and Ser236 are directly phosphorylated by S6K that is subjected to regulation by the nutrient-mTORC1 signaling [36]. In yeast Rps6, the two serine residues (Ser235 and Ser236 in fission yeast and

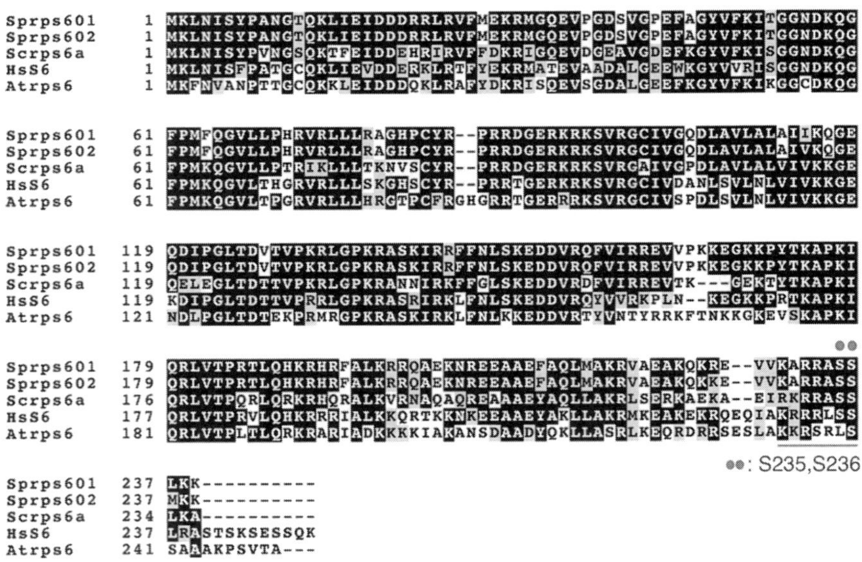

Fig. 8.4. Alignment of predicted amino acid sequences of ribosomal protein S6. Red underline indicates a predicted recognition motif (Arg/Lys-X-Arg/Lys-X-X-Ser/Thr) for the PAS antibody. Circles show phosphorylated serine residues of human S6 by S6Ks. Numbers on the left indicate amino acids. (For interpretation of the references to color in this figure legend, the reader is referred to the Web version of this chapter.)

Ser232 and Ser233 in budding yeast) and their adjacent regions are conserved, whereas the three residues, which extend to the C-terminus in mammalian S6, are missing in the S6 yeast proteins (Figure 8.4). In budding yeast, the conserved serine residues (Ser232 and Ser233) are phosphorylated *in vivo*, and Sch9, which is a functional S6K homolog, can catalyze the phosphorylation of those residues *in vitro* [46, 47]. In fission yeast Ser235, Ser236, and the flanking regions in Rps6 are very similar to the consensus recognition motif of the PAS antibody (Figure 8.4). We have observed in fission yeast that Ser235 and Ser236 are phosphorylated in a nutrient-TORC1-dependent manner, similar to other eukaryotes [8].

It is known that TORC1 signaling controls translation through phosphorylation of S6K and 4E-BP [3], whereas it is possible that Rps6 phosphorylation is dispensable for protein synthesis [46, 48]. In animal model, knock-in mice, in which Rps6 contains alanine substitutions at all five phosphorylatable serine residues, are viable and fertile. In the adult Rps6-mutation mice, interestingly, pancreatic β-cells are smaller than those in normal mice, and the cellular phenotype causes insufficient insulin secretion and impaired glucose homeostasis. Mouse embryo fibroblasts (MEF) generated from the Rps6-mutation knock-in mice are significantly smaller than control cells and unexpectedly exhibit an increased rate of protein synthesis [48]. In fission yeast, although the *rps6* genes themselves are essential for cell survival, its phosphorylation is not necessary for cell proliferation and cell size control [8]. So far, physiological role of Rps6 phosphorylation via TORC1 has not been fully understood, in spite of the conserved event among eukaryotes.

V. S6 Kinase in Fission Yeast

It is predicted that fission yeast possesses a number of potential AGC kinases. In mammalian cells, members in the AGC kinase family include S6K and Akt, which are substrates of mTORC1 and mTORC2, respectively [38]. Based on a search using the BLAST program for p70S6K1 homolog in fission yeast, this microbe has four homologs of S6K: Psk1, Gad8, Sck1, and Sck2 [8, 49–52]. Of these, Gad8 is established as a target of TORC2; therefore, it seems to be a functional homolog of Akt [52, 53]. Our recent analysis has revealed that phosphorylation levels of Psk1, Sck1, and Sck2, but not of Gad8 are altered in response to nitrogen availability [8]. In particular, phosphorylation of Psk1 in response to nitrogen availability is highly dependent on TORC1 activity. A part of Sck1 phosphorylation is also likely to be regulated by TORC1 in response to nitrogen availability (Nakashima and Tamanoi, unpublished results). Importantly, Rps6 phosphorylation in the *psk1* deletion mutant, but not the *sck1* or *sck2* mutants is abolished even in nutrient-rich conditions. Deletion of *psk1* also abolishes Rps6 phosphorylation in a *tor2* active mutant or in a *tsc2* deletion mutant; however, the disruption has no detectable effect on cell proliferation and cell growth, similar to that observed with a strain carrying phosphorylation-defective Rps6 ([49], Nakashima and Tamanoi).

Human S6K is phosphorylated at several sites by different kinases, such as PDK1 and mTOR. For example, mTORC1 phosphorylates serine and threonine residues in the turn motif and the hydrophobic motif, respectively, and PDK1 phosphorylates a threonine in the activation loop. It is known that those phosphorylations are important or critical for the kinase activity of S6K [37, 38]. Our analysis has revealed that a threonine residue in the hydrophobic motif in Psk1 is phosphorylated in response to nitrogen conditions, and TORC1 contributes to its phosphorylation. Analysis of alanine substitutions at phosphorylatable sites has shown that phosphorylation in the hydrophobic motif or the activation loop in Psk1 is critical for its full activation to phosphorylate Rps6 and that phosphorylation in the turn motif is likely to be important in phosphorylation of the hydrophobic motif (Nakashima and Tamanoi). These observations demonstrate that Psk1 is the most functional counterpart of S6K in fission yeast. Thus, the TORC1–S6K–S6 signaling would be conserved in fission yeast like other eukaryotes.

VI. Regulation of the TORC1 Signaling

As described above, TORC1 activity in fission yeast responds to environmental ammonium conditions. We have observed that the TORC1–S6 signaling is inhibited by glucose starvation [8]. When 2-deoxyglucose, a nonmetabolizable glucose analog that blocks glucose–hexokinase activity thereby inhibiting glycolysis, is added to medium instead of glucose, the TORC1 signaling is also abolished (Nakashima and Tamanoi), suggesting that TORC1 might be regulated in response to intracellular energy status, similar to the AMPK–mTORC1 pathway in mammalian cells [54]. Homologs of AMPK are conserved in fission yeast: Ppk9 and Ssp2 are catalytic subunits, Amk2 is a β-subunit, and Cbs2 is a γ-subunit [55]. These AMPKs might mediate energy status to TORC1. On the other hand, cycloheximide, which is a protein synthesis inhibitor, attenuates downregulation of TORC1–S6 signaling upon nitrogen or glucose starvation (Nakashima and Tamanoi).

It has been reported that like other eukaryotes, TORC1 is implicated in the regulation of autophagy and expression of an autophagy-related gene, that is, TORC1 disturbs autophagy induction [30, 31]. In fission yeast, nitrogen depletion becomes a trigger of autophagy induction, whereas glucose starvation does not do so even though both inhibit the TORC1–S6 signaling [8, 56, 57]. Therefore, glucose starvation might interfere with only a part of TORC1 function.

In higher eukaryotes, it has been known that amino acids, especially leucine, acts as a priming cue to activate the mTORC1 signaling, although it has not been established how the amino acid signaling is transduced to mTORC1 [58, 59]. In contrast to mammalian cells requiring supplement of essential amino acids, prototrophic yeast cells do not need any exogenous amino acids for cell proliferation. In budding yeast, it seems that alteration of intracellular glutamine affects TORC1 activity. Furthermore, TORC1 activity in cells carrying a leucine auxotroph is diminished by leucine depletion [44, 47, 60]. In fission yeast, our analysis indicates that refeeding of glutamine following nitrogen starvation can activate the TORC1–S6 signaling instead of ammonium and, conversely, that inhibition of intracellular glutamine synthesis by treatment with L-methionine sulfoximine, a glutamine synthetase inhibitor, results in a decrease in TORC1 activity. Effect of refeeding leucine, proline, or glutamate on the TORC1 signaling is much less than that of glutamine [Nakashima and Tamanoi]. As mentioned above, inhibition of protein synthesis with cycloheximide under nitrogen starvation attenuated inactivation of TORC1. This might result from maintained intracellular amino acid pool. Taken together, TORC1 signaling might monitor intracellular amino acid level.

In mammalian cells and budding yeast, TORC1 signaling is inhibited by several stress conditions in addition to nutrient depletion [47, 61]. We have observed that the TORC1–S6 signaling is inhibited by oxidative and osmotic stresses in fission yeast. In the cases of both stresses, the inhibition occurs slowly (Rps6 phosphorylation disappears by 60 min), compared with the inhibition by nutrient depletion (15 min) [8]. The Wis1–Sty1 (also known as Phh1 and Spc1) signaling in fission yeast is a major stress-activated protein kinases (SAPKs) cascade, which is activated by several stresses, such as nitrogen starvation, oxidative and osmotic stresses, etc., to maintain cell survival under those stress conditions [62–64]. It has not been clarified yet whether the SAPK signaling is involved in the regulation of the TORC1 pathway under the stress conditions.

VII. Effect of Rapamycin on the TORC1 Signaling

A TOR-specific inhibitor, rapamycin, forms an intracellular drug–protein complex with FKBP12, and the complex interacts with the FRB domain in TOR, thereby inhibiting TOR activity [65, 66]. In mammalian cells, this macrolide preferentially and quickly prevent phosphorylation of

S6K and 4E-BP via inhibition of mTORC1 [1]. By contrast, prolonged treatment of the macrolide is needed to reduce the phosphorylation of Akt by mTORC2 [67]. Rapamycin also decreases cell volume through inhibition of mTORC1 signaling [68]. Recently, the existence of rapamycin-resistant mTORC1 function that includes autophagy regulation, a part of regulation of 4E-BP phosphorylation and cell size, has been reported [69–72]. In budding yeast, rapamycin blocks cell proliferation via inhibition of TORC1 function and mimics nitrogen starvation including induction of autophagy, whereas it appears to have no effect on TORC2 activity [73].

In fission yeast, rapamycin treatment has no effect on cell proliferation in normal cells and does not mimic nitrogen starvation [9, 10]. Microarray analysis has revealed that rapamycin has little impact on alteration of gene expression involved in response to nitrogen depletion [31]. On the other hand, the macrolide prevents leucine uptake owing to downregulation of gene expression of amino acid transporters, as well as in a *tor1* deletion mutant; therefore, when cells have a mutation on the *leu1* gene, the cells exhibit growth defect on a minimal medium containing rapamycin and low concentration of leucine [74]. Tor1 is a catalytic subunit of TORC2 in fission yeast and this effect of rapamycin is likely to be caused by the inhibition of Tor1, because deletion of fkh1, a FKBP12 homolog in fission yeast, or expression of a rapamycin-resistant *tor1* allele can suppress the growth defect. Yeast two-hybrid analysis has also shown that the macrolide interacts with the FRB domain of Tor1 [74]. Furthermore, it has been reported that mutations of the *tor* genes in fission yeast confer sensitivity to rapamycin. Deletion of *tor1* results in growth defect in the presence of the macrolide independently of decreased leucine uptake [10], whereas this defect in the *tor1* mutant is suppressed by an *fkh1* disruption [74]. A point mutation (L2048S) in *tor2* gene also causes rapamycin sensitivity [32]. Interestingly, if yeast cells were cultured in medium containing glutamate as sole nitrogen source instead of ammonium, rapamycin induced mitotic onset because of inhibition of Tor1 and facilitated entry into sexual differentiation (until 10 h after addition of rapamycin) [75]. It appears that Tor1 as part of TORC1 regulates mitotic entry through the SAPK following reduced nitrogen availability [76]. However, we observed that the macrolide in an ammonium medium as a nitrogen source did not induce cell cycle arrest in G1 phase and sexual differentiation even when the cells were cultured at least for 17 h (Nakashima and Tamanoi).

We have demonstrated that S6 phosphorylation is inhibited by rapamycin [8]. Phosphorylation of Rps6 is markedly reduced by 30 min and subsequently disappears by 60 min, whereas its inhibition requires high dose of the macrolide (Figure 8.5A). Either disruption of *fkh1* or mutation of the rapamycin-

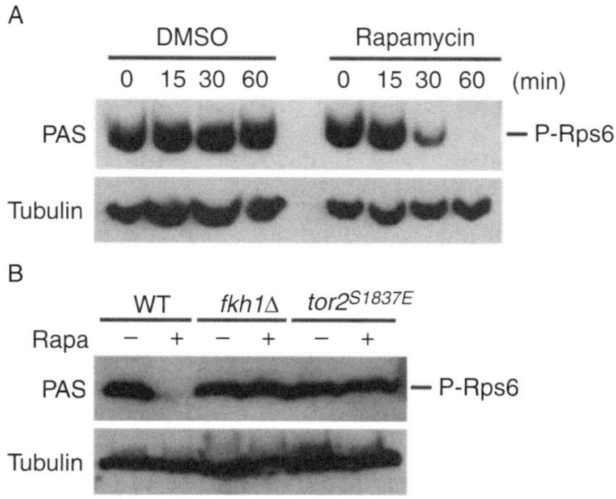

FIG. 8.5. Rapamycin inhibits Rps6 phosphorylation by blocking TORC1 activity. (A) Cells were added with either 200 nM rapamycin or DMSO and incubated for the indicated times. Extracts were subjected to immunoblotting with the PAS antibody. (B) Null mutant of FKBP12 and tor2 mutant resistant to rapamycin were grown and incubated with or without rapamycin. Extracts were immunoblotted with the PAS antibody. Tubulin is shown as a loading control. (From Ref. [8], Fig. 7A and C, p. 782.)

unbinding *tor2* suppresses this inhibition by rapamycin (Figure 8.5B). In addition, phosphorylation of Psk1 is also prevented by rapamycin in a Tor2-dependent manner (Nakashima and Tamanoi), suggesting that like other eukaryotes, rapamycin is targeting the TORC1–S6K–S6 signaling in fission yeast. Furthermore, prolonged rapamycin treatment (14 h) significantly reduces cell size in a TORC1-dependent manner [8]. By contrast, we also observed that the macrolide does not mimic nitrogen starvation, consistent with previous reports. Accordingly, cell behaviors upon rapamycin treatment are markedly different from those in *tor2-ts* mutant at nonpermissive temperature including cell cycle arrest in G1 phase, small round cell, and entry into sexual differentiation; namely the effects of rapamycin are more limited than those of loss of Tor2 function. These observations suggest that rapamycin inhibits a subset of TORC1 activity and that rapamycin-resistant function of TORC1 exists in fission yeast. Consistent with the findings in mutants of Rps6 or Psk1, rapamycin-sensitive functions of TORC1, at least the TORC1–S6K–S6 signaling is not important for cell growth, whereas rapamycin-resistant ones would contribute control of the switch between cell proliferation and cell differentiation. Further investigation is needed to determine why rapamycin prevents only a part of TORC1 functions.

VIII. Future Prospects

Nutrient signaling that senses nutrient cues and then transduces the signal to regulate most cellular processes, such as cell proliferation, differentiation, and metabolism, is the most fundamental machinery among organisms. The TORC1 signaling has been established as one of the major nutrient signaling pathway. We have summarized here the TORC1–S6K–S6 signaling, which is regulated by nutrient stimulations, such as ammonium, amino acid, and glucose, in fission yeast. Therefore, it is established that this signaling is conserved between yeasts and higher eukaryotes. In addition, like in higher eukaryotes, the TSC–Rheb pathway is involved in the regulation of the TORC1 pathway in fission yeast. However, a mechanism for sensing nutrients to regulate the TORC1 signaling has not been elucidated. It is also unknown what the input signaling to regulate the TSC–Rheb pathway in fission yeast is. We have shown that phosphorylation of Rps6 and Psk1 reflects downstream activity of TORC1. Thus, phosphorylation of these proteins should serve as convenient markers to examine the status of TORC1 activity and may provide a powerful tool to identify regulators and mediators of nutrient signaling to TORC1. Furthermore, we have observed that rapamycin inhibits the TORC1–S6K–S6 pathway, suggesting that there exists rapamycin-resistant function of TORC1. In addition to downstream factors in the rapamycin-sensitive TORC1 pathway besides S6K–S6, further investigation into downstreams of rapamycin-resistant TORC1 should provide insights into physiological roles of TORC1 not only in fission yeast, but also in other eukaryotes including animals.

Acknowledgments

We thank Dr. Tatsuhiro Sato for discussion. We also thank members of the Tamanoi laboratory for comments. This work was supported by the NIH grant CA41996.

References

1. Wullschleger, S., Loewith, R., and Hall, M.N. (2006). TOR signaling in growth and metabolism. *Cell* 124:471–484.
2. Chiang, G.G., and Abraham, R.T. (2007). Targeting the mTOR signaling network in cancer. *Trends Mol Med* 13:433–442.
3. Hay, N., and Sonenberg, N. (2004). Upstream and downstream of mTOR. *Genes Dev* 18:1926–1945.
4. Corradetti, M.N., and Guan, K.L. (2006). Upstream of the mammalian target of rapamycin: do all roads pass through mTOR? *Oncogene* 25:6347–6360.

5. Huang, J., and Manning, B.D. (2008). The TSC1-TSC2 complex: a molecular switchboard controlling cell growth. *Biochem J* 412:179–190.
6. Aspuria, P.J., Sato, T., and Tamanoi, F. (2007). The TSC/Rheb/TOR signaling pathway in fission yeast and mammalian cells: temperature sensitive and constitutive active mutants of TOR. *Cell Cycle* 6:1692–1695.
7. Otsubo, Y., and Yamamato, M. (2008). TOR signaling in fission yeast. *Crit Rev Biochem Mol Biol* 43:277–283.
8. Nakashima, A., Sato, T., and Tamanoi, F. (2010). Fission yeast TORC1 regulates phosphorylation of ribosomal S6 proteins in response to nutrients and its activity is inhibited by rapamycin. *J Cell Sci* 123:777–786.
9. Weisman, R., Choder, M., and Koltin, Y. (1997). Rapamycin specifically interferes with the developmental response of fission yeast to starvation. *J Bacteriol* 179:6325–6334.
10. Kawai, M., Nakashima, A., Ueno, M., Ushimaru, T., Aiba, K., Doi, H., and Uritani, M. (2001). Fission yeast Tor1 functions in response to various stresses including nitrogen starvation, high osmolarity, and high temperature. *Curr Genet* 39:166–174.
11. European Chromosome 16 Tuberous Sclerosis Consortium(1993). Identification and characterization of the tuberous sclerosis gene on chromosome16 . *Cell* 75:1305–1315.
12. van Slegtenhorst, M., de Hoogt, R., Hermans, C., Nellist, M., Janssen, B., Verhoef, S., Lindhout, D., van den Ouweland, A., Halley, D., Young, J., Burley, M., Jeremiah, S., *et al.* (1997). Identification of the tuberous sclerosis gene TSC1 on chromosome 9q34. *Science* 277:805–808.
13. Matsumoto, S., Bandyopadhyay, A., Kwiatkowski, D.J., Maitra, U., and Matsumoto, T. (2002). Role of the Tsc1-Tsc2 complex in signaling and transport across the cell membrane in the fission yeast *Schizosaccharomyces pombe*. *Genetics* 161:1053–1063.
14. van Slegtenhorst, M., Carr, E., Stoyanova, R., Kruger, W.D., and Henske, E.P. (2004). *tsc1*$^{+}$ and *tsc2*$^{+}$ regulate arginine uptake and metabolism in *Schizosaccharomyces pombe*. *J Biol Chem* 279:12706–12713.
15. Nakase, Y., Fukuda, K., Chikashige, Y., Tsutsumi, C., Morita, D., Kawamoto, S., Ohnuki, M., Hiraoka, Y., and Matsumoto, T. (2006). A defect in protein farnesylation suppresses a loss of *Schizosaccharomyces pombe tsc2*$^{+}$, a homolog of the human gene predisposing to tuberous sclerosis complex. *Genetics* 173:569–578.
16. Weisman, R., Roitburg, I., Schonbrun, M., Harari, R., and Kupiec, M. (2007). Opposite effects of Tor1 and Tor2 on nitrogen starvation responses in fission yeast. *Genetics* 175:1153–1162.
17. Aspuria, P.J., and Tamanoi, F. (2008). The Tsc/Rheb signaling pathway controls basic amino acid uptake via the Cat1 permease in fission yeast. *Mol Genet Genomics* 279:441–450.
18. Hein, C., Springael, J.Y., Volland, C., Haguenauer-Tsapis, R., and André, B. (1995). NPl1, an essential yeast gene involved in induced degradation of Gap1 and Fur4 permeases, encodes the Rsp5 ubiquitin-protein ligase. *Mol Microbiol* 18:77–87.
19. Soetens, O., De Craene, J.O., and Andre, B. (2001). Ubiquitin is required for sorting to the vacuole of the yeast general amino acid permease, Gap1. *J Biol Chem* 276:43949–43957.
20. Blondel, M.O., Morvan, J., Dupré, S., Urban-Grimal, D., Haguenauer-Tsapis, R., and Volland, C. (2004). Direct sorting of the yeast uracil permease to the endosomal system is controlled by uracil binding and Rsp5p-dependent ubiquitylation. *Mol Biol Cell* 15:883–895.
21. Urano, J., Comiso, M.J., Guo, L., Aspuria, P.J., Deniskin, R., Tabancay, A.P., Kato-Stankiewicz, J., and Tamanoi, F. (2005). Identification of novel single amino acid changes that result in hyperactivation of the unique GTPase, Rheb, in fission yeast. *Mol Microbiol* 58:1074–1086.

22. van Slegtenhorst, M., Mustafa, A., and Henske, E.P. (2005). Pas1, a G1 cyclin, regulates amino acid uptake and rescues a delay in G1 arrest in Tsc1 and Tsc2 mutants in *Schizosaccharomyces pombe*. *Hum Mol Genet* 14:2851–2858.
23. Yang, W., Urano, J., and Tamanoi, F. (2000). Protein farnesylation is critical for maintaining normal cell morphology and canavanine resistance in *Schizosaccharomyces pombe*. *J Biol Chem* 275:429–438.
24. Yang, W., Tabancay, A.P., Urano, J., and Tamanoi, F. (2001). Failure to farnesylate Rheb protein contributes to the enrichment of G0/G1 phase cells in the *Schizosaccharomyces pombe* farnesyltransferase mutant. *Mol Microbiol* 41:1339–1347.
25. Murai, T., Nakase, Y., Fukuda, K., Chikashige, Y., Tsutsumi, C., Hiraoka, Y., and Matsumoto, T. (2009). Distinctive responses to nitrogen starvation in the dominant active mutants of the fission yeast Rheb GTPase. *Genetics* 183:517–527.
26. Urano, J., Tabancay, A.P., Yang, W., and Tamanoi, F. (2000). The *Saccharomyces cerevisiae* Rheb G-protein is involved in regulating canavanine resistance and arginine uptake. *J Biol Chem* 275:11198–11206.
27. Mach, K.E., Furge, K.A., and Albright, C.F. (2000). Loss of Rhb1, a Rheb-related GTPase in fission yeast, causes growth arrest with a terminal phenotype similar to that caused by nitrogen starvation. *Genetics* 155:611–622.
28. Tabancay, A.P., Gau, C.L., Machado, I.M., Uhlmann, E.J., Gutmann, D.H., Guo, L., and Tamanoi, F. (2003). Identification of dominant negative mutants of Rheb GTPase and their use to implicate the involvement of human Rheb in the activation of p70S6K. *J Biol Chem* 278:39921–39930.
29. Alvarez, B., and Moreno, S. (2006). Fission yeast Tor2 promotes cell growth and represses cell differentiation. *J Cell Sci* 119:4475–4485.
30. Uritani, M., Hidaka, H., Hotta, Y., Ueno, M., Ushimaru, T., and Toda, T. (2006). Fission yeast Tor2 links nitrogen signals to cell proliferation and acts downstream of the Rheb GTPase. *Genes Cells* 11:1367–1379.
31. Matsuo, T., Otsubo, Y., Urano, J., Tamanoi, F., and Yamamoto, M. (2007). Loss of the TOR kinase Tor2 mimics nitrogen starvation and activates the sexual development pathway in fission yeast. *Mol Cell Biol* 27:3154–3164.
32. Hayashi, T., Hatanaka, M., Nagao, K., Nakaseko, Y., Kanoh, J., Kokubu, A., Ebe, M., and Yanagida, M. (2007). Rapamycin sensitivity of the *Schizosaccharomyces pombe* tor2 mutant and organization of two highly phosphorylated TOR complexes by specific and common subunits. *Genes Cells* 12:1357–1370.
33. Urano, J., Sato, T., Matsuo, T., Otsubo, Y., Yamamoto, M., and Tamanoi, F. (2007). Point mutations in TOR confer Rheb-independent growth in fission yeast and nutrient-independent mammalian TOR signaling in mammalian cells. *Proc Natl Acad Sci USA* 104:3514–3519.
34. Aspuria, P.J., and Tamanoi, F. (2004). The Rheb family of GTP-binding proteins. *Cell Signal* 16:1105–1112.
35. Manning, B.D., Tee, A.R., Logsdon, M.N., Blenis, J., and Cantley, L.C. (2002). Identification of the tuberous sclerosis complex-2 tumor suppressor gene product tuberin as a target of the phosphoinositide 3-kinase/akt pathway. *Mol Cell* 10:151–162.
36. Ruvinsky, I., and Meyuhas, O. (2006). Ribosomal protein S6 phosphorylation: from protein synthesis to cell size. *Trends Biochem Sci* 31:342–348.
37. Pearce, L.R., Komander, D., and Alessi, D.R. (2010). The nuts and bolts of AGC protein kinases. *Nat Rev Mol Cell Biol* 11:9–22.
38. Jacinto, E., and Lorberg, A. (2008). TOR regulation of AGC kinases in yeast and mammals. *Biochem J* 410:19–37.

39. Kane, S., Sano, H., Liu, S.C., Asara, J.M., Lane, W.S., Garner, C.C., and Lienhard, G.E. (2002). A method to identify serine kinase substrates. Akt phosphorylates a novel adipocyte protein with a Rab GTPase-activating protein (GAP) domain. *J Biol Chem* 277:22115–22118.
40. Smith, E.M., Finn, S.G., Tee, A.R., Browne, G.J., and Proud, C.G. (2005). The tuberous sclerosis protein TSC2 is not required for the regulation of the mammalian target of rapamycin by amino acids and certain cellular stresses. *J Biol Chem* 280:18717–18727.
41. Roccio, M., Bos, J.L., and Zwartkruis, F.J. (2006). Regulation of the small GTPase Rheb by amino acids. *Oncogene* 25:657–664.
42. Sancak, Y., Peterson, T.R., Shaul, Y.D., Lindquist, R.A., Thoreen, C.C., Bar-Peled, L., and Sabatini, D.M. (2008). The Rag GTPases bind raptor and mediate amino acid signaling to mTORC1. *Science* 320:1496–14501.
43. Kim, E., Goraksha-Hicks, P., Li, L., Neufeld, T.P., and Guan, K.L. (2008). Regulation of TORC1 by Rag GTPases in nutrient response. *Nat Cell Biol* 10:935–945.
44. Binda, M., Péli-Gulli, M.P., Bonfils, G., Panchaud, N., Urban, J., Sturgill, T.W., Loewith, R., and De Virgilio, C. (2009). The Vam6 GEF controls TORC1 by activating the EGO complex. *Mol Cell* 35:563–573.
45. Sancak, Y., Bar-Peled, L., Zoncu, R., Markhard, A.L., Nada, S., and Sabatini, D.M. (2010). Ragulator-Rag complex targets mTORC1 to the lysosomal surface and is necessary for its activation by amino acids. *Cell* 141:290–303.
46. Johnson, S.P., and Warner, J.R. (1987). Phosphorylation of the *Saccharomyces cerevisiae* equivalent of ribosomal protein S6 has no detectable effect on growth. *Mol Cell Biol* 7:1338–1345.
47. Urban, J., Soulard, A., Huber, A., Lippman, S., Mukhopadhyay, D., Deloche, O., Wanke, V., Anrather, D., Ammerer, G., Riezman, H., Broach, J.R., De Virgilio, C., *et al.* (2007). Sch9 is a major target of TORC1 in *Saccharomyces cerevisiae*. *Mol Cell* 26:663–674.
48. Ruvinsky, I., Sharon, N., Lerer, T., Cohen, H., Stolovich-Rain, M., Nir, T., Dor, Y., Zisman, P., and Meyuhas, O. (2005). Ribosomal protein S6 phosphorylation is a determinant of cell size and glucose homeostasis. *Genes Dev* 19:2199–2211.
49. Mukai, H., Miyahara, M., Takanaga, H., Kitagawa, M., Shibata, H., Shimakawa, M., and Ono, Y. (1995). Identification of *Schizosaccharomyces pombe* gene *psk1+*, encoding a novel putative serine/threonine protein kinase, whose mutation conferred resistance to phenylarsine oxide. *Gene* 166:155–159.
50. Jin, M., Fujita, M., Culley, B.M., Apolinario, E., Yamamoto, M., Maundrell, K., and Hoffman, C.S. (1995). *sck1*, a high copy number suppressor of defects in the cAMP-dependent protein kinase pathway in fission yeast, encodes a protein homologous to the *Saccharomyces cerevisiae* SCH9 kinase. *Genetics* 140:457–467.
51. Fujita, M., and Yamamoto, M. (1998). *S. pombe sck2+*, a second homologue of *S. cerevisiae* SCH9 in fission yeast, encodes a putative protein kinase closely related to PKA in function. *Curr Genet* 33:248–254.
52. Matsuo, T., Kubo, Y., Watanabe, Y., and Yamamoto, M. (2003). *Schizosaccharomyces pombe* AGC family kinase Gad8p forms a conserved signaling module with TOR and PDK1-like kinases. *EMBO J* 22:3073–3083.
53. Ikeda, K., Morigasaki, S., Tatebe, H., Tamanoi, F., and Shiozaki, K. (2008). Fission yeast TOR complex 2 activates the AGC-family Gad8 kinase essential for stress resistance and cell cycle control. *Cell Cycle* 7:358–364.
54. Kimball, S.R. (2006). Interaction between the AMP-activated protein kinase and mTOR signaling pathways. *Med Sci Sports Exerc* 38:1958–1964.

55. Hanyu, Y., Imai, K.K., Kawasaki, Y., Nakamura, T., Nakaseko, Y., Nagao, K., Kokubu, A., Ebe, M., Fujisawa, A., Hayashi, T., Obuse, C., and Yanagida, M. (2009). *Schizosaccharomyces pombe* cell division cycle under limited glucose requires Ssp1 kinase, the putative CaMKK, and Sds23, a PP2A-related phosphatase inhibitor. *Genes Cells* 14:539–554.
56. Nakashima, A., Hasegawa, T., Mori, S., Ueno, M., Tanaka, S., Ushimaru, T., Sato, S., and Uritani, M. (2006). A starvation-specific serine protease gene, *isp6+*, is involved in both autophagy and sexual development in *Schizosaccharomyces pombe*. *Curr Genet* 49:403–413.
57. Kohda, T.A., Tanaka, K., Konomi, M., Sato, M., Osumi, M., and Yamamoto, M. (2007). Fission yeast autophagy induced by nitrogen starvation generates a nitrogen source that drives adaptation processes. *Genes Cells* 12:155–170.
58. Avruch, J., Hara, K., Lin, Y., Liu, M., Long, X., Ortiz-Vega, S., and Yonezawa, K. (2006). Insulin and amino-acid regulation of mTOR signaling and kinase activity through the Rheb GTPase. *Oncogene* 25:6361–6372.
59. Wang, X., and Proud, C.G. (2009). Nutrient control of TORC1, a cell-cycle regulator. *Trends Cell Biol* 19:260–267.
60. Crespo, J.L., Powers, T., Fowler, B., and Hall, M.N. (2002). The TOR-controlled transcription activators GLN3, RTG1, and RTG3 are regulated in response to intracellular levels of glutamine. *Proc Natl Acad Sci USA* 99:6784–6789.
61. Patel, J., McLeod, L.E., Vries, R.G., Flynn, A., Wang, X., and Proud, C.G. (2002). Cellular stresses profoundly inhibit protein synthesis and modulate the states of phosphorylation of multiple translation factors. *Eur J Biochem* 269:3076–3085.
62. Shiozaki, K., and Russell, P. (1996). Conjugation, meiosis, and the osmotic stress response are regulated by Spc1 kinase through Atf1 transcription factor in fission yeast. *Genes Dev* 10:2276–2288.
63. Kato, T., Jr., Okazaki, K., Murakami, H., Stettler, S., Fantes, P.A., and Okayama, H. (1996). Stress signal, mediated by a Hog1-like MAP kinase, controls sexual development in fission yeast. *FEBS Lett* 378:207–212.
64. Wilkinson, M.G., Samuels, M., Takeda, T., Toone, W.M., Shieh, J.C., Toda, T., Millar, J.B., and Jones, N. (1996). The Atf1 transcription factor is a target for the Sty1 stress-activated MAP kinase pathway in fission yeast. *Genes Dev* 10:2289–2301.
65. Chen, J., Zheng, X.F., Brown, E.J., and Schreiber, S.L. (1995). Identification of an 11-kDa FKBP12-rapamycin-binding domain within the 289-kDa FKBP12-rapamycin-associated protein and characterization of a critical serine residue. *Proc Natl Acad Sci USA* 92:4947–4951.
66. Lorenz, M.C., and Heitman, J. (1995). TOR mutations confer rapamycin resistance by preventing interaction with FKBP12-rapamycin. *J Biol Chem* 270:27531–27537.
67. Sarbassov, D.D., Ali, S.M., Sengupta, S., Sheen, J.H., Hsu, P.P., Bagley, A.F., Markhard, A.L., and Sabatini, D.M. (2006). Prolonged rapamycin treatment inhibits mTORC2 assembly and Akt/PKB. *Mol Cell* 22:159–168.
68. Kim, D.H., Sarbassov, D.D., Ali, S.M., King, J.E., Latek, R.R., Erdjument-Bromage, H., Tempst, P., and Sabatini, D.M. (2002). mTOR interacts with raptor to form a nutrient-sensitive complex that signals to the cell growth machinery. *Cell* 110:163–175.
69. Choo, A.Y., Yoon, S.O., Kim, S.G., Roux, P.P., and Blenis, J. (2008). Rapamycin differentially inhibits S6Ks and 4E-BP1 to mediate cell-type-specific repression of mRNA translation. *Proc Natl Acad Sci USA* 105:17414–17419.
70. Feldman, M.E., Apsel, B., Uotila, A., Loewith, R., Knight, Z.A., Ruggero, D., and Shokat, K.M. (2009). Active-site inhibitors of mTOR target rapamycin-resistant outputs of mTORC1 and mTORC2. *PLoS Biol* 7:e38.

71. García-Martínez, J.M., Moran, J., Clarke, R.G., Gray, A., Cosulich, S.C., Chresta, C.M., and Alessi, D.R. (2009). Ku-0063794 is a specific inhibitor of the mammalian target of rapamycin (mTOR). *Biochem J* 421:29–42.
72. Thoreen, C.C., Kang, S.A., Chang, J.W., Liu, Q., Zhang, J., Gao, Y., Reichling, L.J., Sim, T., Sabatini, D.M., and Gray, N.S. (2009). An ATP-competitive mammalian target of rapamycin inhibitor reveals rapamycin-resistant functions of mTORC1. *J Biol Chem* 284:8023–8032.
73. De Virgilio, C., and Loewith, R. (2006). Cell growth control: little eukaryotes make big contributions. *Oncogene* 25:6392–6415.
74. Weisman, R., Roitburg, I., Nahari, T., and Kupiec, M. (2005). Regulation of leucine uptake by *tor1+* in *Schizosaccharomyces pombe* is sensitive to rapamycin. *Genetics* 169:539–550.
75. Petersen, J., and Nurse, P. (2007). TOR signalling regulates mitotic commitment through the stress MAP kinase pathway and the Polo and Cdc2 kinases. *Nat Cell Biol* 9:1263–1272.
76. Hartmuth, S., and Petersen, J. (2009). Fission yeast Tor1 functions as part of TORC1 to control mitotic entry through the stress MAPK pathway following nutrient stress. *J Cell Sci* 122:1737–1746.

9

The Systemic Control of Growth, Physiology, and Behavior by TOR Signaling in Drosophila

NATHALIE ARQUIER • RENALD DELANOUE •
PIERRE LÉOPOLD

IBDC, University of Nice-Sophia Antipolis
CNRS, Parc Valrose, Nice, France

I. Abstract

One of the critical functions of the TOR complex is to link the nutrient/energy status of cells to metabolic homeostasis and growth. However, in multicellular organisms, the availability of nutrients and energy homeostasis are controlled by endocrine regulation, which brings into question the role of a cell-based sensor. Recent studies using the invertebrate model *Drosophila melanogaster* suggest that the TOR complex can act as a molecular sensor for nutrients in specific endocrine tissues, allowing systemic regulation of metabolism, growth, reproduction, behavior, and ageing.

II. Introduction

Cells, tissues, and organisms all rely on nutrient uptake for growth. In isolated cells, the TOR pathway plays a pivotal role in coupling nutrient and oxygen status with metabolic and biosynthetic regulation as described

ISSN NO: 1874-6047
DOI: 10.1016/S1874-6047(10)28009-5

extensively in several chapters of this book. This is exemplified by the yeast *Saccharomyces cerevisiae*, in which TOR signaling controls a wide variety of cellular processes including transcription, translation, ribosome biogenesis, actin dynamics, protein degradation, mRNA destabilization, and autophagy in response to nutrient availability (see chapters herein). Although it is vital that an isolated cell rapidly senses its direct environment and adapts its behavior accordingly, a cell embedded in a tissue is less prone to abrupt variations in environmental conditions due to hormonal control, which complex organisms have developed to blunt environmental variations. Indeed, in multicellular organisms, hormonal regulation synchronizes the behavior of large numbers of cells and tissues, and nutrients are taken up and transported to tissues using complex processes that maintain homeostatic regulation of consumption and energy balance. When nutrients are scarce, growth is restrained in a homogeneous manner, leading to a proportional reduction of all body parts. These findings highlight the importance of systemic regulation.

Metazoan cells have adapted their growth and proliferation to such indirect energy signals, and tissue/organ growth appears to be mainly controlled by circulating growth/mitotic factors. This is illustrated by the simple fact that providing basic nutrients and therefore energy to metazoan cells kept in culture is not sufficient to promote growth and proliferation. Similar results are obtained in culture of tissue explants and the addition of a complex cocktail of growth factors is necessary to sustain growth and prevent apoptosis.

Therefore, the function of a cell-based nutrient sensor needs to be clarified in the context of a multicellular organism. Are individual cells embedded in a developing organ or tissue still capable of regulating their metabolism and growth through direct assessment of local availability of nutrients? Alternatively, is nutrient sensing taking place in specialized organs capable of relaying the information to the rest of the organism by sending signals that control growth and metabolism in a systemic manner? Several recent reports have addressed this point by using genetics to decipher the physiology of nutrient response in the context of a whole organism. This chapter focuses on advances made using invertebrate models in which our understanding of the control of organismal growth and physiology has made considerable progress. The general message emerging from these studies is that TOR signaling serves as a nutrition sensor in a limited number of endocrine organs that couple important physiological processes like metabolism, growth, behavior, reproduction, and ageing, to the energy status of the whole organism (Figure 9.1).

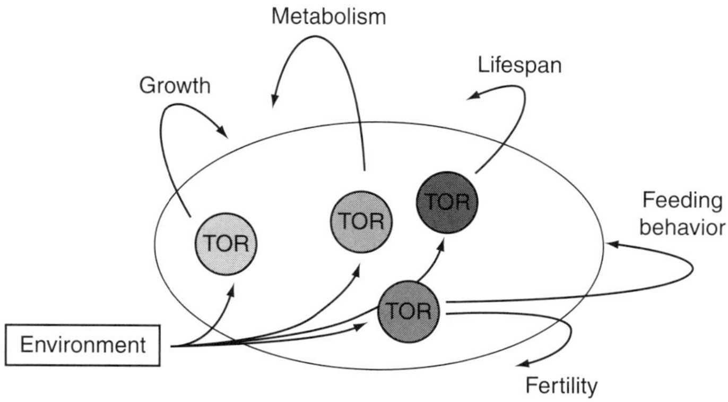

FIG. 9.1. Pleiotropic actions of TOR signaling in *Drosophila*. Environmental cues, mainly nutrient availability, control the level of TORC1 activity in most tissues. A limited number of endocrine "sentinelle organs" (color circles) use this information to systemically control physiological processes at the level of the whole organism. (For interpretation of the references to color in this figure legend, the reader is referred to the Web version of this chapter.)

III. Growth Rate

In many species, including the invertebrate and vertebrate phyla, growth is restricted to the juvenile period of development. In *Drosophila*, most of the programming of growth takes place during larval development, a 4-day period marked by active feeding [1]. Precise coupling between feeding and growth is set up during this period, which results in a spectacular increase in the body mass. This coupling involves the production, release, and circulation of specific *Drosophila* insulin-like peptides called Dilps, which promote systemic growth of all larval tissues [2]. In a similar way to growth induction by mammalian Insulin-like Growth Factors (IGFs), the binding of Dilps to their unique receptor, InR, activates a conserved downstream pathway leading to cell and tissue growth. Nevertheless, this pathway does not respond directly to nutrient levels, as the activity of downstream components like PI3-kinase and AKT are not modified upon amino acid withdrawal [3]. This indicates that a separate sensory mechanism independent of insulin/IGF signaling (IIS) must operate to assess the energy levels of the growing organism and to control the general levels of IIS. Genetic experiments in which amino acid import is decreased in a tissue-directed manner have allowed screening for larval tissues in which decreased amino acid levels lead to a systemic growth defect. These experiments indicate that the fat body, an important endocrine organ of the larva, which has functional similarities to both vertebrate liver and white fat, plays a pivotal role in sensing amino acids and controlling larval growth [4]. When the amino acid transporter Slimfast (Slif) is shutdown in this tissue (referred to as *Slif*^{A} animals), amino acid import is reduced, creating local starvation. Interestingly, this local effect results in general growth inhibition, producing adults with a markedly reduced body size. The knock-down of *Slif* in fat cells is accompanied by a shutdown in S6-kinase activity, suggesting that TORC1 inhibition is taking place locally. Indeed, the growth defects observed in *Slif*^{A} animals are partially rescued by fat-specific expression of an activated form of S6-kinase. Conversely, the downregulation of TORC1 signaling in fat cells by expression of either a dominant-negative form of the kinase (TOR^{TED}) or the TORC1 inhibitors TSC1 and TSC2 also leads to nonautonomous growth defects. Therefore, TORC1 is used in the fat body as a molecular sensor for amino acid levels, and links systemic growth with the nutrient status [4].

In addition, TORC1 inhibition in fat cells induces peripheral inhibition of IIS, since a reduction in the PI3kinase activity is observed in the epidermis of *Slif*^{A} animals. Therefore, a relay mechanism conveys the nutrition signal from the fat body where it is sensed, to IIS in peripheral tissues. What is the nature of this relay? A recent report indicates that TOR signaling in fat cells is both necessary and sufficient to control the secretion of insulin-like peptides from the brain, and therefore controls remotely IIS in response to changes in nutritional information [5]. Several Dilps are produced in specific brain neurosecretory cells called insulin-producing cells (IPCs). In scarce nutrient conditions, a drastic accumulation of Dilp2 and Dilp5 is observed in brain IPCs, suggesting that Dilp secretion is subjected to control by nutrient conditions. These observations were suggested by earlier work performed on adult flies showing that refeeding after protein fasting was accompanied by rapid activation of IIS without any transcriptional activation of *DILP* genes [6]. Experiments on larvae indicate that dietary amino acids are the main factor controlling Dilp accumulation, since feeding of nutrient-deficient animals with casein is sufficient to provoke brain Dilp release. More specifically, branched amino acids like leucine or isoleucine are particularly potent at triggering Dilp release, a result in line with the prominent role of leucine in the activation of TORC1 (see chapters herein). Interestingly, a similar accumulation of Dilps is observed in brain IPCs of *Slif*^{A} animals, suggesting that the fat body can remotely control the secretion of brain Dilps according to an amino acid signal [5]. Further experiments established that TORC1 in adipose tissue plays a major role in this regulation: inhibition of TORC1 by silencing different components of the complex in fat cells is sufficient to block brain Dilp release. Conversely, the retention of brain Dilps observed upon starvation can be efficiently overridden by activating TORC1 activity in fat cells. These results establish that amino acids control brain Dilp release through the modulation of TORC1 levels in the fat body [5].

While this constitutes a step forward in our understanding of the mechanisms linking nutrient and systemic growth control, further work will be necessary to elucidate the nature of the link between the fat body and the brain. In this respect, coculture experiments of dissected brain and fat tissue indicate that this remote regulation is neither due to a physical connection existing between the two tissues nor to an indirect consequence of changes in general metabolism. Rather, a humoral factor produced by fat cells is released in the hemolymph and acts on brain to modulate insulin secretion. The identification of this factor by biochemical and/or genetic approaches will constitute an important goal for future experiments. In addition, it will be of primary importance to identify which cells in the brain constitute the primary target for such a factor. IPCs are suspected to be a target for such a signal. In addition, one group of serotonergic neurons was recently shown to control Dilp function upstream of IPCs. Flies mutant for the *nucleostemin 3* (*NS3*) gene have a reduced body size and accumulation of Dilp2 in IPCs, suggestive of a defect in Dilp secretion. *NS3* is expressed in a subgroup of serotonergic neurons that project axons in the close vicinity of the IPC axons [7]. At present, implication of these neurons in controlling Dilp secretion as a function of nutrient availability remains to be established, but if this turns out to be true, these cells could constitute a bona fide target for the fat body hormone.

Another aspect of systemic growth regulation by IIS involves insulin/IGF-binding partners. In vertebrates, IGF-I circulates in the blood in the form of a trimeric complex in association with an IGF-binding protein (IGFBP) and with a scaffold partner called Acid Labile Subunit (ALS) [8]. This complex protects IGF-I from proteolytic degradation and allows the formation of a circulating reservoir of IGF-I in the blood. The proteolytic degradation of the non-IGF partners is required to allow the binding of IGF-I to its receptor, therefore, certain IGFBPs play important inhibitory functions on IGF signaling [9]. In flies, it was recently shown that similar partners bind to circulating Dilps and modulate their function. The *Imp-L2* gene encodes a homolog of the vertebrate IGFBP 7, a far related member of the IGFBP family. The Imp-L2 protein binds Dilp2 *in vitro* and *in vivo*, and inhibits growth in a nonautonomous manner. Animals lacking Imp-L2 function are viable and larger than controls, whereas overexpression of the *Imp-L2* gene leads to tiny flies [10]. Interestingly, the Imp-L2 protein accumulates in fat cells and is required for IIS inhibition and survival in starvation conditions [10]. Similar, although weaker, phenotypes are obtained by knocking down the *dALS* gene that encodes an ALS homolog in the fly. The dALS protein forms a complex with Imp-L2 and Dilp2 that could represent a functional fly homolog of the trimeric complexes found in vertebrate blood. Interestingly, dALS expression is restricted to the fat body and the IPCs. It is severely inhibited under starvation, as well as in $slif^{A}$ animals, and the knock-down of *dALS* upon starvation aggravates growth inhibition [11]. This series of results leads to a model whereby in response to starvation, increased accumulation of Imp-L2 leads to the inactivation of circulating Dilps and, concomitantly, inhibition of *dALS* expression reduces the stability of circulating Dilp complexes. Together, these mechanisms could lead to a reduction in circulating active Dilps and contribute to the reduction of IIS in response to low nutrients. The link with TOR signaling in the fat body remains to be fully established, but it is noteworthy that regulation takes place in fat cells, which use TOR to sense global levels of amino acids.

The fat body orchestrates two other types of responses that contribute to the adaptation of growth following changes in nutrient availability. Fat body cells constitute an important site for the storage of high energy nutrients, mainly glycogen and triacylglycerides. Under limiting nutrient conditions, these stores are efficiently mobilized by an active autophagic process that provokes the shrinkage of fat cells and the release of a pool of nutrients directly available to peripheral larval tissues. Starvation induces a drop in TOR activity and triggers autophagy in fat cells. Conversely, in fed conditions, TOR activity prevents autophagy and high levels of IIS induce fat cells to accumulate nutrients [12, 13]. Therefore, this efficient mechanism allows the switching in fat cells from an anabolic to a catabolic state, providing a nutrients buffer to peripheral tissues in adverse food conditions.

Then the question is: how do animals use this additional pool of nutrient when they switch to a catabolic state and no longer produce/release brain Dilps? A partial answer to this question could come from recent reports indicating that nonfeeding states and low amino acid diets induce the production of Dilp6, a specific insulin-like peptide, in fat cells. Removal of the *DILP6* gene by mutation leads to smaller adult flies with no metabolic phenotype, indicating that Dilp6 is a strict growth inducer. Starvation induces *DILP6* expression and this relies on the dFoxO transcription factor, therefore constituting a regulation loop of IIS on itself [14]. Whether TOR signaling participates directly in the transcriptional regulation of *DILP6* is not yet known, but local inhibition of amino acid import in fat cells is sufficient to induce *DILP6* expression (M. S., R. D., and P. L., unpublished data). In conditions of limiting nutrients, the growth of some larval tissues, like the fat body and other endoreduplicating tissues (ERTs), is strongly affected. By contrast, the imaginal tissues (ITs), which contribute adult structures, and the brain are somehow protected from the effect of starvation and continue to grow, although at a slower pace. One attractive hypothesis proposes that the combined induction of autophagy and *DILP6* expression by starvation in fat cells could provide both a source of nutrients and the necessary growth inducers to ensure the development of adult tissues in adverse nutrient conditions.

As an interesting parallel to these studies, experiments in mice have demonstrated that knocking out TORC1 or TORC2 activities in white adipose tissues also lead to systemic effects on animal mass. The involved mechanisms appear different though, since adipose-specific knockout of Raptor induces enhanced energy expenditure in the fat and leanness, whereas adipose-knockout of Rictor leads to elevation of circulating IGF-I levels and mass increase [15, 16].

IV. Developmental Timing

A second important aspect of the establishment of adult target size is to determine when it is necessary to stop growth. A recent report shows that during fly larval development, TOR signaling in a specific endocrine tissue called the ring gland plays a key role in coupling the duration of the growth period with variations in nutritional information (Figure 9.2).

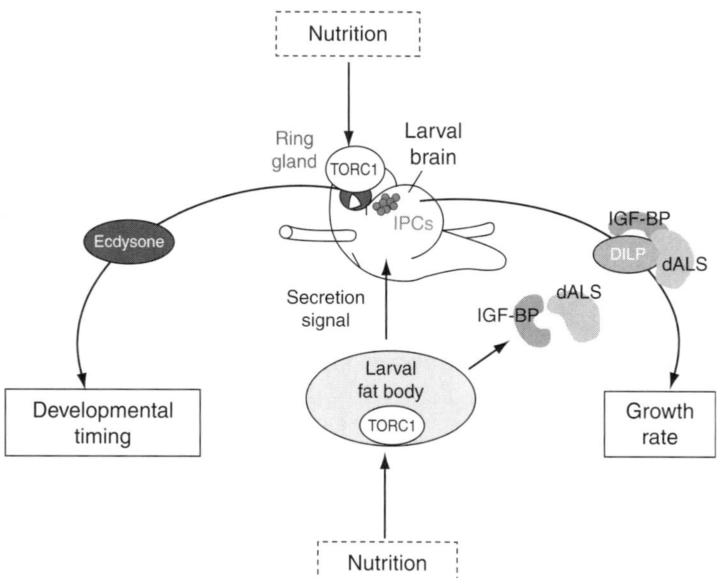

FIG. 9.2. The systemic control of growth by TOR signaling in *Drosophila*. The fat body and the ring gland are two important endocrine organs using TOR signaling to couple growth with nutritional information. In the fat body, the activation of TORC1 complex by amino acids modulates the animal's growth rate via the remote regulation of insulin-like peptide secretion in the brain IPCs (see text for abbreviations). In the ring gland, TORC1 activity controls the production of the molting hormone ecdysone to regulate the cessation of growth and entry into the maturation stage.

In holometabolous insects, commitment to metamorphosis is preceded by several checkpoints controlling animal size that ensure that the larva reach a size or weight allowing successful metamorphosis and adult emergence. For instance, two successive checkpoints have been defined for the large insect *Manduca sexta*. When starved at the "minimal viable weight" (MVW), animals die before reaching the pupal stage. When starved between the MVW and the "critical weight" (CW), animals enter the pupal stage with delayed development and do not reach adulthood. When starved passed the CW, animals form small pupae without delayed development [17]. This description does not fully apply to *Drosophila*. Nevertheless, as for *Manduca* larvae, *Drosophila* larvae pass a critical size or weight after which they are committed to pupal development independently of nutrition [18, 19]. A molecular definition of the CW is still lacking and would definitely help to understand this important physiological transition. The interval between the attainment of critical size and final size at the end of larval development could correspond to the time the cascade of endocrine regulation, which controls larval/pupal transition, takes place. In *Manduca*, this cascade has been relatively well described [17]. At the time animals reach a critical size, juvenile hormone (JH), a sesquiterpenoid that maintains animals in the juvenile or larval state, starts to be degraded. Degradation allows the production of the prothoracicotropic hormone (PTTH) by two pairs of neurosecretory brain cells. In turn, PTTH activates the massive production of ecdysone in the prothoracic gland. Ecdysone, the main steroid in insects, is the ultimate signal that triggers entry into pupal stage. The duration of the larval period affects critically the size at pupariation, which in turn determines final adult size. This was demonstrated in recent experiments, where the timing of ecdysone production at the larval/pupal transition is modified. *Drosophila* larvae in which PTTH-producing cells are genetically ablated grow for a prolonged time [20]. Some eventually pupariate giving rise to adults 50–70% heavier than control animals. Conversely, ectopic induction of the PTTH gene induces precocious pupariation and leads to dwarf adults. Examination of the animal's developmental timing reveals that nutrition affects the time between critical weight and pupariation (also called the Terminal Growth Period, TGP). This suggests a mechanism capable of coupling the production of ecdysone in the ring gland to nutrition. Indeed, the downregulation of TORC1 activity in the prothoracic gland (PG, the part of the ring gland that produces ecdysone) delays ecdysone production, larval/pupal transition, and leads to bigger pupae and adults [21]. Conversely, activation of TORC1 by overexpression of Rheb in the PG is sufficient to reduce the developmental delay induced by a low amino acid diet. This indicates that TORC1 activity in the PG is both necessary and sufficient to control the duration of larval development as a function of nutrient availability [21]. By contrast, inhibition of TORC1 activity in the PTTH-producing neurons has no effect on the timing of pupariation and the gene encoding PTTH is still expressed at normal levels under low nutrient conditions. This suggests that nutrition, via TORC1, determines the capacity of the ring gland to induce ecdysone production in response to the PTTH signal.

Therefore, a common message emerges from studies into the growth rate and timing control: in both cases, TORC1 serves as a molecular sensor and couples endocrine activity in sensor tissues to environmental factors, contributing to systemic control of animal growth (Figures 9.1 and 9.2).

V. Feeding Behavior

The relationship between energy expenditure and caloric intake has been intensively studied in mammalian systems. The control of caloric consumption by the brain involves a hypothalamic region, also known as "feeding center," and the pituitary gland axis to regulate eating and metabolism [22]. Pharmacological studies have implicated a variety of chemicals of the brain, including bioactive peptides in promoting or inhibiting food intake. Several hypothalamic neuropeptides, including the agouti-related protein and neuropeptide Y, are thought to be potent feeding stimulants. On the other hand, peptides such as leptin, melanocortins, or insulin suppress food intake [22]. However, the elucidation of the physiological roles of these molecules and their sites of action has been difficult, largely because of the complexity of vertebrate models. In addition, the signaling pathways that are involved remain unknown.

Functional similarities have been evoked between the insect intercerebralis/corpora cardiaca (CC)/corpora allata (CA) axis and the vertebrate hypothalamus–pituitary axis [23, 24]. Consequently, *Drosophila* has been used as a relatively simple system to elucidate molecular and neural mechanisms underlying regulation of hunger. A study on fly larvae suggested that changes in TOR-dependent activity, namely S6-kinase function, in a specific set of neurons affects feeding behavior. Indeed, ectopic expression of either active or dominant-negative forms of S6K in the brain IPCs is sufficient to affect hunger-driven behavior: fed larvae with neural S6K loss of function have an increased food intake. Conversely, fasted larvae with neural S6K overexpression show an attenuated feeding response [25]. These observations suggest that neuronal S6K controls the hunger-driven response. However, the contribution of the full TORC1 axis to this process remains to be evaluated.

In other invertebrates, such as *Caenorhabditis elegans*, the contribution of TOR to feeding has been recently unveiled. RNAi against *rictor*, a component of the TORC2 complex, leads to a decrease in food consumption [26]. In addition, *rictor* mutant animals show decreased lawn residence, indicating that TORC2 is required in worms for proper feeding behavior. In mammals, TOR signaling is known to play a central role in this process. Indeed, intracerebroventricular administration of rapamycin significantly increases food intake of presatiated rats, suggesting that hypothalamic mTOR regulates food intake [27]. In conclusion, recent studies in different model systems reveal that the activity of the TOR complex has a strong impact on feeding behavior and constitutes a primary signaling pathway to control the hunger-driven response to environmental conditions. Having major function in nutrient sensing, TOR controls a homeostatic cross talk between organs determining consumption and energy balance. However, this aspect of TOR function remains largely unknown and constitutes a vast field for future investigation. This will be of primary importance to help understand human pathologies associated with feeding disorders such as obesity, diabetes, or cardiovascular disease.

VI. Fertility

In *Drosophila* as in many insects, egg laying is directly linked to nutrient availability [28]. Under normal conditions, mated females deposit up to 90 eggs per day, a total amount of material corresponding to half their body weight. However, upon nutrient shortage, the females respond rapidly and egg laying drops to 1.5 eggs per day [29]. Egg production in the ovarioles is very sensitive to nutritional resources: upon nutrient limitation, the rate of division in the germline lineage is reduced and death of the cysts is increased [29]. Additionally, young egg chambers do not proceed to vitellogenesis and are often arrested at mid oogenesis, while egg chambers already in early vitellogenesis undergo apoptosis [30]. Finally, mature eggs are frequently retained by the starving female [31]. These phenotypes are efficiently rescued within a few hours after refeeding [29, 30].

The nutrient sensing mechanism and the starvation signal that is sent to arrest oogenesis are not known yet, even if both TOR and insulin signaling pathways appear to be of primary importance during oogenesis. IIS has a critical role in the control of germline stem cell division and cyst growth. Key players in the pathway such as the Insulin receptor (dInR) and the adapter protein Chico are required to control oogenesis in response to nutrient levels [29, 32]. In addition, the production of Dilps in brain neurosecretory cells (IPCs) is important for egg production and vitellogenesis [33], and IPC-ablated females lose their ability to upregulate follicular cell growth in response to a rich diet [34]. Recent data have shown that during larval development, Dilp secretion is remotely regulated by a TORC1-dependent signal emitted from the fat body [5] (see above). A similar mechanism could be envisaged for coupling systemic regulation of egg production with nutrient availability in the adult female. At present, the role of the TOR pathway in coupling oogenesis to nutrition has not been clarified in *Drosophila*.

In other insects such as the mosquito *Aedes aegypti*, the situation seems clearer. In this species, a blood meal is required to trigger oogenesis and the production of yolk proteins by the fat cells. This process requires TORC1 activity, as both rapamycin and TOR silencing in cultured fat bodies effectively blocks the vitellogenic response. Conversely, inhibition of *TSC2* is sufficient to promote the production of yolk proteins in the absence of a blood meal. Moreover, mosquito females injected with double-strand RNA against TOR or S6K before the blood meal show a dramatic reduction in the number and size of ovarioles, and a significant reduction in the number of deposited eggs [35, 36]. These studies suggest that, in addition to its role in regulating systemic growth during larval development, adipose TOR activity controls yolk protein production and vitellogenesis in the adult. A recent study on another insect, *Blattella germanica*, proposes an alternative scenario, in which TOR activity couples nutrition to JH synthesis, vitellogenesis, and ovary development. In this insect, vitellogenesis is controlled by the production of JH in the CA, and RNAi-mediated TOR knock-down leads to a reduction in JH synthesis, inhibition of vitellogenin transcription and aborted ovarial growth, as observed under starvation [37].

In addition to systemic regulation of egg production, some data in the literature suggest that a local response to nutrients could also play an important role. *TOR* expression and S6K phosphorylation are induced in mosquito ovaries after a blood meal [36], whereas *TSC1* expression increases in ovaries of starved flies [38] and *Dilp5* expression in the oocyte is thought to be important for oogenesis [33, 34]. However, the role of these components in coupling the nutritional status of the egg chamber to oogenesis remains to be evaluated.

In insects with diphenic development, TOR signaling has an unforeseen role in directing developmental fate. In the honeybee, *Apis meliffera*, a diphenism in female development is associated with feeding conditions: excess feeding of female larvae results in queens (large reproductive females), whereas a moderate diet yields workers (small helpers). Recent findings show that TOR signaling regulates the dimorphism of queen-workers. Queen-destined larvae have an approximately twofold higher *TOR* mRNA level than worker-destined larvae. Moreover, rapamycin and *TOR*-RNAi silencing of larvae prevent queen fate and favor the emergence of workers [39]. These observations also underline the importance of TOR signaling in controlling egg production, as it is critical to generate reproductive adult females. More importantly, it shows that the level of TOR signaling determines the important morphological changes differentiating queen and worker individuals. The exact mechanism of this caste differentiation remains unknown, but JH is a key component of the developmental determination of queen-like characters [40]. Therefore, like in the case of *B. germanica* adult females, JH synthesis might be controlled by TOR signaling during the late larval stages in the honeybee fed a rich diet.

In summary, TOR plays a critical role in insect reproduction and especially in egg production. However, the requirement of TOR and the mode of action are not fully established and need further study. Alternatively, the nutrient quality also affects the male reproductive ability by modifying testis mass, sperm production, or male courtship behavior, but the inference of TOR signaling in these processes has not been assessed.

VII. Control of Lifespan

Understanding the regulation of lifespan is a fascinating goal, which has a very high social and economic impact. The first molecular connection between lifespan extension and the TOR pathway was obtained from *C. elegans*. Animals bearing mutations in the *C. elegans* orthologs of *TOR* and *Raptor* live longer [41, 42]. In *Drosophila*, the ubiquitous expression of *dTSC1*, *dTSC2* or dominant-negative forms of dTOR and dS6K is sufficient to extend lifespan. Similarly, a viable hypomorphic mutant for dTOR presents a lifespan extension [44]. In addition, the modulation of TOR signaling in specific tissues like the fat body also influences global lifespan [43]. Molecular connections are now emerging, providing clues as to how this works. In *C. elegans* like in *Drosophila*, accumulating genetic evidence demonstrate that protein translation is the main output from TOR signaling that controls lifespan [45–49]. More recently, rapamycin fed to flies was shown to greatly extend lifespan. Interestingly, this extension is not associated with an alteration in IIS or AMP-kinase activity, but is blocked by altering autophagy or by increasing the protein translation efficiency [50]. Both in worm and in flies, inhibition of autophagy affects lifespan *per se* and blocks lifespan extension provoked by TORC1 inactivation or dietary restriction (DR) [51–55]. Taken together, experimental data on invertebrate models suggest that the effects of TOR on longevity involve at least two regulatory outputs, namely protein translation and autophagy.

Since the pioneering experiments on rats performed in the 1930s [56], it has been established that DR extends lifespan in most animal models. The recent contribution of invertebrate models such as worms and flies has been key to understanding the molecular basis of lifespan extension by DR [57]. In particular, the fact that DR does not further increase lifespan in the context of TOR loss of function in worms or flies strongly suggests that TOR inhibition could constitute an important part of the mechanisms by which DR acts on lifespan [58].

Many studies performed on *Drosophila* have shown that IIS is involved in the control of lifespan. Hypomorphic dInR mutants, homozygous *chico* (IRS) mutants, or flies with dFOXO overexpressed in the head fat body, all present lifespan extension in a dose-dependent manner [59–62]. All these animals show downregulation of *dilp2* expression in brain IPCs [61]. Nevertheless, no clear causality between these events is established since the silencing of *dilp2* in the IPCs does not lead to extended lifespan [63]. Moreover, despite the numerous connections existing between IIS and the TOR pathway, lifespan extensions observed upon inactivation of each pathways could have separate molecular origins. Indeed, rapamycin treatment of long-lived heterozygous *chico* mutants can extend lifespan beyond that induced by rapamycin alone, indicating that IIS might act through additional mechanisms on aging [50].

References

1. Edgar, B.A. (2006). How flies get their size: genetics meets physiology. *Nat Rev Genet* 7:907–916.
2. Geminard, C., *et al.* (2006). Control of metabolism and growth through insulin-like peptides in *Drosophila*. *Diabetes* 55(Suppl. 2):S5–S8.
3. Radimerski, T., *et al.* (2002). dS6K-regulated cell growth is dPKB/dPI(3)K-independent, but requires dPDK1. *Nat Cell Biol* 4:251–255.
4. Colombani, J., *et al.* (2003). A nutrient sensor mechanism controls *Drosophila* growth. *Cell* 114:739–749.
5. Geminard, C., Rulifson, E.J., and Leopold, P. (2009). Remote control of insulin secretion by fat cells in *Drosophila*. *Cell Metab* 10:199–207.
6. Gershman, B., *et al.* (2007). High-resolution dynamics of the transcriptional response to nutrition in *Drosophila*: a key role for dFOXO. *Physiol Genomics* 29:24–34.
7. Kaplan, D.D., *et al.* (2008). A nucleostemin family GTPase, *NS3*, acts in serotonergic neurons to regulate insulin signaling and control body size. *Genes Dev* 22:1877–1893.
8. Boisclair, Y.R., *et al.* (2001). The acid-labile subunit (ALS) of the 150 kDa IGF-binding protein complex: an important but forgotten component of the circulating IGF system. *J Endocrinol* 170:63–70.
9. Firth, S.M., and Baxter, R.C. (2002). Cellular actions of the insulin-like growth factor binding proteins. *Endocr Rev* 23:824–854.

10. Honegger, B., *et al.* (2008). Imp-L2, a putative homolog of vertebrate IGF-binding protein 7, counteracts insulin signaling in *Drosophila* and is essential for starvation resistance. *J Biol* 7:10.
11. Arquier, N., *et al.* (2008). *Drosophila* ALS regulates growth and metabolism through functional interaction with insulin-like peptides. *Cell Metab* 8:446.
12. Rusten, T.E., *et al.* (2004). Programmed autophagy in the *Drosophila* fat body is induced by ecdysone through regulation of the PI3K pathway. *Dev Cell* 7:179–192.
13. Scott, R.C., Schuldiner, O., and Neufeld, T.P. (2004). Role and regulation of starvation-induced autophagy in the *Drosophila* fat body. *Dev Cell* 7:167–178.
14. Slaidina, M., *et al.* (2009). A *Drosophila* insulin-like peptide promotes growth during non-feeding states in *Drosophila. Dev Cell* 17:874–884.
15. Cybulski, N., *et al.* (2009). mTOR complex 2 in adipose tissue negatively controls whole-body growth. *Proc Natl Acad Sci USA* 106:9902–9907.
16. Polak, P., *et al.* (2008). Adipose-specific knockout of raptor results in lean mice with enhanced mitochondrial respiration. *Cell Metab* 8:399–410.
17. Nijhout, H.F. (2003). The control of body size in insects. *Dev Biol* 261:1–9.
18. Mirth, C.K., and Riddiford, L.M. (2007). Size assessment and growth control: how adult size is determined in insects. *Bioessays* 29:344–355.
19. Stieper, B.C., *et al.* (2008). Imaginal discs regulate developmental timing in *Drosophila melanogaster*. *Dev Biol* 321:18–26.
20. McBrayer, Z., *et al.* (2007). Prothoracicotropic hormone regulates developmental timing and body size in *Drosophila. Dev Cell* 13:857–871.
21. Layalle, S., Arquier, N., and Leopold, P. (2008). The TOR pathway couples nutrition and developmental timing in *Drosophila. Dev Cell* 15:568–577.
22. Dietrich, M.O., and Horvath, T.L. (2009). Feeding signals and brain circuitry. *Eur J Neurosci* 30:1688–1696.
23. Buch, S., and Pankratz, M.J. (2009). Making metabolic decisions in *Drosophila. Fly (Austin)* 3:74–77.
24. Wang, S., *et al.* (2007). The origin of islet-like cells in *Drosophila* identifies parallels to the vertebrate endocrine axis. *Proc Natl Acad Sci USA* 104:19873–19878.
25. Wu, Q., *et al.* (2005). Regulation of hunger-driven behaviors by neural ribosomal S6 kinase in *Drosophila. Proc Natl Acad Sci USA* 102:13289–13294.
26. Soukas, A.A., *et al.* (2009). Rictor/TORC2 regulates fat metabolism, feeding, growth, and life span in *Caenorhabditis elegans*. *Genes Dev* 23:496–511.
27. Cota, D., *et al.* (2006). Hypothalamic mTOR signaling regulates food intake. *Science* 312:927–930.
28. Rivero, A., Giron, D., and Casas, J. (2001). Lifetime allocation of juvenile and adult nutritional resources to egg production in a holometabolous insect. *Proc Biol Sci* 268:1231–1237.
29. Drummond-Barbosa, D., and Spradling, A.C. (2001). Stem cells and their progeny respond to nutritional changes during *Drosophila* oogenesis. *Dev Biol* 231:265–278.
30. Terashima, J., *et al.* (2005). Nutritional status affects 20-hydroxyecdysone concentration and progression of oogenesis in *Drosophila melanogaster*. *J Endocrinol* 187:69–79.
31. Terashima, J., and Bownes, M. (2004). Translating available food into the number of eggs laid by *Drosophila melanogaster*. *Genetics* 167:1711–1719.
32. Hsu, H.J., LaFever, L., and Drummond-Barbosa, D. (2008). Diet controls normal and tumorous germline stem cells via insulin-dependent and -independent mechanisms in *Drosophila. Dev Biol* 313:700–712.

33. Ikeya, T., *et al.* (2002). Nutrient-dependent expression of insulin-like peptides from neuroendocrine cells in the CNS contributes to growth regulation in *Drosophila. Curr Biol* 12:1293–1300.
34. LaFever, L., and Drummond-Barbosa, D. (2005). Direct control of germline stem cell division and cyst growth by neural insulin in *Drosophila. Science* 309:1071–1073.
35. Hansen, I.A., *et al.* (2004). Target of rapamycin-mediated amino acid signaling in mosquito anautogeny. *Proc Natl Acad Sci USA* 101:10626–10631.
36. Hansen, I.A., *et al.* (2005). Target of rapamycin-dependent activation of S6 kinase is a central step in the transduction of nutritional signals during egg development in a mosquito. *J Biol Chem* 280:20565–20572.
37. Maestro, J.L., Cobo, J., and Belles, X. (2009). Target of rapamycin (TOR) mediates the transduction of nutritional signals into juvenile hormone production. *J Biol Chem* 284:5506–5513.
38. Terashima, J., and Bownes, M. (2005). A microarray analysis of genes involved in relating egg production to nutritional intake in *Drosophila melanogaster. Cell Death Differ* 12:429–440.
39. Patel, A., *et al.* (2007). The making of a queen: TOR pathway is a key player in diphenic caste development. *PLoS ONE* 2:e509.
40. Barchuk, A.R., *et al.* (2007). Molecular determinants of caste differentiation in the highly eusocial honeybee *Apis mellifera. BMC Dev Biol* 7:70.
41. Vellai, T., *et al.* (2003). Genetics: influence of TOR kinase on lifespan in *C. elegans. Nature* 426:620.
42. Jia, K., Chen, D., and Riddle, D.L. (2004). The TOR pathway interacts with the insulin signaling pathway to regulate *C. elegans* larval development, metabolism and life span. *Development* 131:3897–3906.
43. Kapahi, P., *et al.* (2004). Regulation of lifespan in *Drosophila* by modulation of genes in the TOR signaling pathway. *Curr Biol* 14:885–890.
44. Luong, N., *et al.* (2006). Activated FOXO-mediated insulin resistance is blocked by reduction of TOR activity. *Cell Metab* 4:133–142.
45. Hansen, M., *et al.* (2007). Lifespan extension by conditions that inhibit translation in *Caenorhabditis elegans. Aging Cell* 6:95–110.
46. Syntichaki, P., Troulinaki, K., and Tavernarakis, N. (2007). eIF4E function in somatic cells modulates ageing in *Caenorhabditis elegans. Nature* 445:922–926.
47. Pan, K.Z., *et al.* (2007). Inhibition of mRNA translation extends lifespan in *Caenorhabditis elegans. Aging Cell* 6:111–119.
48. Zid, B.M., *et al.* (2009). 4E-BP extends lifespan upon dietary restriction by enhancing mitochondrial activity in *Drosophila. Cell* 139:149–160.
49. Kaeberlein, M., and Kennedy, B.K. (2008). Protein translation, 2008. *Aging Cell* 7:777–782.
50. Bjedov, I., *et al.* (2010). Mechanisms of life span extension by rapamycin in the fruit fly *Drosophila melanogaster. Cell Metab* 11:35–46.
51. Hansen, M., *et al.* (2008). A role for autophagy in the extension of lifespan by dietary restriction in *C. elegans. PLoS Genet* 4:e24.
52. Jia, K., and Levine, B. (2007). Autophagy is required for dietary restriction-mediated life span extension in *C. elegans. Autophagy* 3:597–599.
53. Toth, M.L., *et al.* (2008). Longevity pathways converge on autophagy genes to regulate life span in *Caenorhabditis elegans. Autophagy* 4:330–338.
54. Simonsen, A., *et al.* (2008). Promoting basal levels of autophagy in the nervous system enhances longevity and oxidant resistance in adult *Drosophila. Autophagy* 4:176–184.

55. Juhasz, G., *et al.* (2007). Atg7-dependent autophagy promotes neuronal health, stress tolerance, and longevity but is dispensable for metamorphosis in *Drosophila*. *Genes Dev* 21:3061–3066.
56. McCay, C., Crowell, M., and Maynard, L. (1935). The effect of retarded growth upon the length of life span and upon the ultimate body size. *J Nutr* 10:63.
57. Masoro, E.J. (2005). Overview of caloric restriction and ageing. *Mech Ageing Dev* 126:913–922.
58. Grandison, R.C., Piper, M.D., and Partridge, L. (2009). Amino-acid imbalance explains extension of lifespan by dietary restriction in *Drosophila*. *Nature* 462:1061–1064.
59. Tatar, M., *et al.* (2001). A mutant *Drosophila* insulin receptor homolog that extends life-span and impairs neuroendocrine function. *Science* 292:107–110.
60. Clancy, D.J., *et al.* (2001). Extension of life-span by loss of CHICO, a *Drosophila* insulin receptor substrate protein. *Science* 292:104–106.
61. Hwangbo, D.S., *et al.* (2004). *Drosophila* dFOXO controls lifespan and regulates insulin signalling in brain and fat body. *Nature* 429:562–566.
62. Giannakou, M.E., *et al.* (2004). Long-lived *Drosophila* with overexpressed dFOXO in adult fat body. *Science* 305:361.
63. Broughton, S., *et al.* (2008). Reduction of DILP2 in *Drosophila* triages a metabolic phenotype from lifespan revealing redundancy and compensation among DILPs. *PLoS ONE* 3:e3721.

10

Cell-Intrinsic Functions and Regulation of TOR Signaling in Drosophila

THOMAS P. NEUFELD
Department of Genetics, Cell Biology and Development
University of Minnesota
Minneapolis, Minnesota, USA

I. Abstract

The fruit fly *Drosophila melanogaster* has provided a valuable animal model system for a multitude of experimental questions, and has proven to be particularly amenable to studies of signaling pathways. The full repertoire of technical approaches available in this system have been applied to the study of TOR signaling, including innovative genetic screens, high-throughput expression profiling, detailed cell biological analyses, and sensitive genetic interaction studies. Results from these diverse lines of study have provided the field with a wealth of new information and insights into the TOR pathway, from identification of novel signaling factors to elucidation of regulatory mechanisms and physiologically relevant targets. This chapter highlights the progress in our understanding of the regulation and cellular functions of TOR stemming from work in the *Drosophila* system.

THE ENZYMES, Vol. XXVIII

ISSN NO: 1874-6047

DOI: 10.1016/S1874-6047(10)28010-1

II. Introduction

Conservation of cellular signaling networks across the broad range of eukaryotic life has allowed researchers working with widely disparate cell types and organisms to contribute to and draw from a common body of knowledge. Discoveries made in one experimental system have been quickly tested and expanded upon in others, allowing a clear picture of the generalized network and informative exceptions to emerge. A remarkable degree of similarity can be found in the primary structure and enzymatic activity of orthologous proteins in such networks, their association into interaction pairs and multiprotein complexes, and the organization and logic of the signaling pathways themselves. This conservation has facilitated the rapid evolution of the TOR field from the discovery of *TOR1* and *TOR2* in *Saccharomyces cerevisiae* to the development, both potential and realized, of rapamycin and other TOR inhibitors for a variety of medical applications.

Drosophila has served as a useful intermediate model system in this field, with a physiological complexity and experimental facility between that of yeast and vertebrate animal systems. The powerful genetics of the fly system has made possible the discovery of several of the key players in the TOR network, and has supported a detailed analysis of their functions *in vivo*. This chapter discusses the contributions that studies in *Drosophila* have made to the field of TOR signaling, emphasizing how a variety of genetics-based approaches have been used to dissect different aspects of this pathway.

III. Genetic Screens: Identification of Network Components and Their Relationships

Almost from its inception as a model organism, the potential of *Drosophila* for research into the mechanisms of cell growth control has been evident. Mutations that interfere with the normal programs of growth in this organism can give rise to reduced or increased size of the adult fly body or of individual structures. Large numbers of such spontaneous or induced mutations were identified and collected beginning in the 1910s and 1920s in the fly labs of Morgan, Bridges, and Sturtevant, and were described in the first comprehensive catalog of *Drosophila* mutants of Lindsey and Grell [1]. Although many of these lines have been lost long ago, others have been maintained over time and some are in use currently. For example, mutations in the *diminutive* gene were identified in 1933 based on their reduced body size, and were shown in 1996 to be allelic to the *Drosophila* ortholog of the Myc oncogene [2].

Despite this long history, the identification of TOR pathway components was more directly influenced by the development of modern genetic mosaic techniques allowing conditional disruption genes in spatially limited clones of cells during fly development [3, 4]. This approach is especially useful for the analysis of genes that are required for viability or for development to adulthood, which is the case for most of the genes in the TOR pathway. Hafen and colleagues developed a cleverly designed screen based on this technique, inducing mitotic recombination of randomly mutagenized chromosomes specifically in the developing eye and head capsule, and selecting for individuals with reduced eye and head size, so-called pinhead flies. Large-scale saturating mutagenesis screens based on this approach led to the identification of several components of the TOR pathway, including mutations in the *Drosophila* TOR gene itself [5]. Concurrently, a more standard genetic knockout strategy was used to generate null alleles of TOR [6], and hypomorphic point mutants were subsequently isolated through noncomplementation screening [7]. Together, these fly lines allowed the first analyses of the *in vivo* functions of TOR. The reduced growth of tissues and organs mutant for TOR was shown to be due largely to a marked reduction in cell size, one of the hallmarks of this pathway. TOR mutant cells displayed a number of phenotypes reminiscent of starvation, including an extended G1 phase of the cell cycle, reduced nucleolar mass, and mobilization of nutrient stores. These studies also demonstrated that S6-kinase functions as physiologically relevant target of TOR *in vivo*, confirming studies from transformed mammalian cell lines.

Similar screens also led to the identification of negative regulators of the TOR pathway. Mutations in the *Drosophila* orthologs of *Tsc1* and *Tsc2*, genes responsible for the tuberous sclerosis complex in human patients, were identified based on their causing a striking overgrowth when targeted to the developing eye [8–11]. Genetic epistasis experiments placed these genes as functioning downstream of the insulin/PI3K/Akt pathway and upstream of S6K, consistent with a role in TOR regulation. Subsequent studies directly demonstrated that the *Tsc1*/*Tsc2* complex functions as a specific antagonist of TOR signaling [12, 13], a finding confirmed in mammalian systems. The observations that reduction of TOR signaling can suppress the overgrowth and lethality of Tsc mutants in *Drosophila* was a critical early step in the development of rational therapies aimed at tuberous sclerosis.

The identification of the Ras-related GTPase Rheb as the key link between Tsc proteins and TOR was also first established in *Drosophila*. Rheb was identified using an approach complementary to the mosaic loss-of-function screen described above. In this case, random integration of a transposable element allowing targeted overexpression of neighboring genes was used to identify genes capable of driving overgrowth of the fly eye. Rheb was among the genes captured in these screens, and again through order of action experiments were assigned to this pathway directly upstream of TOR [14–16]. Together with biochemical studies demonstrating that *Tsc2* possesses GAP activity toward Rheb, these genetic studies helped to establish the core pathway leading from insulin signaling to TOR activation.

In contrast to the rapid pace of discovery into the Tsc–Rheb–Tor arm of this pathway, insight into how TOR activity is regulated by nutrients has progressed more slowly, despite the apparent conservation of this regulation from yeast to mammals. Only in recent years have potential regulators of this branch of TOR signaling been identified and evaluated. Among the more compelling candidates are the GTR/Rag proteins, members of another family of small GTPases. Studies in yeast, *Drosophila*, and mammalian cells have found that these proteins function as obligate components in sensing or conveying amino acid sufficiency to TOR [17–20]. In *Drosophila*, RagA and RagC were identified through a selective RNAi screen in cultured S2 cells, in which all *Drosophila* GTPases were individually depleted and assayed for their effect on S6K phosphorylation. Standard loss-of-function mutants in RagC were also characterized *in vivo*, and displayed effects on cell growth and autophagy consistent with reduced TOR signaling. Studies in cultured mammalian cells have demonstrated that Rag proteins stimulate TOR activity by controlling the recruitment of TOR complexes to the surface of the late endosomal/lysosomal compartment, thus facilitating interaction with Rheb, a constitutive resident of these organelles [19, 21]. The class III PI(3)-kinase Vps34, a central regulator of endocytic trafficking, has also been identified as an amino acid-dependent activator of TOR in mammalian cells [22–24]. Whether Vps34 influences the localization of Rheb or TOR complexes and interacts with GTR/Rag-dependent control has not yet been determined. Interestingly, despite the conservation of Vps34's functions in endocytosis and autophagy, it does not appear to have an appreciable role in TOR signaling in *Drosophila* (Figure 10.1), thus representing a rare case in which these pathways appear to differ between flies and mammals [25]. Cells that are homozygous for null alleles of *Drosophila* Vps34 are normal in size and do not show defects in S6K activation. Vps34 mutant cells also have normal rates of proliferation, and do not display increased sensitivity to altered levels of TOR signaling. This apparent difference in the functions of fly and mammalian Vps34 may reflect real differences in the architecture of these pathways, or this function of Vps34 may be masked in *Drosophila* by the presence of redundant mechanisms of PI(3)P generation. Alternatively, Vps34 may play substantially different roles in cell culture and *in vivo* settings.

Studies in *Drosophila* have also pointed toward additional factors potentially involved in nutrient signaling upstream of TOR. A class of amino acid transporter-related proteins with high substrate affinity but limited or

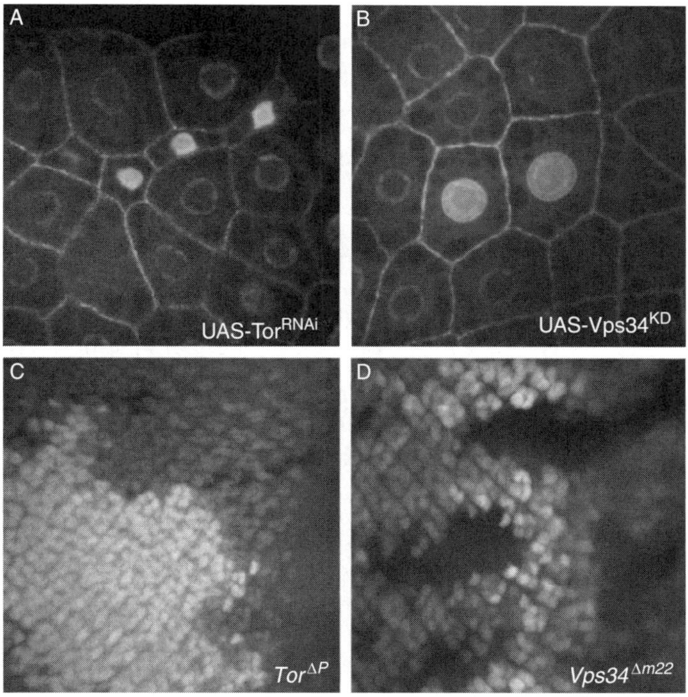

FIG. 10.1. Loss of Vps34 function does not cause reduced TOR activity in *Drosophila*. Analysis of cell size and proliferation in the fat body (A, B) and wing imaginal disc (C, D) in cells with decreased TOR or Vps34 function. Reduction of TOR activity through clonal induction of RNAi against TOR (A, GFP-positive cells) causes a decrease in cell size, whereas inhibition of Vps34 activity does not affect cell size (B). Null alleles of TOR (C) but not Vps34 (D) in clones of wing disc cells lead to a decrease in cell proliferation. Clones of homozygous mutant cells are marked by lack of GFP expression, and their rate of proliferation can be compared to that of their twin clones which carry two copies of GFP.

absent transporter activity has been suggested to function in nutrient sensing [26]. In *Drosophila*, the proton-assisted amino acid transporter family member *pathetic* displays characteristics of such a protein acting in the TOR pathway [27]. *Path* mutants display cell growth defects and interact genetically with TOR, whereas the Path protein has little capacity for amino acid transport. Finally, an RNAi screen in *Drosophila* S2 cells for protein kinases that regulate S6K and 4EBP phosphorylation identified the Ste20 family member MAP4K3 as an upstream element in TOR signaling [28]. This and a subsequent study [29] demonstrated that amino acid levels regulate MAP4K3 and can influence its association with the Rag GTPases, and that loss of MAP4K3 in flies leads to multiple starvation-like phenotypes. Further investigation into the relationships among these various components acting upstream of TOR will help to clarify whether and how nutrients act through single or multiple independent mechanisms to control TOR activity.

IV. Identification and Analysis of TOR-Dependent Cellular Functions in *Drosophila*

Given the dramatic impact of TOR signaling on cellular growth, considerable effort has been devoted to identifying the effector molecules and cellular processes responsible for TOR's effects. In yeast, TOR has a well-established effect on gene expression, and is known to regulate a number of processes in part through transcription-dependent mechanisms. More recently, this possibility has been studied in the *Drosophila* system through a number of complementary approaches. Sabatini and coworkers assayed changes in gene expression in *Drosophila* S2 cells in response to short-term (0–2 h) treatment with rapamycin [30]. 84 genes with significant, greater than twofold changes in expression were identified, with an overrepresentation of downregulated genes involved in ribosome biogenesis. These authors went on to individually test the potential role of each of these genes in cell growth, through RNAi-mediated depletion in S2 cells. Over half of the genes tested caused a significant change in cell size, generally in accordance with the direction of the transcriptional regulation in response to rapamycin, suggesting that coordinated changes in large sets of genes likely contribute to the growth effects of TOR. Interestingly, the rapamycin-responsive gene set shows significant overlap with genes identified as responsive to levels of the transcription factor Myc in S2 cells [31], implicating Myc as a potential mediator of TOR-dependent transcription. This idea was tested directly by the Cohen group, which found that depletion of Myc blunted the effect of rapamycin on a panel of TOR-regulated genes [32]. A number of rapamycin-responsive genes contain Myc-binding E boxes in their promoter regions, and it was found that the protein levels of Myc and its association with these elements is decreased in cells treated with rapamycin. Recently, a more comprehensive comparison of the transcriptional responses to a number of starvation protocols and Myc or Tor mutation in *Drosophila* larvae has been described [33]. Consistent with the results using cultured cells and rapamycin treatment, the changes in gene expression in response to genetic ablation of Myc or Tor in whole animals showed significant overlap with each other and with those caused by starvation for amino acids. Together, these results strongly implicate Myc as an important downstream target of TOR, and point toward future studies for identifying the mechanisms underlying this connection.

Genetic interaction screens represent another powerful, unbiased method for identifying novel targets and processes regulated by signaling pathways. In general, such screens work by establishing a sensitized genetic system in which dosage-dependent effects of functionally related genes can be observed. A common approach in *Drosophila* is to overexpress a gene of interest, and identify interacting mutations based on their dominant (typically haploinsufficient) suppression or enhancement of the overexpression phenotype. In the case of TOR, overexpression of the wild-type protein was found to cause a counterintuitive reduction in signaling and resultant decrease in cell and organ size [34], presumably resulting from titration and dilution of effective signaling complexes. The reduced eye phenotype caused by eye-specific TOR overexpression was used as the basis for a genetic modifier screen in our lab, and led to the identification of the clathrin-uncoating ATPase Hsc4 as a dominant enhancer [35]. This protein plays an essential role in the early recycling steps of clathrin-mediated endocytosis, and its interaction with TOR thus implicated endocytosis as a TOR-regulated process. This was confirmed experimentally, as disruption of TOR was found to cause a marked decrease in nonselective endocytic uptake, whereas TOR activation stimulated bulk endocytosis (Figure 10.2). Interestingly, a converse effect of TOR signaling was observed on the selective endocytic degradation of specific cell surface proteins, such as amino acid transporters. Related findings have been reported in cultured mammalian cells, with clathrin-mediated endocytosis found to require mTOR, and nutrient transporter turnover inhibited by growth factor

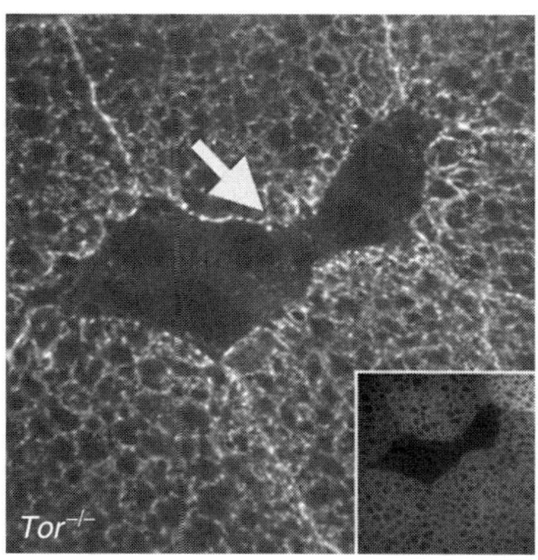

FIG. 10.2. TOR mutant cells are defective for endocytosis. Arrow marks a clone of cells homozygous for a null allele of TOR in the larval fat body, which has been incubated with the endocytic tracer Texas Red avidin. Uptake of tracer can be observed in wild-type cells (indicated by GFP expression in the inset) but not TOR mutant cells.

signaling [36, 37]. Although these dual endocytic effects of TOR would be expected to stimulate nutrient uptake in cells with active TOR signaling, this has not been borne out by direct transport assays in S2 cells [38]. While fast growing cells by definition must import and incorporate mass at an accelerated rate, such differences may be too slight to be appreciable and may not be a significant driving force in cell growth. Overall, endocytosis is regulated by TOR and appears to contribute to the growth effects of this pathway, as disruption of endocytosis compromises cell growth *in vivo*.

The regulation of endocytic processes by TOR and its localization and regulation on intracellular vesicles have placed increased attention on the cell biological functions of TOR related to membrane trafficking. Perhaps the best understood of these functions is the role of TOR in regulating autophagy. This process involves the engulfment of cytoplasmic components within a novel class of double membraned vesicles called autophagosomes, which fuse with the endolysosomal compartment to effect degradation of the sequestered material (reviewed in Ref. [39]). This process proceeds at a basal rate in nearly all eukaryotic cells, playing a critical role in the homeostatic turnover of old or damaged organelles and protein complexes. This process can be significantly upregulated in response to nutrient withdrawal and reduced TOR signaling (Figure 10.3), generating an internal supply of nutrients that can be used to sustain cell viability under starvation conditions. The growing appreciation of the role of autophagy in cancer, neurodegeneration, and aging has motivated the effort to better understand and exploit the regulatory pathways involved in controlling autophagy.

Many of the core components involved in autophagosome formation, a group of about 30 so-called ATG proteins, were identified in genetic screens in *S. cerevisiae*. Elegant genetic studies in yeast have delineated a hierarchy of action of these factors, and have identified the Ser/Thr kinase Atg1 as functioning at or near the initial steps of this process. Importantly, these studies have shown that Atg1 activity is regulated by TOR, which phosphorylates the Atg1-associated protein Atg13, thereby preventing the formation of active Atg1–Atg13 complexes under growth conditions [39]. Recent studies have tested this model in both *Drosophila* and cultured mammalian cells [40–43]. In *Drosophila*, autophagy can be readily assayed in the large cells of the larval fat body, which display a robust induction of autophagy in response to starvation. Mutations in the fly orthologs of Atg1 and Atg13 disrupt autophagy induction by starvation or rapamycin, consistent with a requirement for these factors downstream of TOR [43, 44]. In addition, overexpression of Atg1 is sufficient to induce autophagy under feeding conditions, implying a central role for this kinase [45]. In contrast to the behavior of these proteins in yeast, TOR appears to regulate the phosphorylation of Atg1 rather than Atg13 in *Drosophila*, and does not

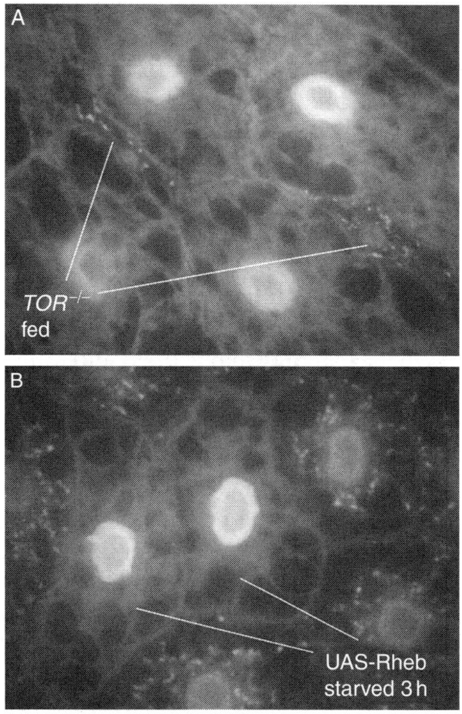

FIG. 10.3. TOR activity regulates autophagy in the larval fat body. Clones of cells with inactivation (A) or activation (B) of TOR signaling display inverse effects on formation of autophagic vesicles, marked by the punctate appearance of mCherry-Atg8a (red). (For interpretation of the references to color in this figure legend, the reader is referred to the Web version of this chapter.).

have a significant effect on the interaction between these proteins [43]. Similar results were reported for the mammalian orthologs of Atg1 and Atg13. Thus, although the general function of these proteins in promoting autophagy has been conserved between yeast and higher eukaryotes, their specific relationships have considerably diverged.

Although autophagy generally functions as a survival mechanism under stress conditions, high levels of autophagy can be damaging to the cell and in some circumstances can play an active role in cell death. Recent work in *Drosophila* has identified a number of feedback and self-limiting mechanisms that serve to restrain and fine-tune the effects of autophagy under prolonged starvation conditions. For example, whereas the TOR pathway overall acts to inhibit autophagy, S6K plays an essential activating role in this process [46]. Thus, inactivation of TOR both induces autophagy and simultaneously downregulates the activity of an essential autophagy component, resulting in a self-buffering effect. Similarly, the increased intracellular concentration of amino acids resulting from autophagic degradation of protein can also act to limit the rate of autophagy through reactivation of TOR. In contrast, this pathway also contains feed-forward mechanisms of signal amplification that may function to override these buffering effects. The best characterized of these results from the reciprocal antagonism between TOR and Atg1; just as TOR inhibits Atg1 activity, Atg1 has been shown to downregulate TOR signaling [45, 47]. Intriguingly, this involves an Atg1-dependent disruption of TOR localization to a premature endocytic compartment [43], suggesting that the autophagic membrane trafficking functions of Atg1 may have been incorporated into its own regulation.

An important goal of TOR research in the fly system and elsewhere will be to investigate how these various cellular functions and regulatory mechanisms of TOR are integrated to produce a coordinated cellular response. For example, it is poorly understood how the regulation of TOR on endomembranes and its functions in vesicle trafficking relate to the important roles of this pathway in translational and transcriptional regulation. Whether distinct pools of TOR are found in different cellular compartments to regulate these functions is presently unclear. To what extent the transcriptional output of TOR influences TOR's various cellular functions will also need to be addressed, as well as the role each of these functions plays in the effects of TOR on cell growth. Genetically accessible models such as *Drosophila* will continue to play an important role in these discoveries.

References

1. Lindsey, D.L., and Grell, E.H. (1968). Genetic Variations of *Drosophila melanogaster*. Carnegie Institute, Washington, DC.
2. Gallant, P., Shiio, Y., Cheng, P.F., Parkhurst, S.M., and Eisenman, R.N. (1996). Myc and Max homologs in *Drosophila*. *Science* 274:1523–1527.
3. Brand, A.H., and Perrimon, N. (1993). Targeted gene expression as a means of altering cell fates and generating dominant phenotypes. *Development* 118:401–415.
4. Xu, T., and Rubin, G.M. (1993). Analysis of genetic mosaics in developing and adult *Drosophila* tissues. *Development* 117:1223–1237.
5. Oldham, S., Montagne, J., Radimerski, T., Thomas, G., and Hafen, E. (2000). Genetic and biochemical characterization of dTOR, the *Drosophila* homolog of the target of rapamycin. *Genes Dev* 14:2689–2694.
6. Zhang, H., Stallock, J.P., Ng, J.C., Reinhard, C., and Neufeld, T.P. (2000). Regulation of cellular growth by the *Drosophila* target of rapamycin dTOR. *Genes Dev* 14:2712–2724.
7. Zhang, Y., Billington, C.J., Jr., Pan, D., and Neufeld, T.P. (2006). *Drosophila* target of rapamycin kinase functions as a multimer. *Genetics* 172:355–362.

8. Ito, N., and Rubin, G.M. (1999). gigas, a *Drosophila* homolog of tuberous sclerosis gene product-2, regulates the cell cycle. *Cell* 96:529–539.
9. Gao, X., and Pan, D. (2001). TSC1 and TSC2 tumor suppressors antagonize insulin signaling in cell growth. *Genes Dev* 15:1383–1392.
10. Tapon, N., Ito, N., Dickson, B.J., Treisman, J.E., and Hariharan, I.K. (2001). The *Drosophila* tuberous sclerosis complex gene homologs restrict cell growth and cell proliferation. *Cell* 105:345–355.
11. Potter, C.J., Huang, H., and Xu, T. (2001). *Drosophila* Tsc1 functions with Tsc2 to antagonize insulin signaling in regulating cell growth, cell proliferation, and organ size. *Cell* 105:357–368.
12. Radimerski, T., Montagne, J., Hemmings-Mieszczak, M., and Thomas, G. (2002). Lethality of *Drosophila* lacking TSC tumor suppressor function rescued by reducing dS6K signaling. *Genes Dev* 16:2627–2632.
13. Gao, X., *et al.* (2002). Tsc tumour suppressor proteins antagonize amino-acid-TOR signalling. *Nat Cell Biol* 4:699–704.
14. Saucedo, L.J., Gao, X., Chiarelli, D.A., Li, L., Pan, D., and Edgar, B.A. (2003). Rheb promotes cell growth as a component of the insulin/TOR signalling network. *Nat Cell Biol* 5:566–571.
15. Stocker, H., *et al.* (2003). Rheb is an essential regulator of S6K in controlling cell growth in *Drosophila*. *Nat Cell Biol* 5:559–566.
16. Zhang, Y., Gao, X., Saucedo, L.J., Ru, B., Edgar, B.A., and Pan, D. (2003). Rheb is a direct target of the tuberous sclerosis tumour suppressor proteins. *Nat Cell Biol* 5:578–581.
17. Dubouloz, F., Deloche, O., Wanke, V., Cameroni, E., and De Virgilio, C. (2005). The TOR and EGO protein complexes orchestrate microautophagy in yeast. *Mol Cell* 19:15–26.
18. Kim, E., Goraksha-Hicks, P., Li, L., Neufeld, T.P., and Guan, K.L. (2008). Regulation of TORC1 by Rag GTPases in nutrient response. *Nat Cell Biol* 10:935–945.
19. Sancak, Y., *et al.* (2008). The Rag GTPases bind raptor and mediate amino acid signaling to mTORC1. *Science* 320:1496–1501.
20. Binda, M., *et al.* (2009). The Vam6 GEF controls TORC1 by activating the EGO complex. *Mol Cell* 35:563–573.
21. Sancak, Y., Bar-Peled, L., Zoncu, R., Markhard, A.L., Nada, S., and Sabatini, D.M. (2010). Ragulator-Rag complex targets mTORC1 to the lysosomal surface and is necessary for its activation by amino acids. *Cell* 141:290–303.
22. Byfield, M.P., Murray, J.T., and Backer, J.M. (2005). hVps34 is a nutrient-regulated lipid kinase required for activation of p70 S6 kinase. *J Biol Chem* 280:33076–33082.
23. Nobukuni, T., *et al.* (2005). Amino acids mediate mTOR/raptor signaling through activation of class 3 phosphatidylinositol 3OH-kinase. *Proc Natl Acad Sci USA* 102:14238–14243.
24. Gulati, P., *et al.* (2008). Amino acids activate mTOR complex 1 via Ca2+/CaM signaling to hVps34. *Cell Metab* 7:456–465.
25. Juhász, G., *et al.* (2008). The class III PI(3)K Vps34 promotes autophagy and endocytosis but not TOR signaling in *Drosophila*. *J Cell Biol* 181:655–666.
26. Hundal, H.S., and Taylor, P.M. (2009). Amino acid transceptors: gate keepers of nutrient exchange and regulators of nutrient signaling. *Am J Physiol Endocrinol Metab* 296: E603–E613.
27. Goberdhan, D.C., Meredith, D., Boyd, C.A., and Wilson, C. (2005). PAT-related amino acid transporters regulate growth via a novel mechanism that does not require bulk transport of amino acids. *Development* 132:2365–2375.

28. Findlay, G.M., Yan, L., Procter, J., Mieulet, V., and Lamb, R.F. (2007). A MAP4 kinase related to Ste20 is a nutrient-sensitive regulator of mTOR signalling. *Biochem J* 403:13–20.
29. Bryk, B., Hahn, K., Cohen, S.M., and Teleman, A.A. (2010). MAP4K3 regulates body size and metabolism in *Drosophila*. *Dev Biol* 344:150–157.
30. Guertin, D.A., Guntur, K.V., Bell, G.W., Thoreen, C.C., and Sabatini, D.M. (2006). Functional genomics identifies TOR-regulated genes that control growth and division. *Curr Biol* 16:958–970.
31. Hulf, T., *et al.* (2005). Whole-genome analysis reveals a strong positional bias of conserved dMyc-dependent E-boxes. *Mol Cell Biol* 25:3401–3410.
32. Teleman, A.A., Hietakangas, V., Sayadian, A.C., and Cohen, S.M. (2008). Nutritional control of protein biosynthetic capacity by insulin via Myc in *Drosophila*. *Cell Metab* 7:21–32.
33. Li, L., Edgar, B.A., and Grewal, S.S. (2010). Nutritional control of gene expression in *Drosophila* larvae via TOR, Myc and a novel cis-regulatory element. *BMC Cell Biol* 11:7.
34. Hennig, K.M., and Neufeld, T.P. (2002). Inhibition of cellular growth and proliferation by dTOR overexpression in *Drosophila*. *Genesis* 34:107–110.
35. Hennig, K.M., Colombani, J., and Neufeld, T.P. (2006). TOR coordinates bulk and targeted endocytosis in the *Drosophila melanogaster* fat body to regulate cell growth. *J Cell Biol* 173:963–974.
36. Pelkmans, L., *et al.* (2005). Genome-wide analysis of human kinases in clathrin- and caveolae/raft-mediated endocytosis. *Nature* 436:78–86.
37. Edinger, A.L. (2007). Controlling cell growth and survival through regulated nutrient transporter expression. *Biochem J* 406:1–12.
38. Hall, D.J., Grewal, S.S., de la Cruz, A.F., and Edgar, B.A. (2007). Rheb-TOR signaling promotes protein synthesis, but not glucose or amino acid import, in *Drosophila*. *BMC Biol* 5:10.
39. He, C., and Klionsky, D.J. (2009). Regulation mechanisms and signaling pathways of autophagy. *Annu Rev Genet* 43:67–93.
40. Jung, C.H., *et al.* (2009). ULK-Atg13-FIP200 complexes mediate mTOR signaling to the autophagy machinery. *Mol Biol Cell* 20:1992–2003.
41. Hosokawa, N., *et al.* (2009). Nutrient-dependent mTORC1 association with the ULK1-Atg13-FIP200 complex required for autophagy. *Mol Biol Cell* 20:1981–1991.
42. Ganley, I.G., Lam du, H., Wang, J., Ding, X., Chen, S., and Jiang, X. (2009). ULK1. ATG13.FIP200 complex mediates mTOR signaling and is essential for autophagy. *J Biol Chem* 284:12297–12305.
43. Chang, Y.Y., and Neufeld, T.P. (2009). An Atg1/Atg13 complex with multiple roles in TOR-mediated autophagy regulation. *Mol Biol Cell* 20:2004–2014.
44. Reggiori, F., Tucker, K.A., Stromhaug, P.E., and Klionsky, D.J. (2004). The Atg1–Atg13 complex regulates Atg9 and Atg23 retrieval transport from the pre-autophagosomal structure. *Dev Cell* 6:79–90.
45. Scott, R.C., Juhasz, G., and Neufeld, T.P. (2007). Direct induction of autophagy by Atg1 inhibits cell growth and induces apoptotic cell death. *Curr Biol* 17:1–11.
46. Scott, R.C., Schuldiner, O., and Neufeld, T.P. (2004). Role and regulation of starvation-induced autophagy in the *Drosophila* fat body. *Dev Cell* 7:167–178.
47. Lee, S.B., *et al.* (2007). ATG1, an autophagy regulator, inhibits cell growth by negatively regulating S6 kinase. *EMBO Rep* 8:360–365.

11

TOR Signaling and Cell Death

TAO WANG[a] • BRUCE A. EDGAR[a,b]

[a]*Division of Basic Sciences*
Fred Hutchinson Cancer Research Center
Seattle, Washington, USA

[b]*DKFZ-ZMBH Alliance*
Im Neuenheimer Feld 282
Heidelberg, Germany

I. Abstract

The evolutionarily conserved serine/threonine protein kinase, Target of Rapamycin (TOR), acts as a central effector of cell growth by controlling diverse metabolic processes, including translation initiation, ribosome biosynthesis, lipid synthesis, endocytosis, nutrient uptake, cytoskeleton organization, and autophagy. In metazoans, the coupling of cell growth and cell death together controls tissue homeostasis. As a growth regulator, TOR signaling affects cell death as well, in a bidirectional manner. Consistent with its cell growth functions, the TOR pathway has a cell-protective role that counteracts multiple cell death signals. On the other hand, hyperactive TOR signaling can also enhance the induction of cell death by various signals. Moreover, inhibition of TOR signaling can protect cells against a variety of insults that promote cell death, especially in neuronal degeneration disorders. Therefore, manipulation of the TOR pathway or its downstream targets has emerged as a promising approach for the therapy of diseases, including cancers and neurodegenerative diseases. In this review, we summarize recent findings regarding the modification of cell death signals by TOR signaling, suggest possible mechanisms, and explore the potential implications of these findings for disease therapy.

ISSN NO: 1874-6047
DOI: 10.1016/S1874-6047(10)28011-3

II. Introduction: Overview of the TOR Signaling Pathway

The Target of Rapamycin (TOR) signaling pathway has central roles in controlling the growth of cells, tissues, and organisms. TOR is an evolutionarily conserved Ser/Thr kinase that nucleates two distinct multiprotein complexes, TOR complex 1 (TORC1) and TOR complex 2 (TORC2), which have been respectively characterized as the rapamycin-sensitive and rapamycin-insensitive complexes.

TORC1 containing regulatory associated protein of mTOR (raptor) is sensitive to rapamycin inhibition. It is now clear that rapamycin binds to FK506-binding protein of 12 kDa (FKBP12), which interacts with the FKBP12-rapamycin-binding domain of TOR and inhibits TORC1 activity. TORC1 functions as a central growth regulator with diverse substrates involved in many growth-related events such as translation initiation, ribosome biosynthesis, lipid synthesis, endocytosis, nutrient uptake, and autophagy. However, some functions of the TOR signaling (TOS) pathway might vary according to cellular type and context. For example, although it has been demonstrated that nutrient uptake is prompted by TORC1 in a few cell types [1], a recent report showed that the TOS does not have direct effects on the import of amino acids or glucose in *Drosophila* S2 cells [2].

TORC1 is regulated by various intracellular and extracellular signals, including growth factors, energy status, oxygen, and amino acids. TORC1 is activated by the small GTPase protein Rheb (Ras homolog enriched in brain) in a GTP-dependent manner [3–5]. Tuberous sclerosis 1 (TSC1) and TSC2 form a complex, which negatively regulates TORC1 and cell growth by functioning as GTPase-activating protein (GAP) for Rheb (Figure 11.1) [6, 7]. Growth factors such as insulin or insulin-like growth factors (IGFs) activate TORC1 signaling by phosphorylation of TSC2 and inactivation of TSC complex through insulin receptor (InR)/phosphoinositide 3-kinases (PI3K)/protein kinase B (PKB or AKT) signaling cascade (Figure 11.1) [8, 9].

As a major growth control pathway, TORC1 is regulated by the availability of nutrients. Energy sources such as glucose change the energy status of the cell and increase the intracellular ATP/AMP ratio, which activates TORC1 by inactivation of AMP-activated protein kinase (AMPK). AMPK is activated by the conjunction of both AMP and the tumor suppressor LKB1, and serves as a negative regulator of TORC1 by phosphorylating Raptor and TSC2, which both lead to reduction of TORC1 activation (Figure 11.1) [10–13]. TORC1 also senses amino acid availability through mechanisms independent/downstream of TSC1/2. Although the mechanisms of regulation of TORC1 by amino acids have not been well understood, the Rag small GTPase proteins have been found to interact with TOR and promote the relocalization of TORC1 from discrete locations

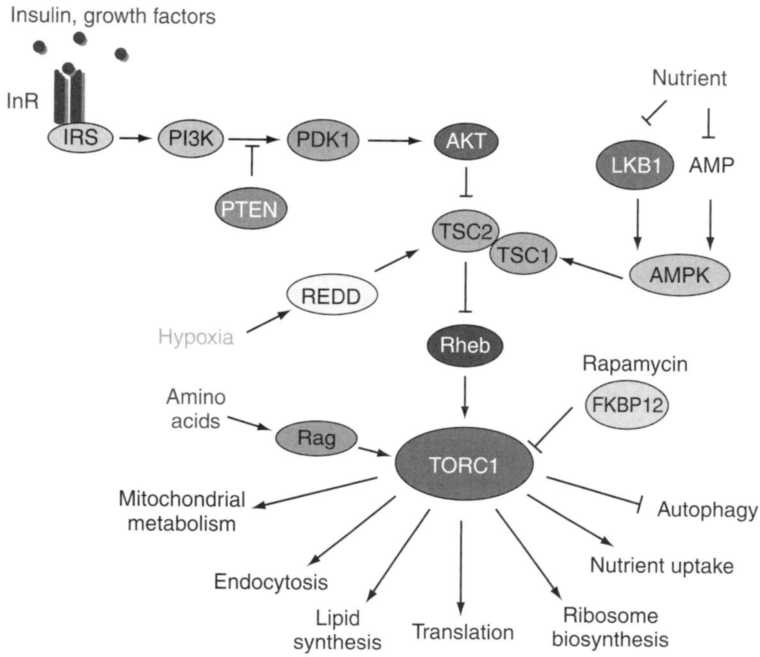

FIG. 11.1. Signaling pathway of TORC1. TORC1 plays a key role in cell growth by regulating translation initiation, ribosome biosynthesis, lipid synthesis, endocytosis, nutrient uptake, and autophagy. Rapamycin inhibits TORC1 through binding with FKBP12. Growth factors activate TORC1 through inactivation of TSC1/2 complex via InR/PI3K/Akt pathway. TSC1/2 complex negatively regulates TORC1 by functioning as GAP of Rheb, the direct activator of TORC1. TORC1 senses the cellular energy status through AMPK, the negative regulator of TORC1, which is activated by LKB and AMP. Amino acids control TORC1 activity through small GTPase Rag. REDD1 is induced by hypoxia and negatively regulates TOR through TSC1/2.

throughout the cytoplasm to a perinuclear region that contains its activator Rheb, in an amino acid-sensitive manner. Therefore, Rag is necessary for the activation of the TORC1 pathway by amino acids [14, 15]. Hypoxia can also block TORC1 activity through reduction of ATP/AMP ratio and through transcriptional activation of the DNA damage response 1 (REDD1), which suppresses TORC1 signaling by releasing TSC2 from its growth-factor-induced association with 14-3-3 proteins (Figure 11.1) [16].

Dysregulation of the TORC1 singling has been found in many growth-related diseases, including cancers. For example, the mutation of *pten*, the negative regulator of PI3K and TOR pathway, is one of the most common causes of human cancers. More directly, mutation of either TSC1 or TSC2 gene causes a benign tumor syndrome, tuberous sclerosis. Therefore, TOR inhibitors such as rapamycin have been proposed as promising drugs for cancer therapies. On the other hand, TORC2 containing rapamycin-insensitive companion of TOR (rictor) is not inhibited by rapamycin. TORC2 regulates the phosphorylation and activation of Akt, conventional protein kinase Cα (cPKCα), and serum/glucocorticoid-regulated kinase (SGK), and thereby plays roles in various processes, including cell survival, cell growth, and cytoskeletal organization. In contrast to TORC1, for which many upstream signals and cellular functions have been defined, little is known about TORC2 regulation.

III. Anti-Cell Death Functions of TOR

Cell growth and survival signals are often coupled in metazoans. TOR, as a central regulator of cell growth, also regulates cell survival in both normal and cancer cells. Constitutively activated TOR can support cell survival in the absence of growth factor signals [1], whereas inhibition of TORC1 with rapamycin induces or facilitates cell death in several cancer cell lines [17, 18]. For instance, cells expressing an active form of TOR were protected from many of the sequelae of growth factor deprivation, including amino acid transporter degradation, reduction of the glycolytic rate, cellular atrophy, decreased mitochondrial membrane potential, and Bax activation [1]. This increased growth factor-independent, nutrient-dependent cell survival and enhanced the ability of p53 deficient mouse embryonic fibroblasts (MEFs) to form colonies in soft agar.

A. Prosurvival Function of S6K

Among the direct targets of TORC1, ribosomal protein p70 S6 kinase (S6K) is a crucial effector of TORC1 signaling, and TOR's effects on cell growth to produce a genetically predetermined cell size, organ shape, or body plan are greatly mediated by S6K [19]. The serine/threonine protein kinase S6K regulates protein synthesis by phosphorylating multiple targets, including 40S ribosomal protein S6, protein synthesis initiation factor 4B, and elongation factor 2 kinase [20]. Besides regulation of protein synthesis, it has also been suggested that S6K, a direct target of TORC1, serves an important function in delivering survival signals in different cell types by phosphorylating and inactivating the proapoptotic molecule BAD (Figure 11.2) [21, 22]. S6K1 has been shown to directly phosphorylate BAD on Ser-136 both *in vitro* and *in vivo*, and to thereby inactivate its proapoptotic function. Conversely, inhibition of TORC1 by rapamycin or

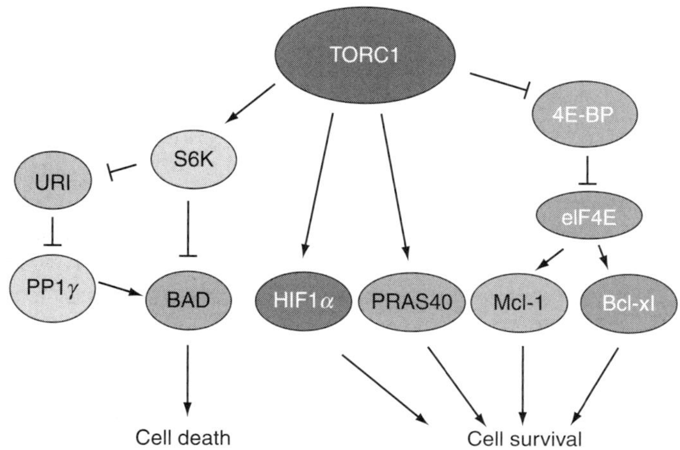

FIG. 11.2. Cell survival function of TORC1. TORC1 activates S6K, which phosphorylates and inactivates pro-apoptotic protein BAD. S6K also activates BAD through regulation of URI/PP1γ complexes. TORC1 phosphorylates 4E-BP, and leads to activation of eIF4E-dependent translation, which increases express ion of anti-apoptotic protein Mcl-1 and Bcl-xl. HIF1α and PRAS40, which promote cell survival in different conditions, are also activated through TORC1-dependent phosphorylation.

S6K1 knockout prevented IGF-1-induced phosphorylation of BAD and blocked IGF-1-induced cell survival.

The phosphorylation of BAD by S6K1 and other kinases, including protein kinase A (PKA), Akt, and mitogen-activated protein kinase (MAPK), inactivates the cell death function of BAD and promotes cell survival. This kinase-mediated inactivation of BAD is counterbalanced by phosphatases, including mitochondria-based BAD-bound phosphatase 1 (PP1). Mitochondrial prefoldin chaperone URI (unconventional prefoldin *R*PB5 interactor) was found to form stable complexes with PP1γ in growth factor-deprived or rapamycin-treated cells, which inhibited the antiapoptosis function of PP1γ. On the other hand, URI was phosphorylated by S6K1 at Ser-371, and phosphorylation of URI by S6K1 induced disassembly of URI/PP1γ complexes, leading to activation of PP1γ and BAD, thus inducing cell death (Figure 11.2) [21]. Therefore, S6K1-dependent phosphorylation of URI opposes sustained S6K1 survival signaling of phosphorylation and inactivation of BAD, thus setting up a threshold for cell death in response to different nutrient and growth factor conditions.

In experiments using ischemia conditions, oxygen and glucose deprivation were found to decrease S6K1 mRNA levels in astrocyte cultures, leading to a reduction of S6K1 protein and activity. This loss of S6K1 function led to an imbalance between mitochondrial pro- and antiapoptotic factors, a rise in ROS levels, and increased astrocyte death during ischemia [23]. In chronic myeloid leukemia cells, S6K1 suppressed expression of programmed cell death 4 (PDCD4), a tumor suppressor protein with a role in apoptosis, and thereby constitutively activated TOR pathway promoted cell growth and survival of leukemic cells [24]. S6K2 has also been observed to mediate cell survival by forming a mitogen-stimulated complex with B-Raf and protein kinase C ε (PKCε). In lung cancer cells, RNAi-mediated downregulation of S6K2 abolished fibroblast growth factor-2 (FGF-2)-mediated cell survival, while increased S6K2 kinase activity triggered prosurvival effects through upregulation of the antiapoptotic proteins XIAP (X-chromosome-linked inhibitor of apoptosis) and Bcl-XL (B-cell lymphoma-extra large) [25].

B. Translational Control of Cell Death Pathway by 4E-BP

Upon activation, TORC1 phosphorylates eIF4E-binding proteins (4E-BP), leading to the release of eIF4E and subsequent assembly of the eIF4F complex [26]. Besides its growth control function by regulation of translation, eIF4E also plays a role in preventing cell death, which is best documented in cancer cells. For example, eIF4E can recapitulate Akt action on lymphomas as an antiapoptotic oncogene, which provides chemotherapeutic agent resistance to cancer cells [27]. eIF4E is a rate-limiting factor for formation of the 48S complex containing mRNA, a major point of regulation mediated by the eIF-4 group of factors, in directing particular mRNAs from the untranslated pool to the ribosomes for translation. The eIF4E-sensitive mRNAs are characterized by long and complex 5′-UTR regions, and encode several proliferation and survival promoting proteins such as cyclins, c-Myc, and Bcl-XL [28].

Recently, the antiapoptotic protein Mcl-1, a Bcl-2 family protein, has been demonstrated as an eIF4E-regulated factor. In this context, eIF4E has been proposed to promote tumorigenesis primarily by suppressing apoptosis through enhancing the translation of Mcl-1 [29].

C. Other Anti-Cell Death Function of TORC1

TORC1 potentially functions as a cell death suppressor through HIF1α in hypoxia-induced cell death. HIF1α, a transcriptional factor that upregulates several genes to promote survival in low-oxygen conditions, is a potential downstream target of TORC1 (Figure 11.2). Raptor interacts with HIF1α through a TOS motif located in the N terminus of HIF1α. During hypoxia, activation of TORC1 potently enhanced the activity of

HIF1α and vascular endothelial growth factor (VEGF)-A secretion, which was reversed by treatment with rapamycin [30].

PRAS40 (proline-rich Akt substrate of 40 kDa) is also phosphorylated by TORC1, and inhibition of TORC1 by either withdrawal of amino acids or treatment with rapamycin alters the phosphorylation of PRAS40 (Figure 11.2). Binding of PRAS40 to 14-3-3 proteins, an effector of survival signaling pathways, was inhibited by expression of TSC1/2, the negative regulators of TORC1, and stimulated by expression of Rheb in a rapamycin-sensitive manner [31]. PRAS40 itself is a proapoptotic molecule, and PRAS40 knockdown protects cells against the induction of apoptosis by various signals, including TNFα (tumor necrosis factor α) and cycloheximide. However, the fact that rapamycin failed to prevent the reduction in apoptosis caused by PRAS40 knockdown casts doubt upon the hypothesis that TOR mediates cell survival through PRAS40 [32].

TOR has an antiapoptotic role against DNA damage-induced cell death, and TOR's antiapoptosis function following DNA damage is p53-dependent. RAD001, a rapamycin derivative, dramatically enhanced apoptosis following treatment with the DNA-damaging agent cisplatin in tumor cells with wild-type p53, but not in p53 mutant or deficient tumor cells. Further evidence shows that RAD001 inhibits cisplatin-induced induction of p21, a major antiapoptotic target of p53 in the DNA damage response. This appears to occur by global translation inhibition, and downregulation of p21 by inactivation of TORC1 results in enhanced apoptosis upon DNA damage [17].

Many cancer cells, for example, human melanoma cell lines, respond poorly to doxorubicin-induced apoptosis. Except inducing cell death, anthracycline compounds such as doxorubicin also activated NF-κB nuclear translocation, which made cells resistant to apoptotic stimuli. Accordingly, disruption of the NF-κB signaling pathway increased the efficacy of apoptogenic stimuli. In this context, rapamycin inhibited doxorubicin-induced NF-κB activation in bone marrow samples of childhood acute lymphoblastic leukemia and increased doxorubicin-induced apoptosis [18]. Moreover, this occurred in spite of the fact that Akt was activated. Therefore, activation of NF-κB may be another anti-cell death pathway relevant to TOR, which can overcome negative feedback of TOR on Akt signaling.

D. Suppression of TORC2 Leads to Inactivation of Akt

The TOR protein is a central component of two functionally distinct multiprotein complexes: TORC1 and TORC2. In complex with FKBP12, rapamycin binds and inhibits TOR exclusively in TORC1, whereas FKBP12–rapamycin does not bind TOR in TORC2, making TORC2 insensitive to rapamycin. Although it has been reported that long-term treatment with rapamycin prevented TORC2 assembly and thereby indirectly inhibited TORC2 activity in a few cell types, TORC2 is well recognized as rapamycin-insensitive complex [33]. TORC2 is believed to control cell survival, cell growth, and cytoskeletal organization by phosphorylating several AGC kinases, including SGK, cPKCα, and Akt. As the first identified direct substrate of mTORC2, Akt was found to be phosphorylated at Ser473 by TORC2 [34]. Early evidence indicated that Ser473 phosphorylation of Akt primed it for Thr308 phosphorylation, leading to its full activation [34, 35]. However, later reports suggested that TORC2-mediated Akt phosphorylation does not determine absolute activity, but instead determines substrate specificity. Akt lacking TORC2-mediated Ser473 phosphorylation appeared to remain largely active against many substrates such as glycogen synthase kinase 3 (GSK3) and TSC2, but showed decreased activity toward Forkhead box O 1/3a (FOXO1/3a) and BAD [35, 36]. Given TORC2's ability to control the activity of FOXO and BAD, two major cell survival substrates of Akt, TORC2 potentially has a key function in cell survival. Indeed, it has been reported that disruption of TORC2 by mutation of SIN1, one of its components, increased cells' sensitivity to stress-induced apoptosis following exposure to hydrogen peroxide (H_2O_2) or the DNA damage-inducing agent, etoposide (ETOP) [35].

E. TOR Activation can Make Cancer Cells Resistant to Chemotherapeutic Agents

Chemotherapy using cytotoxic chemicals to induce cell death is a frontline defense against malignant tumors, and resistance to cell death is a major contributing cause of chemoresistance in hard-to-treat cancers. Numerous studies have demonstrated that most of the chemotherapeutic agents induce cell death by activating the apoptotic pathway, and that resistance to apoptosis occurs because of high intracellular levels of antiapoptotic factors. The TOR pathway is constitutively activated at a high frequency in many tumors, and the cell-protective role of TOR through its inhibition of proapoptotic proteins and pathways suggests that TOR activation can affect the sensitivity of some tumors' chemotherapy. Indeed, a number of studies have suggested a correlation between tumor chemoresistance and upregulation of TOR. Activated TOR can drive resistance to common chemotherapeutics in cell and mouse tumor models [27]. TOR has been demonstrated as an important determinant of cancer cell response to a wide range of clinically important chemotherapeutic agents. TOR inhibitors, such as rapamycin and its analogs, have been found to be additive or synergistic with many chemotherapeutic agents, such as paclitaxel, cyclophosphamide, cisplatin, vinorelbine, doxorubicin, and campthotecin, to enhance cancer cell death during the chemotherapy [27, 37, 38]. Compared with single agent therapy, the combination of rapamycin with other chemotherapeutics can restore sensitivity of cancer cells to chemotherapies and enhance antitumor efficacy in cell lines and animal models. Ongoing clinical trials are currently evaluating the efficacy of rapamycin and its analogs in combination with a broad spectrum of chemotherapeutic agents [39].

F. Cancer and TOR

Consistent with its role in cell proliferation, the TOR pathway is frequently hyperactivated in a number of human malignant tumor cells, and is thus considered to be an attractive target for anticancer therapy. Activated TOS is associated with human cancers, including breast cancer, endometrial cancer, thyroid cancer, prostate cancer, melanoma, and glioblastoma. It is also a causal factor in several tumor-predisposition syndromes: Cowden's syndrome (*PTEN* mutations), Peutz–Jeghers syndrome (*LKB1* mutations), tuberous sclerosis (*TSC1/2* mutations), and neurofibromatosis (*NF1* mutations) [39].

The ability of rapamycin to suppress both cell proliferation and growth through inhibition of TORC1 has suggested that rapamycin and its analogs might be effective as anticancer agents. Indeed, it has been reported that rapamycin and its analogs effectively inhibit cell proliferation and angiogenesis in some human tumors with relatively minor toxicity [40]. The rapamycin analogs temsirolimus (CCI-779) and everolimus (RAD001) have recently been approved by the USFDA (U.S. Food and Drug Administration) for the treatment of advanced stage renal cell carcinomas and sarcomas, respectively [41]. However, the efficacy of rapamycin and its analogs as broad-based monotherapies for cancer has not proved as promising as initially expected [42]. Several aspects of the molecular biology of TOS need to be considered in assessing the shortcomings of rapamycin analogs as anticancer drugs.

During prolonged application, rapamycin has distinct effects on inhibition of S6K versus 4E-BP. While rapamycin potently inhibits S6K activity throughout a treatment, 4E-BP1 can recover to the phosphorylated, inactive state within a short period of time, despite initial inhibition. Nevertheless, it has been suggested that rapamycin blocks only some TOR-dependent activities, as it has relatively little effect on the inhibition of phosphorylation of 4E-BP [43]. Since the TOR/4E-BP axis that regulates eIF4E availability and cap-dependent translation plays an important role in TOR-dependent cell growth and survival [27, 44], the poor inhibition of 4E-BP during long-term treatment with rapamycin analogs renders these TOR inhibitors ineffective. Moreover, the emerging role of rapamycin-resistant TORC2 complexes on tumorigenesis and cell survival underscores the importance of developing comprehensive inhibitors of both TORC1 and TORC2 to efficiently target TOS in cancer. A number of active-site TOR inhibitors have now been developed, which specifically suppress TOS by competing with ATP for binding to the kinase domain of TOR [45–49]. These compounds, such as PP242, Torin1, WYE-354, Ku-0063794, and AZD8055, suppress both TORC1 and TORC2 activity with significant specificity, and display more dramatic effects on inhibition of cell growth and proliferation than rapamycin. Moreover, these compounds may be able to inhibit TOR activity more effectively than rapamycin. For example, Torin1, unlike rapamycin, potently blocked 4E-BP1 phosphorylation during long-term application [48].

Finally, since TOR inhibition activates functions that promote cell survival (e.g., autophagy, AKT activation), TOR inhibitors have the potential to make cells resistant to death signals. In the context of carcinogenesis and cancer chemotherapy, this might make cancer cells resistant to toxic chemicals and stress conditions, and thereby be counterproductive. This possibility should be considered when TOR inhibitors are used as anticancer chemotherapeutics.

IV. Cell Death Associated with the Upregulation of TOR

A. Cells are Susceptible to Death Signals When TOR Signaling is Hyperactivated

Although TOS can promote cell survival in some contexts, increasing evidence suggests that inappropriate activation of TOR can also make cells sensitive to death signals. This is especially true in quiescent cells such as neurons. Notably, activated caspase 8 and TUNEL-labeled cells have been detected in neural tuber sections from patients with TSC diseases, and some of these *Tsc* mutant tumor cells exhibited morphological features of apoptosis [50]. In addition, in a mouse model of TSC disease, conditional *Tsc1* loss induced cell death in the hippocampus and neocortex [51]. Furthermore, TSC1 or TSC2-deficient MEFs exhibit caspase 3 activation and compromised cell survival following exposure to death signals such as DNA damage, TNFα stimulation, or treatment with the chemotherapeutic agents camptothecin (CMPT) and ETOP, whereas reconstitution of TSC2 expression or rapamycin treatment protected these cells from the cell death signals [52, 53]. Consistently, a recent study in *Drosophila* showed that activation of TORC1 by either overexpression of Rheb or mutation of the *Tsc1* gene resulted, in the long term, to neuronal cell death [54]. Thus, constraining TOR activity can be crucial for cell survival, especially in neurons.

B. Multiple Pathways may Mediate TOR-Induced Cell Death

Besides regulation of protein synthesis, recent evidence suggests that S6K also elicits a selective inhibitory effect on Akt activation at a point downstream of the InR [55]. Insulin receptor substrate (IRS) is directly phosphorylated by S6K, leading to inhibition of Akt signaling and insulin resistance. Moreover, S6K also controls expression of IRS, as mRNA levels of IRS were also reduced in *Tsc2*-deficient cells and restored by both rapamycin treatment and S6K suppression. Therefore, TORC1 suppresses InR/PI3K/Akt signaling via the S6K-dependent control of IRS [55, 56]. PI3K/Akt pathway appears to provide a universal survival signal, as Akt phosphorylates a number of substrates important for the regulation of apoptosis, including Bad, FKHR, and MDM2, and also phosphorylates FOXO transcription factors, resulting in their export from the nucleus and the downregulation of FOXO target genes, including the death gene FasL and proapoptotic Bcl-2 family member Bim [57, 58]. The increased sensitivity of *Tsc2*-deficient MEFs to DNA-damaging reagents has been ascribed to TOR, because rapamycin treatment can restore insulin-stimulated Akt activation and chemoresistance in these cells. Inappropriate activation of the TORC1 is sufficient to downregulate Akt, and thus attenuated cell survival [52].

In MEFs, *Tsc1*- and *Tsc2*-deficient cells are more sensitive to DNA damage and TNFα-induced cell death with reduced NF-κB activation through suppression of ERK1/2 and Akt signaling [53]. ERK1/2 signaling mediates NF-κB activation and cell survival in response to DNA damage, while Akt functions in TNFα-dependent NF-κB activation. The inhibition of TOR by rapamycin restored NF-κB activation and survival in *Tsc2*-deficient MEFs in response to both DNA damage reagents and TNF treatments, whereas downregulation of NF-κB or inhibition of ERK1/2 MAP kinase or PI3K pathways reversed this rapamycin-mediated cell survival effect. NF-κB has prosurvival activity in a variety of cell types, by virtue of its ability to induce expression of antiapoptotic genes, including several members of the Bcl-2 family, most notably, Bcl-XL and A1/Bfl-1, c-IAPs, XIAP, and TRAF [59]. In situations of impropriate activation of TOR, cross talk between TOS and NF-κB may suppress NF-κB-dependent cell survival, leading to cell death.

In tauopathies and Alzheimer diseases (AD), neurodegenerative diseases characterized by abnormal accumulations of hyperphosphorylated tau, several positive regulators of the cell-cycle have been found to be aberrantly expressed or localized, including cyclins, cyclin-dependent kinases (Cdks), the S-phase marker proliferating cell nuclear antigen (PCNA), and the M-phase marker phosphohistone-3 (PH-3) [60, 61]. TOR kinase pathway was aberrantly upregulated in AD, as both phosphorylated S6K and activated eIF4E was elevated in AD models [62, 63]. In a *Drosophila* tauopathy model, upregulation of TOR activity was linked to neurodegeneration, such that TOR activation was shown to enhance tau-induced neurodegeneration in a cell cycle-dependent manner. In that study, transgenic expression of cell cycle inhibitors such as the Retinoblastoma protein (Rbf) or the cyclin-dependent kinase 2 (Cdk2) inhibitor Dacapo (Dap) reduced tau-induced cell death, whereas ectopic cell-cycle activation driven by ectopic E2F led to apoptotic neurodegeneration. Similarly, blocking TOR activity pharmacologically or genetically reduced tau-induced neurodegeneration, whereas TOR activation enhanced tau-induced neurotoxicity. Interestingly, this enhancement could be blocked by concomitant inhibition of the cell-cycle gene activity. These results suggest that ectopically activated TOR may drive cell-cycle activation, and as a result induce apoptosis in postmitotic neurons [64]. However, it should be noted that a direct linkage between TOR activation and cell cycle progression has not been clearly demonstrated, except several observations indicating that TOR is required for cell cycle progression.

TOR is also likely to affect cell viability through its profound effects on mitochondrial metabolism and biogenesis. Mitochondrial integrity is very important for maintaining cell viability and suppressing apoptosis. This is because several proapoptotic factors, such as cytochrome *c*, apoptosis-inducing factor (AIF), and Smac (or Diablo), are released from mitochondria through the mitochondrial permeability transition pore in response to mitochondrial dysfunction or apoptotic stimuli, and also because mitochondria are major sites of reactive oxygen species (ROS) production. These factors are affected by TOS because it plays a significant role in determining both resting oxygen consumption and oxidative capacity. Inhibition of TORC1 by rapamycin lowers mitochondrial membrane potential, oxygen consumption, and cellular ATP levels and profoundly alters the mitochondrial phosphoproteome [65]. TORC1 activity also regulates mitochondrial biogenesis and mitochondrial gene expression, and thus regulates mitochondrial oxidative function, in part by targeting the transcription factor *yin-yang 1* [66]. In hematopoietic stem cells, upregulation of TOR by loss of *Tsc1* was shown to cause increased mitochondrial biogenesis and accumulation of intracellular ROS. Since ROS has harmful effects, such as damage of DNA, oxidation of polyunsaturated fatty acids in lipids, and oxidation of amino acids in proteins, which induces cell death, the TOR-dependent accumulation of intracellular ROS is another factor that may account for TOR-related cell death.

C. Suppression of Autophagy as a TOR-Dependent Death Signal

The word "autophagy" is derived from the Greek "to eat" ("phagy") oneself ("auto"). There are three forms of autophagy: chaperone-mediated autophagy, microautophagy, and macroautophagy, which are separated according to their mechanisms, physiological functions and cargo specificity. The major form of autophagy, macroautophagy (herein referred to as autophagy) is a catabolic process whereby cytoplasmic materials are engulfed by autophagosomes and shuttled to lysosomes for degradation (Figure 11.3A) [67]. Recent studies have suggested conflicting roles of autophagy in cell death and survival. Excessive autophagy can deplete organelles and critical proteins, thus inducing a form of nonapoptotic cell death, which has been implicated in the death of a variety of cell types during development and tissue homeostasis [68, 69].

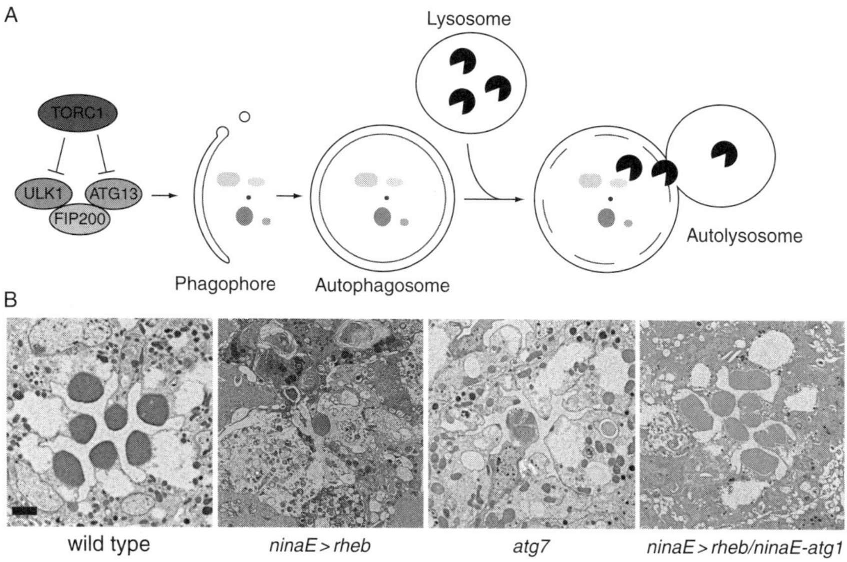

FIG. 11.3. TORC1 suppresses autophagy, making cells vulnerable to death signals. Autophagy is a bulk degradation pathway for clearance of damaged organelles or proteins. (A) Autophagy involves formation of double-membrane structures phagophores, followed by formation of autophagosomes through elongation and engulfment of phagophores. Autophagosomes then fuse with lysosomes to form autolysosomes, where their contents are degraded by acidic lysosomal hydrolases. TORC1 negatively regulates autophagy by phosphorylation of ATG13 and ULK1/ATG1 and inhibition of ULK1-ATG13-FIP200 complex. (B) Upregulation of TOR signals drives cell death in *Drosophila* photoreceptor cells by suppression of autophagy. Transmission EM images of cross sections from: wild type, *ninaE > rheb* (hyperactivation of TORC1), $atg7^{d77}$ (disruption of autophagy), and *ninaE > rheb/ninaE-atg1* (induction of autophagy with hyperactivation of TORC1) [54].

On the other hand, autophagy functions at basal levels in every cell and is beneficial in that it promotes cellular homeostasis by regulating the turnover of organelles and proteins, thus recycling nonessential, redundant, or damaged organelles and macromolecular components. Hence, the absence of autophagy genes results in cell death in many cell types under stress conditions. Recent studies in both mice and fly models found that defective autophagy in neurons could disrupt neural function by the induction of neural cell death. Mice deficient for *Atg5* (autophagy-related 5) or *Atg7* specifically in neural cells accumulated cytoplasmic inclusion bodies, ultimately leading to neurodegeneration [70, 106]. In flies, mutation of *Atg7* also resulted in accumulation of ubiquitin-positive aggregates in degenerating neurons [71]. This genetic evidence highlights an important cell survival role for autophagy.

In most types of cells, autophagy is profoundly regulated by TOS. Stimulation of TORC1 strongly suppresses autophagy, whereas inhibition of TORC1 increases the process. It has recently become clear that TOR controls autophagy through the regulation of a protein complex composed of unc-51-like kinase 1 (ULK1)/ATG1, autophagy-related gene 13 (ATG13) and focal adhesion kinase family-interacting protein of 200 kDa (FIP200) by phosphorylating and thereby repressing ULK1 and ATG13 [72–75]. The Ser/Thr kinase ULK1 functions as a key regulator of autophagy via multiple pathways, including TOR, AMPK, and Ras/PKA [72, 76, 77]. Direct activation of ULK1 is sufficient to drive full induction of autophagy [69]. Our recent studies in *Drosophila* suggested that the suppression of autophagy is the major cause of cell death induced by hyperactive TOS [54]. Suppression of autophagy by disruption of *Atg7* phenocopied the activity-dependent photoreceptor neuron degeneration phenotype caused by overexpressed Rheb. Importantly, induction of autophagy by overexpression of ULK1/ATG1 largely suppressed the Rheb-induced photoreceptor cell death (Figure 11.3B) [54]. This evidence strongly supports the idea that the continuous clearance of diffuse cytosolic proteins through basal autophagy is important for preventing the accumulation of toxic proteins, which can disrupt neural function and ultimately lead to cell death.

D. Activated TOR Modulates but Does Not Directly Cause Cell Death

While hyperactivation of TOR can induce cell death, it is not always sufficient to induce cell death. However, cells with hyperactive TOR are susceptible to a variety of death signals. The link between TOR's cell death function and autophagy may explain why TOR hyperactivity does not drive cell death under normal conditions in many cells. Only in abnormal conditions, such as stress, injury and disease, the toxic molecules which can induce death are accumulated in cells. Under such conditions, autophagy may be required to relieve the accumulation of toxic molecules or damaged organelles.

Different types of cells have different sensitivity to TOR activation, and neurons have been shown to be the most vulnerable tissues to TOR-induced cell death. Consistently, autophagy is particularly important in neural cells, and loss of autophagy has been found to induce cell death exclusively in neural cells. Recent evidence suggested a relationship between TOR-induced neuronal degeneration and neuronal activity, as suppression of neuronal activity significantly suppressed the cell death caused by TOR upregulation in neurons [54]. The activation of neurons leads to generation of toxic products or damaged proteins. For example, continuous exposure to light leads to photoreceptor neuron damage and blindness in wild-type animals [78]. Therefore, it was suggested that continuous clearance of damaged proteins through autophagy is only important for cell survival in neurons with high activity.

V. Autophagy Protects Cells from Neurodegenerative Diseases

A. Suppression of Neuronal Degeneration by Inhibition of TOR

As increasing evidence has revealed that hyperactivation of TOR can promote cell death in quiescent cells such as neurons, it has been suggested that TOR might regulate cell death pathways in neuronal degenerative diseases. Indeed, altered TOR activity has been detected in neurodegenerative diseases involving the formation of intracellular aggregates including Parkinson's disease (PD), Huntington's disease (HD), and AD. In cellular HD models, as well as transgenic mice and brain samples from HD patients, the TOR pathway has been reported to be downregulated, apparently due to the sequestration of TOR in polyglutamine aggregates or polyglutamine-expanded huntington fragments [79]. Similarly, TOS has also been reported to be suppressed in PDs. This may be because RTP801/REDD1, a suppressor of TORC1, was induced in multiple cellular models and an animal model of PD, and in affected neurons of PD patients [80]. In contrast, upregulation of TOR has been indicated ADs. Both upregulation of phosphorylated, active S6K and elevated eIF4E activity were detected in AD brain samples [62, 63], and upregulation of S6K phosphorylation was induced in *Drosophila* tauopathy models [64]. These observations suggest potential roles of TOS pathway in the pathology of neurodegenerative diseases.

Inhibition of TOR by rapamycin and its analogs has been shown to attenuate cell death in several neurodegenerative diseases. Rapamycin and its analog CCI-779 attenuated huntingtin aggregation and cell death in both cell and fly models of HD [79, 81]. In addition, rapamycin has been shown to reduce the toxicity of different proteins with long polyglutamine repeats, and inhibition of TOR was able to attenuate neuronal cell death in several polyglutamine expansion disorders [82]. Rapamycin also reduced the toxicity of wild-type or mutant forms of tau in *Drosophila* tauopathy models, by reducing levels of insoluble tau [64, 82]. Furthermore (and in contrast to the observation that TOR is suppressed in PD noted earlier), rapamycin also protected neurons from death in both cellular and animal models of PD [83]. Consistent with these pharmacological studies, genetic experiments have verified the protective role of inactivation TOR in neurodegenerative diseases involving the formation of intracellular aggregates. In a *Drosophila* model of HD, genetic inactivation of TORC1 pathway by coexpression of TSC1/2 reduced cell death, and in a fly model of tauopathy, suppression of TOR by expression of the constitutively active form of TSC2 reduced tau-induced cell death [54, 64]. Thus, despite some conflicting reports, the bulk of available evidence is consistent with the notion that suppression of TOR activity can counteract neurodegenerative disease.

B. Autophagy Alleviates the Toxicity of Aggregate-Prone Proteins

As the major cell-protective target of TOR, autophagy is a critical factor in the development of a number of diseases, especially neurodegenerative diseases involving formation of intracellular aggregates. Autophagy is highly involved in the clearance of cellular aggregates formed by aggregate-prone proteins such as multiple polyglutamine expansion proteins, including mutant huntingtin, α-synuclein, and tau [82, 84, 85]. Disruption of autophagy-lysosome degradation pathway by treatment with lysosome inhibitor lactacystin strongly increased soluble protein levels of the polyglutamine-expended proteins and α-synuclein. Similar effects were detected when autophagy was inhibited by 3-methyl adenine (3-MA) and bafilomycin A1 (Baf). In both cell and animal models of PD induced by overexpression of α-synuclein, expression of beclin 1, a positive regulator of the autophagy pathway, reduced accumulation of α-synuclein and ameliorated associated neurodegeneration [86]. Expression of beclin 1 was also decreased in affected brain regions of patients with AD early in the disease process, and the restoration of beclin 1 was able to reduce both intracellular and extracellular amyloid pathology in a mouse model of AD driven by the expression of human amyloid precursor protein [87].

Consistent with these findings, recent studies in cell and fly models have suggested that rapamycin can attenuate cell death in many intracytosolic proteinopathies, including HD, principally through the upregulation of autophagy [79, 82]. In addition to rapamycin, treatment with lithium also stimulated a novel TOR-independent pathway that regulates autophagy, and promoted aggregate clearance in cells and in *Drosophila* models of HD, again through the induction of autophagy [88]. Other autophagy inducers that act independent of TOS, for example, minoxidil and clonidine, which induce autophagy through regulation of IP3 and cAMP pathway, alleviated cell death in HD using mammalian cell, fly and zebrafish models without affecting the TOS pathway [88, 89]. In a *Drosophila* model of HD, genetic induction of autophagy through overexpression of ULK1/ATG1 suppressed neuronal degeneration to a similar extent as shutting down the TOR pathway through overexpression of TSC1/2 [54]. Given that autophagy, regardless of whether it is TOR-dependent or TOR-independent, can protect cells against pathogenic neurodegeneration, it has been suggested that rapamycin protects cells from neuronal degeneration through induction of autophagy.

C. Inhibition of TOR Generally Suppresses Cell Death

Besides protecting cells from aggregation-prone proteins, rapamycin can also protect cells against a range of proapoptotic insults, including staurosporine (STA), 3-nitro-propionic acid (3-NP), and Bax expression. Loss-of-function mutations in ULK1/ATG1 and the autophagy inhibitor 3-MA have been shown to abolish this beneficial effect of rapamycin, suggesting that rapamycin achieves its cytoprotective effects through autophagy. This protection can be accounted for by enhanced clearance of mitochondria by autophagy, thereby reducing cytosolic cytochrome *c* release and downstream caspase activation after proapoptotic insults [90]. By induction of TOR-dependent or independent autophagy, toxic rhodopsin–arrestin complexes generated in a fly retinal degeneration model using *norpA* mutants were removed, alleviating the consequent degeneration. Given that rhodopsin–arrestin complexes are not large aggregates, the model that autophagy can remove toxic soluble proteins, and thereby protect cells against degeneration cues may be generally applied to neuronal degenerative diseases [54].

Autophagy can also prevent ischemia-associated cell death. For instance, Beclin 1 was significantly increased at short times after neonatal hypoxia–ischemia both in the hippocampus and in the cerebral cortex of mice, leading to increased autophagy. Inhibition of autophagy by methyladenine and wortmannin reduced beclin 1 expression followed by necrotic cell death.

Conversely, rapamycin reduced this necrotic cell death, decreasing brain injury [91]. In another example, ischemia/reperfusion induced cell death in cardiac myocytes by impairing both the formation and downstream lysosomal degradation of autophagosomes. In this case, overexpression of Beclin1 enhanced autophagic flux after ischemia/reperfusion and significantly reduced activation of proapoptotic protein Bax, whereas knocking down of Beclin1 increased Bax activation and apoptosis. Genetic inhibition of autophagy by expression of a dominant-negative mutant of ATG5 also increased cellular injury during Ischemia/reperfusion [92]. Therefore, induction of autophagy by drugs such as rapamycin might be useful to treat certain diseases involving increased cell death, including various neurodegenerative diseases, even if they are not associated with overt protein aggregate formation.

D. Suppression of Neuronal Degeneration by Other Branches Downstream of TOR Independent of Autophagy

Although it is currently well recognized that induction of autophagy is the major cause of rapamycin-induced antineurodegeneration function, this is still challenged by some evidence. From genetic studies in a *Drosophila* model of polyglutamine disease, it was found that reduction of another TORC1 target, S6K, suppressed the progression of neuron degeneration. In addition, this was achieved by comprehensive inactivation of TORC1 signaling by knocking down the upstream regulator PDK1. The authors therefore suggested that inactivation of the S6K branch of the TOS pathway might play a role in suppression of neuronal degeneration such as in polyglutamine disease [93].

In PD, induction of RTP801/REDD is both sufficient and necessary for cell death. Rapamycin blocked RTP801/REDD induction by suppression of RTP801/REDD translation in PC12 cells and a mouse model of PD induced by 6-OHDA, and protected those cells from PD toxins. However, Torin1, a comprehensive inhibitor of TOR which blocks all actions of TORC1 and TORC2, blocked RTP801/REDD protein induction by 6-OHDA without inhibiting cell death promoted by 6-OHDA. Given the similar abilities of rapamycin and Torin to induce autophagy [48], it is unlikely that rapamycin suppresses cell death in PD through induction of autophagy [83].

E. Induction of Autophagy can also Adversely Affect Cell Survival

Autophagy generally protects cells against toxic insults, especially aggregate-prone proteins, and upregulating autophagy has been proposed as a therapy for certain neuronal degenerative diseases. However, it should be noted that the over-induction of autophagy can induce autophagic cell death, as characterized by morphological changes such as shrinkage of the cell, contraction of the cytoplasm, and condensation of chromatin [94]. Moreover, autophagy also plays a role in facilitating necrotic cell death. Recent studies in *C. elegans* indicated a causative role of autophagy in necrotic cell death [95, 96]. In this system, excessive autophagy was induced early during necrotic cell death, and impairment of autophagy by downregulation of the autophagy genes *bec-1*, *unc-51*, and *lgg-1* or pharmacological treatment with 3-MA partially suppressed necrotic neuronal death induced by hyperactive MEC-4, DEG-1, and DEG-3 ion channels. In contrast, upregulation of autophagy by knockdown of TOR, or by starvation, further promoted neuron necrosis.

In ADs, it has been suggested that autophagy contributes to the formation of Aβ, the main constituent of amyloid plaques. In a mouse model of AD, neuronal autophagy was induced early on, and presenilin, Aβ precursor protein and γ-secretase activity were detected in autophagic vacuoles. Moreover, rapamycin induced Aβ production and secretion, concomitant with an increase in autophagic vacuole formation [97]. Dependence of autophagy on the formation of Aβ raises the possibility of a deleterious role of autophagy in the pathogenesis of AD. Indeed, a recent report, using a fly model of AD overexpressing Aβ(1–42), demonstrated an age-dependent increase in autophagic dysfunction. In this case, downregulating autophagy by silencing autophagy genes in selective neurons resulted in a protective effect against Aβ toxicity, whereas induction of autophagy by treatment with rapamycin increased the toxicity of Aβ [98]. However, opposite results demonstrating the protective effects of autophagy on Aβ(1–42) toxicity have been achieved in a neuroblastoma cell model [99]. Cells treated with rapamycin exhibited a decrease in ROS accumulation and increased survival, suggesting that defects in autophagic regulation may impair the clearance of Aβ and enhance neuronal death [99]. Clearly, further investigation into the detailed roles of altered autophagy on Aβ formation is required to determine the true role of autophagy on the pathogensis of AD.

F. Selectivity and Regulation of Autophagy

Autophagy used to be considered as a nonselective bulk protein degradation system. However, this idea has been challenged recently. In contrast to wild-type proteins, aggregate-prone mutant proteins are highly dependent on autophagy for their clearance. For example, the inhibition of autophagy by 3-MA largely reduced the total levels of polyglutamine-expanded huntington protein, which did not affect the amount of wild-type huntington. The ubiquitin–proteasome system (UPS) is the main selective degradative pathway for ubiquitinated proteins. Interestingly, cross talk between UPS and autophagy systems has been demonstrated recently. Histone deacetylase 6 (HDAC6), a microtubule-associated deacetylase that interacts with polyubiquitinated proteins, is required in dynein-mediated transport of autophagosomes, suggesting an essential mechanistic link between UPS and autophagy systems. The toxicity resulting from proteasome inhibition, mediated by a dominant-negative proteasome subunit, was largely rescued by induction of autophagy-mediated degradation by either overexpression of HDAC6 or treatment with rapamycin [100]. Later evidence showed that the binding of ubiquitinated substrates to the ubiquitin-binding protein p62 might target them to autophagosomes. The interaction between UPS and autophagy systems suggests a selective feature of autophagy-mediated degradation for some ubiquitinated proteins [107]. Another adaptor protein for selective autophagy, yeast ATG32, was found to be involved in the delivery of the mitochondria to autophagic vacuole by binding with ATG11, which represents a selective mitochondrial degradation by autophagy [108, 109]. These recent studies have been shedding light on the selective autophagy modes that degrade aggregated proteins or damaged organelles.

Due to the multiple targets and cellular processes controlled by TOR, the use of rapamycin and other TOR inhibitors as autophagy induction drugs potentially has many unpredictable and adverse side effects. Another consideration is that inhibition of TOR might not affect continuous autophagy upregulation. Therefore, TOR-independent autophagy inducers, or a combination of TOR-dependent and -independent drugs, may be more suitable for long-term use. Indeed, in a fly model of HD, greater protection against neurodegeneration was reported with rapamycin and the TOR-independent autophagy inducer lithium, than with either inhibitor alone [101].

Autophagy is negatively regulated by levels of inositol and inositol trisphosphate (IP_3), and lithium has been shown to induce autophagy by inhibition of inositol monophosphatase, which prevents inositol recycling and leads to depletion of cellular inositol and IP_3 (Figure 11.4) [88]. Via this mechanism, pharmacological inhibition of the IP_3 receptor (IP_3R) by xestospongin B can also induce autophagy. It was further shown that xestospongin B induced autophagy by disrupting the IP_3R-beclin 1 complex and that Bcl2 might also be involved in this process [103]. Lithium has also been shown to reduce mutant huntingtin aggregation in HD, protecting against neurodegeneration in cell and *Drosophila* models of HD. Furthermore, lithium treatment in patients and mouse models of Amyotrophic lateral sclerosis, a motor neuron degenerative disease, was found to

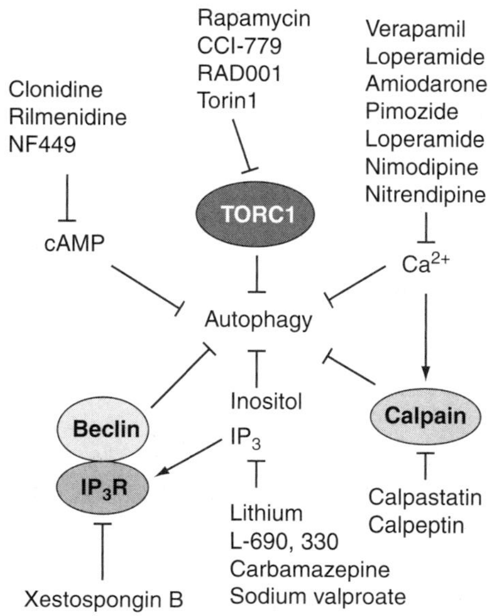

FIG. 11.4. Regulation of autophagy pathway. Autophagy is negatively regulated by TOR, cAMP, Ca^{2+}, calpain, Inostiol/IP3, and IP3R/Beclin. Inhibition of TOR (rapamycin, CCI-779, RAD001, and Torin1), downregulation of cAMP levels (Clonidine, Rilmenidine, and NF449), blocking of Ca^{2+} channels (Verapamil, Loperamide, Amiodarone, Pimozide, Loperamide, Nimodipine, and nitrendipine), inhibition of calpain (Calpastatin and Calpeptin), prevention of inositol recycling (Lithium, L-690,330, Carbamazepine, and Sodium valproate), and disruption of IP_3R/beclin 1 interaction—all lead to increased autophagy [102].

increase survival and attenuate disease progression [88, 104]. A few chemical screens have also been reported to identify novel chemicals and pathways regulating autophagy. For instance, a screen of FDA-approved drugs for new autophagy inducers revealed that L-type Ca^{2+} channel antagonists, the K^+ATP channel opener minoxidil, and the Gi signaling activator clonidine, all induced autophagy through regulation of the cAMP pathway. As expected, these drugs alleviated HD using mammalian cell, fly, and zebrafish models, independent of the TOS pathway [89]. Ca^{2+} channel blockers, including verapamil, loperamide, amiodarone, pimozide, loperamide, nimodipine, and nitrendipine, have also been identified as TOR-independent autophagy inducers in two independent screens for regulators of autophagy and polyglutamine aggregates (Figure 11.4) [89, 105]. Consistent with the role of Ca^{2+}, inhibition of the calcium-dependent cysteine protease calpain by calpastatin or calpeptin can induce autophagy and reduce mutant huntingtin aggregation (Figure 11.4) [89].

Finally, another chemical screen for inhibitors of polyglutamine-mediated protein aggregation *in vitro* identified trehalose as an inhibitor of mutant huntingtin aggregation. Trehalose also reduced toxicity in cell models and a mouse model of HD. Follow-up experiments showed that trehalose could enhance the autophagic pathway independently of TOR, thereby increasing the clearance of mutant aggregate-prone proteins. Before pharmaceutical use of these new autophagy inducers can commence, however, questions about their specificity must be addressed. For example, it is well known that lithium is an inhibitor of GSK3β.

VI. Conclusions and Prospectives

The TOR pathway has been well recognized as a central growth control pathway, and upregualtion of TOR has been found in many growth-related diseases. Consistent with this central role in cell growth, TOR has anti-cell death functions that can ensure the growth of cells, tissues, and organisms. However, on the opposite side, in some contexts, TOR also has a pro-cell death function, and inhibition of TOR is cytoprotective. The impact of TOR on multiple cell death pathways may potentially explain the relatively weak malignancy in *TSC*-derived tumors in TSC diseases and the limited efficacy of TOR inhibitors such as rapamycin as cancer therapeutics. The obvious question that arises concerns how TOR's conflicting roles in cell death and survival are integrated.

TOR might achieve its dual roles in cell death and survival by having distinct outputs in different cellular contexts, such that its effect on survival or death might depend on the cell type, timing, and additional cell-extrinsic conditions. In fact, neurons are the cell type most vulnerable to TOR-induced cell death. Thus, it is not illogical that inhibition of TOR has been proposed as an effective treatment for neurodegenerative diseases. However, even in neurons, the dual roles of TOR make the effects of TOR inhibitors unpredictable. For example in ADs, totally opposite results were observed in different stages or different models after treatment with TOR inhibitors. Differential TOR activity might also account for the different effects of TOR on cell death. Many cellular processes are finely balanced, such that activity in excess of certain thresholds might have opposite effects. For instance, moderate levels of autophagy have a cytoprotective role against cell death signals, but excessive autophagy results in cell death. In this vein, it has been demonstrated that the partial inhibition of TOR by rapamycin provided neuroprotection in PD models, whereas full inhibition of TOR by Torin1 was not protective and even induced neuronal cell death [83]. Future efforts aimed at understanding the differential cytoprotective or cytotoxic effects of TOR on different cell types and conditions could provide the background and rationale needed for the application of TOR inhibitors as therapy for cancers and neuronal degenerative diseases.

References

1. Edinger, A.L., and Thompson, C.B. (2004). An activated mTOR mutant supports growth factor-independent, nutrient-dependent cell survival. *Oncogene* 23:5654–5663.
2. Hall, D.J., Grewal, S.S., de la Cruz, A.F., and Edgar, B.A. (2007). Rheb-TOR signaling promotes protein synthesis, but not glucose or amino acid import, in *Drosophila. BMC Biol* 5:10.
3. Long, X., Lin, Y., Ortiz-Vega, S., Yonezawa, K., and Avruch, J. (2005). Rheb binds and regulates the mTOR kinase. *Curr Biol* 15:702–713.
4. Stocker, H., *et al.* (2003). Rheb is an essential regulator of S6K in controlling cell growth in *Drosophila. Nat Cell Biol* 5:559–565.
5. Saucedo, L.J., Gao, X., Chiarelli, D.A., Li, L., Pan, D., and Edgar, B.A. (2003). Rheb promotes cell growth as a component of the insulin/TOR signalling network. *Nat Cell Biol* 5:566–571.
6. Potter, C.J., Huang, H., and Xu, T. (2001). *Drosophila* Tscl functions with Tsc2 to antagonize insulin signaling in regulating cell growth, cell proliferation, and organ size. *Cell* 105:357–368.
7. Zhang, Y., Gao, X., Saucedo, L.J., Ru, B., Edgar, B.A., and Pan, D. (2003). Rheb is a direct target of the tuberous sclerosis tumour suppressor proteins. *Nat Cell Biol* 5:578–581.
8. Potter, C.J., Pedraza, L.G., and Xu, T. (2002). Akt regulates growth by directly phosphorylating Tsc2. *Nat Cell Biol* 4:658–665.
9. Inoki, K., Li, Y., Zhu, T., Wu, J., and Guan, K.L. (2002). TSC2 is phosphorylated and inhibited by Akt and suppresses mTOR signalling. *Nat Cell Biol* 4:648–657.
10. Inoki, K., Zhu, T., and Guan, K.L. (2003). TSC2 mediates cellular energy response to control cell growth and survival. *Cell* 115:577–590.
11. Gwinn, D.M., *et al.* (2008). AMPK phosphorylation of raptor mediates a metabolic checkpoint. *Mol Cell* 30:214–226.
12. Corradetti, M.N., Inoki, K., Bardeesy, N., DePinho, R.A., and Guan, K.L. (2004). Regulation of the TSC pathway by LKB1: evidence of a molecular link between tuberous sclerosis complex and Peutz-Jeghers syndrome. *Genes Dev* 18:1533–1538.
13. Shaw, R.J., *et al.* (2004). The LKB1 tumor suppressor negatively regulates mTOR signaling. *Cancer Cell* 6:91–99.
14. Sancak, Y., *et al.* (2008). The Rag GTPases bind raptor and mediate amino acid signaling to mTORC1. *Science* 320:1496–1501.
15. Kim, E., Goraksha-Hicks, P., Li, L., Neufeld, T.P., and Guan, K.L. (2008). Regulation of TORC1 by Rag GTPases in nutrient response. *Nat Cell Biol* 10:935–945.
16. DeYoung, M.P., Horak, P., Sofer, A., Sgroi, D., and Ellisen, L.W. (2008). Hypoxia regulates TSC1/2-mTOR signaling and tumor suppression through REDD1-mediated 14-3-3 shuttling. *Genes Dev* 22:239–251.
17. Beuvink, I., *et al.* (2005). The mTOR inhibitor RAD001 sensitizes tumor cells to DNA-damaged induced apoptosis through inhibition of p21 translation. *Cell* 120:747–759.
18. Avellino, R., *et al.* (2005). Rapamycin stimulates apoptosis of childhood acute lymphoblastic leukemia cells. *Blood* 106:1400–1406.

19. Um, S.H., D'Alessio, D., and Thomas, G. (2006). Nutrient overload, insulin resistance, and ribosomal protein S6 kinase 1, S6K1. *Cell Metab* 3:393–402.
20. Ruvinsky, I., and Meyuhas, O. (2006). Ribosomal protein S6 phosphorylation: from protein synthesis to cell size. *Trends Biochem Sci* 31:342–348.
21. Djouder, N., *et al.* (2007). S6K1-mediated disassembly of mitochondrial URI/PP1gamma complexes activates a negative feedback program that counters S6K1 survival signaling. *Mol Cell* 28:28–40.
22. Harada, H., Andersen, J.S., Mann, M., Terada, N., and Korsmeyer, S.J. (2001). p70S6 kinase signals cell survival as well as growth, inactivating the pro-apoptotic molecule BAD. *Proc Natl Acad Sci USA* 98:9666–9670.
23. Pastor, M.D., *et al.* (2009). mTOR/S6 kinase pathway contributes to astrocyte survival during ischemia. *J Biol Chem* 284:22067–22078.
24. Carayol, N., Katsoulidis, E., Sassano, A., Altman, J.K., Druker, B.J., and Platanias, L.C. (2008). Suppression of programmed cell death 4 (PDCD4) protein expression by BCR-ABL-regulated engagement of the mTOR/p70 S6 kinase pathway. *J Biol Chem* 283:8601–8610.
25. Pardo, O.E., *et al.* (2006). FGF-2 protects small cell lung cancer cells from apoptosis through a complex involving PKCepsilon, B-Raf and S6K2. *EMBO J* 25:3078–3088.
26. Gingras, A.C., *et al.* (1999). Regulation of 4E-BP1 phosphorylation: a novel two-step mechanism. *Genes Dev* 13:1422–1437.
27. Wendel, H.G., *et al.* (2004). Survival signalling by Akt and eIF4E in oncogenesis and cancer therapy. *Nature* 428:332–337.
28. De Benedetti, A., and Graff, J.R. (2004). eIF-4E expression and its role in malignancies and metastases. *Oncogene* 23:3189–3199.
29. Wendel, H.G., *et al.* (2007). Dissecting eIF4E action in tumorigenesis. *Genes Dev* 21:3232–3237.
30. Land, S.C., and Tee, A.R. (2007). Hypoxia-inducible factor 1alpha is regulated by the mammalian target of rapamycin (mTOR) via an mTOR signaling motif. *J Biol Chem* 282:20534–20543.
31. Fonseca, B.D., Smith, E.M., Lee, V.H., MacKintosh, C., and Proud, C.G. (2007). PRAS40 is a target for mammalian target of rapamycin complex 1 and is required for signaling downstream of this complex. *J Biol Chem* 282:24514–24524.
32. Thedieck, K., *et al.* (2007). PRAS40 and PRR5-like protein are new mTOR interactors that regulate apoptosis. *PLoS ONE* 2:e1217.
33. Sarbassov, D.D., *et al.* (2006). Prolonged rapamycin treatment inhibits mTORC2 assembly and Akt/PKB. *Mol Cell* 22:159–168.
34. Sarbassov, D.D., Guertin, D.A., Ali, S.M., and Sabatini, D.M. (2005). Phosphorylation and regulation of Akt/PKB by the rictor–mTOR complex. *Science* 307:1098–1101.
35. Jacinto, E., *et al.* (2006). SIN1/MIP1 maintains rictor–mTOR complex integrity and regulates Akt phosphorylation and substrate specificity. *Cell* 127:125–137.
36. Yang, Q., Inoki, K., Ikenoue, T., and Guan, K.L. (2006). Identification of Sin1 as an essential TORC2 component required for complex formation and kinase activity. *Genes Dev* 20:2820–2832.
37. Mondesire, W.H., *et al.* (2004). Targeting mammalian target of rapamycin synergistically enhances chemotherapy-induced cytotoxicity in breast cancer cells. *Clin Cancer Res* 10:7031–7042.
38. Steelman, L.S., *et al.* (2008). Suppression of PTEN function increases breast cancer chemotherapeutic drug resistance while conferring sensitivity to mTOR inhibitors. *Oncogene* 27:4086–4095.

39. Meric-Bernstam, F., and Gonzalez-Angulo, A.M. (2009). Targeting the mTOR signaling network for cancer therapy. *J Clin Oncol* 27:2278–2287.
40. Liu, P., Cheng, H., Roberts, T.M., and Zhao, J.J. (2009). Targeting the phosphoinositide 3-kinase pathway in cancer. *Nat Rev Drug Discov* 8:627–644.
41. Kapoor, A., and Figlin, R.A. (2009). Targeted inhibition of mammalian target of rapamycin for the treatment of advanced renal cell carcinoma. *Cancer* 115:3618–3630.
42. Guertin, D.A., and Sabatini, D.M. (2009). The pharmacology of mTOR inhibition. *Sci Signal* 2:pe24.
43. Choo, A.Y., Yoon, S.O., Kim, S.G., Roux, P.P., and Blenis, J. (2008). Rapamycin differentially inhibits S6Ks and 4E-BP1 to mediate cell-type-specific repression of mRNA translation. *Proc Natl Acad Sci USA* 105:17414–17419.
44. Graff, J.R., *et al.* (2007). Therapeutic suppression of translation initiation factor eIF4E expression reduces tumor growth without toxicity. *J Clin Invest* 117:2638–2648.
45. Feldman, M.E., *et al.* (2009). Active-site inhibitors of mTOR target rapamycin-resistant outputs of mTORC1 and mTORC2. *PLoS Biol* 7:e38.
46. Garcia-Martinez, J.M., *et al.* (2009). Ku-0063794 is a specific inhibitor of the mammalian target of rapamycin (mTOR). *Biochem J* 421:29–42.
47. Chresta, C.M., *et al.* (2010). AZD8055 is a potent, selective, and orally bioavailable ATP-competitive mammalian target of rapamycin kinase inhibitor with in vitro and in vivo antitumor activity. *Cancer Res* 70:288–298.
48. Thoreen, C.C., *et al.* (2009). An ATP-competitive mammalian target of rapamycin inhibitor reveals rapamycin-resistant functions of mTORC1. *J Biol Chem* 284:8023–8032.
49. Yu, K., *et al.* (2009). Biochemical, cellular, and in vivo activity of novel ATP-competitive and selective inhibitors of the mammalian target of rapamycin. *Cancer Res* 69:6232–6240.
50. Maldonado, M., *et al.* (2003). Expression of ICAM-1, TNF-alpha, NF kappa B, and MAP kinase in tubers of the tuberous sclerosis complex. *Neurobiol Dis* 14:279–290.
51. Zeng, L.H., *et al.* (2007). Abnormal glutamate homeostasis and impaired synaptic plasticity and learning in a mouse model of tuberous sclerosis complex. *Neurobiol Dis* 28:184–196.
52. Shah, O.J., Wang, Z., and Hunter, T. (2004). Inappropriate activation of the TSC/Rheb/mTOR/S6K cassette induces IRS1/2 depletion, insulin resistance, and cell survival deficiencies. *Curr Biol* 14:1650–1656.
53. Ghosh, S., *et al.* (2006). Essential role of tuberous sclerosis genes TSC1 and TSC2 in NF-kappaB activation and cell survival. *Cancer Cell* 10:215–226.
54. Wang, T., Lao, U., and Edgar, B.A. (2009). TOR-mediated autophagy regulates cell death in *Drosophila* neurodegenerative disease. *J Cell Biol* 186:703–711.
55. Um, S.H., *et al.* (2004). Absence of S6K1 protects against age- and diet-induced obesity while enhancing insulin sensitivity. *Nature* 431:200–205.
56. Harrington, L.S., *et al.* (2004). The TSC1-2 tumor suppressor controls insulin-PI3K signaling via regulation of IRS proteins. *J Cell Biol* 166:213–223.
57. Franke, T.F., Hornik, C.P., Segev, L., Shostak, G.A., and Sugimoto, C. (2003). PI3K/Akt and apoptosis: size matters. *Oncogene* 22:8983–8998.
58. Parcellier, A., Tintignac, L.A., Zhuravleva, E., and Hemmings, B.A. (2008). PKB and the mitochondria: AKTing on apoptosis. *Cell Signal* 20:21–30.
59. Luo, J.L., Kamata, H., and Karin, M. (2005). The anti-death machinery in IKK/NF-kappaB signaling. *J Clin Immunol* 25:541–550.
60. Busser, J., Geldmacher, D.S., and Herrup, K. (1998). Ectopic cell cycle proteins predict the sites of neuronal cell death in Alzheimer's disease brain. *J Neurosci* 18:2801–2807.
61. Ogawa, O., *et al.* (2003). Ectopic localization of phosphorylated histone H3 in Alzheimer's disease: a mitotic catastrophe? *Acta Neuropathol* 105:524–528.

62. An, W.L., *et al.* (2003). Up-regulation of phosphorylated/activated p70 S6 kinase and its relationship to neurofibrillary pathology in Alzheimer's disease. *Am J Pathol* 163:591–607.
63. Li, X., An, W.L., Alafuzoff, I., Soininen, H., Winblad, B., and Pei, J.J. (2004). Phosphorylated eukaryotic translation factor 4E is elevated in Alzheimer brain. *NeuroReport* 15:2237–2240.
64. Khurana, V., Lu, Y., Steinhilb, M.L., Oldham, S., Shulman, J.M., and Feany, M.B. (2006). TOR-mediated cell-cycle activation causes neurodegeneration in a *Drosophila* tauopathy model. *Curr Biol* 16:230–241.
65. Schieke, S.M., *et al.* (2006). The mammalian target of rapamycin (mTOR) pathway regulates mitochondrial oxygen consumption and oxidative capacity. *J Biol Chem* 281:27643–27652.
66. Cunningham, J.T., Rodgers, J.T., Arlow, D.H., Vazquez, F., Mootha, V.K., and Puigserver, P. (2007). mTOR controls mitochondrial oxidative function through a YY1-PGC-1alpha transcriptional complex. *Nature* 450:736–740.
67. Klionsky, D.J. (2007). Autophagy: from phenomenology to molecular understanding in less than a decade. *Nat Rev Mol Cell Biol* 8:931–937.
68. Berry, D.L., and Baehrecke, E.H. (2007). Growth arrest and autophagy are required for salivary gland cell degradation in *Drosophila*. *Cell* 131:1137–1148.
69. Scott, R.C., Juhasz, G., and Neufeld, T.P. (2007). Direct induction of autophagy by Atg1 inhibits cell growth and induces apoptotic cell death. *Curr Biol* 17:1–11.
70. Hara, T., *et al.* (2006). Suppression of basal autophagy in neural cells causes neurodegenerative disease in mice. *Nature* 441:885–889.
71. Juhasz, G., Erdi, B., Sass, M., and Neufeld, T.P. (2007). Atg7-dependent autophagy promotes neuronal health, stress tolerance, and longevity but is dispensable for metamorphosis in *Drosophila*. *Genes Dev* 21:3061–3066.
72. Kamada, Y., Funakoshi, T., Shintani, T., Nagano, K., Ohsumi, M., and Ohsumi, Y. (2000). Tor-mediated induction of autophagy via an Apg1 protein kinase complex. *J Cell Biol* 150:1507–1513.
73. Ganley, I.G., Lam du, H., Wang, J., Ding, X., Chen, S., and Jiang, X. (2009). ULK1. ATG13.FIP200 complex mediates mTOR signaling and is essential for autophagy. *J Biol Chem* 284:12297–12305.
74. Hosokawa, N., *et al.* (2009). Nutrient-dependent mTORC1 association with the ULK1-Atg13-FIP200 complex required for autophagy. *Mol Biol Cell* 20:1981–1991.
75. Jung, C.H., *et al.* (2009). ULK-Atg13-FIP200 complexes mediate mTOR signaling to the autophagy machinery. *Mol Biol Cell* 20:1992–2003.
76. Wang, Z., Wilson, W.A., Fujino, M.A., and Roach, P.J. (2001). Antagonistic controls of autophagy and glycogen accumulation by Snf1p, the yeast homolog of AMP-activated protein kinase, and the cyclin-dependent kinase Pho85p. *Mol Cell Biol* 21:5742–5752.
77. Budovskaya, Y.V., Stephan, J.S., Deminoff, S.J., and Herman, P.K. (2005). An evolutionary proteomics approach identifies substrates of the cAMP-dependent protein kinase. *Proc Natl Acad Sci USA* 102:13933–13938.
78. Lee, S.J., and Montell, C. (2004). Suppression of constant-light-induced blindness but not retinal degeneration by inhibition of the rhodopsin degradation pathway. *Curr Biol* 14:2076–2085.
79. Ravikumar, B., *et al.* (2004). Inhibition of mTOR induces autophagy and reduces toxicity of polyglutamine expansions in fly and mouse models of Huntington disease. *Nat Genet* 36:585–595.
80. Malagelada, C., Ryu, E.J., Biswas, S.C., Jackson-Lewis, V., and Greene, L.A. (2006). RTP801 is elevated in Parkinson brain substantia nigral neurons and mediates death in

cellular models of Parkinson's disease by a mechanism involving mammalian target of rapamycin inactivation. *J Neurosci* 26:9996–10005.
81. Shibata, M., *et al.* (2006). Regulation of intracellular accumulation of mutant Huntingtin by Beclin 1. *J Biol Chem* 281:14474–14485.
82. Berger, Z., *et al.* (2006). Rapamycin alleviates toxicity of different aggregate-prone proteins. *Hum Mol Genet* 15:433–442.
83. Malagelada, C., Jin, Z.H., Jackson-Lewis, V., Przedborski, S., and Greene, L.A. (2010). Rapamycin protects against neuron death in in vitro and in vivo models of Parkinson's disease. *J Neurosci* 30:1166–1175.
84. Ravikumar, B., Duden, R., and Rubinsztein, D.C. (2002). Aggregate-prone proteins with polyglutamine and polyalanine expansions are degraded by autophagy. *Hum Mol Genet* 11:1107–1117.
85. Webb, J.L., Ravikumar, B., Atkins, J., Skepper, J.N., and Rubinsztein, D.C. (2003). Alpha-Synuclein is degraded by both autophagy and the proteasome. *J Biol Chem* 278:25009–25013.
86. Spencer, B., *et al.* (2009). Beclin 1 gene transfer activates autophagy and ameliorates the neurodegenerative pathology in alpha-synuclein models of Parkinson's and Lewy body diseases. *J Neurosci* 29:13578–13588.
87. Pickford, F., *et al.* (2008). The autophagy-related protein beclin 1 shows reduced expression in early Alzheimer disease and regulates amyloid beta accumulation in mice. *J Clin Invest* 118:2190–2199.
88. Sarkar, S., *et al.* (2005). Lithium induces autophagy by inhibiting inositol monophosphatase. *J Cell Biol* 170:1101–1111.
89. Williams, A., *et al.* (2008). Novel targets for Huntington's disease in an mTOR-independent autophagy pathway. *Nat Chem Biol* 4:295–305.
90. Ravikumar, B., Berger, Z., Vacher, C., O'Kane, C.J., and Rubinsztein, D.C. (2006). Rapamycin pre-treatment protects against apoptosis. *Hum Mol Genet* 15:1209–1216.
91. Carloni, S., Buonocore, G., and Balduini, W. (2008). Protective role of autophagy in neonatal hypoxia-ischemia induced brain injury. *Neurobiol Dis* 32:329–339.
92. Hamacher-Brady, A., Brady, N.R., and Gottlieb, R.A. (2006). Enhancing macroautophagy protects against ischemia/reperfusion injury in cardiac myocytes. *J Biol Chem* 281:29776–29787.
93. Nelson, B., *et al.* (2005). Isolation of gene sets affected specifically by polyglutamine expression: implication of the TOR signaling pathway in neurodegeneration. *Cell Death Differ* 12:1115–1123.
94. Baehrecke, E.H. (2005). Autophagy: dual roles in life and death? *Nat Rev Mol Cell Biol* 6:505–510.
95. Samara, C., Syntichaki, P., and Tavernarakis, N. (2008). Autophagy is required for necrotic cell death in *Caenorhabditis elegans*. *Cell Death Differ* 15:105–112.
96. Toth, M.L., Simon, P., Kovacs, A.L., and Vellai, T. (2007). Influence of autophagy genes on ion-channel-dependent neuronal degeneration in *Caenorhabditis elegans*. *J Cell Sci* 120:1134–1141.
97. Yu, W.H., *et al.* (2005). Macroautophagy—a novel beta-amyloid peptide-generating pathway activated in Alzheimer's disease. *J Cell Biol* 171:87–98.
98. Ling, D., Song, H.J., Garza, D., Neufeld, T.P., and Salvaterra, P.M. (2009). Abeta42-induced neurodegeneration via an age-dependent autophagic-lysosomal injury in *Drosophila*. *PLoS ONE* 4:e4201.
99. Hung, S.Y., Huang, W.P., Liou, H.C., and Fu, W.M. (2009). Autophagy protects neuron from Abeta-induced cytotoxicity. *Autophagy* 5:502–510.

100. Pandey, U.B., *et al.* (2007). HDAC6 rescues neurodegeneration and provides an essential link between autophagy and the UPS. *Nature* 447:859–863.
101. Sarkar, S., Krishna, G., Imarisio, S., Saiki, S., O'Kane, C.J., and Rubinsztein, D.C. (2008). A rational mechanism for combination treatment of Huntington's disease using lithium and rapamycin. *Hum Mol Genet* 17:170–178.
102. Renna, M., Jimenez-Sanchez, M., Sarkar, S., and Rubinsztein, D.C. (2010). Chemical inducers of autophagy that enhance the clearance of mutant proteins in neurodegenerative diseases. *J Biol Chem*.
103. Vicencio, J.M., *et al.* (2009). The inositol 1, 4, 5-trisphosphate receptor regulates autophagy through its interaction with Beclin 1. *Cell Death Differ* 16:1006–1017.
104. Fornai, F., *et al.* (2008). Lithium delays progression of amyotrophic lateral sclerosis. *Proc Natl Acad Sci USA* 105:2052–2057.
105. Zhang, L., *et al.* (2007). Small molecule regulators of autophagy identified by an image-based high-throughput screen. *Proc Natl Acad Sci USA* 104:19023–19028.
106. Komatsu, M., *et al.* (2006). Loss of autophagy in the central nervous system causes neurodegeneration in mice. *Nature* 441:880–884.
107. Zheng, Y.T., Shahnazari, S., Brech, A., Lamark, T., Johansen, T., and Brumell, J.H. (2009). The adaptor protein p62/SQSTM1 targets invading bacteria to the autophagy pathway. *J Immunol* 183:5909–5916.
108. Okamoto, K., Kondo-Okamoto, N., and Ohsumi, Y. (2009). Mitochondria-anchored receptor Atg32 mediates degradation of mitochondria via selective autophagy. *Dev Cell* 17:87–97.
109. Kanki, T., Wang, K., Cao, Y., Baba, M., and Klionsky, D.J. (2009). Atg32 is a mitochondrial protein that confers selectivity during mitophagy. *Dev Cell* 17:98–109.

12

Elucidating TOR Signaling in Chlamydomonas reinhardtii

MARÍA ESTHER PÉREZ-PÉREZ • JOSÉ L. CRESPO

Instituto de Bioquímica Vegetal y Fotosíntesis
Consejo Superior de Investigaciones Científicas (CSIC)-Universidad de Sevilla
Sevilla, Spain

I. Abstract

Cell growth is tightly coupled to favorable nutrient conditions. The target of rapamycin (TOR) kinase transmits nutritional and environmental cues to the cellular growth machinery. TOR functions in two distinct multiprotein complexes termed TORC1 and TORC2, and core components of these complexes are, like TOR, evolutionarily conserved. The TOR pathway controls cell growth by promoting anabolic processes, including protein synthesis and ribosome biogenesis, and inhibiting catabolic processes such as autophagy. The TOR kinase has been identified in plants and microalgae, indicating that this signaling pathway is conserved in photosynthetic eukaryotes. The central role that these organisms, particularly microalgae, play in global biomass production, together with their high biotechnological potential in biofuel generation, has recently boosted basic research focused on cell growth improvement. Here, we review recent advance in the study of the growth-controller TOR pathway from the unicellular green alga *Chlamydomonas reinhardtii*.

THE ENZYMES, Vol. XXVIII

ISSN NO: 1874-6047
DOI: 10.1016/S1874-6047(10)28012-5

II. Introduction

Since its identification in the budding yeast *S. cerevisiae* [1], the study of the TOR signaling pathway has been extended to other organisms, including fission yeasts, pathogenic fungi, protozoa, and insects, although the most significant advance in TOR research has taken place in mammals, mostly due to the participation of the TOR pathway and TOR-regulatory proteins in human diseases [2–5]. In both lower and higher eukaryotes, the TOR kinase plays an essential and conserved role in integrating nutritional and energy inputs for the proper regulation of cell growth. TOR controls cell growth by promoting a number of anabolic processes, including translation, transcription, and ribosome biogenesis, and by antagonizing catabolic processes such as mRNA degradation and autophagy [6]. Genetic and biochemical studies in yeast demonstrated that TOR regulates these and other important cellular processes by two separate signaling pathways defined by TORC1 and TORC2 protein complexes [7–10]. The basic architecture of TORC1 and TORC2 appears to be structurally and functionally conserved in most organisms analyzed so far, which highlights the importance of these signaling complexes in cell growth control [11].

The TOR kinase has been described in plants and algae, indicating that this signaling pathway is also conserved in photosynthetic organisms [12, 13]. In both systems, TOR has been involved in cell growth control, since disruption of the *AtTOR* gene in Arabidopsis is lethal [13], and rapamycin-mediated inactivation of CrTOR in Chlamydomonas cells results in strong inhibition of cell growth [12]. Our current knowledge about the TOR pathway in photosynthetic organisms is still limited, although recent findings have contributed in understanding how TOR controls cell growth in plants and algae. Components of TORC1 signaling have been identified in these systems [14–17], and the study of TOR inactivation by gene silencing or rapamycin treatment has revealed that TOR promotes protein synthesis and inhibits autophagy to control cell growth [17–20]. In yeasts and metazoans, TOR controls these and other cellular processes in response to nitrogen (amino acids) and carbon availability [6]. Since plants and algae are able to photo-autotrophically obtain certain nutrients such as carbohydrates which are essential for the cell, our challenge is to determine whether TOR responds to these nutrients to control photosynthetic cell growth. Unlike plants, growth of the unicellular green alga *Chlamydomonas reinhardtii* is sensitive to rapamycin [12]. This particular feature of Chlamydomonas, together with the easy manipulation of this organism, its simple life cycle, and a growing array of genetic and molecular tools, have positioned this green alga as a useful model system to investigate the TOR pathway in photosynthetic organisms.

III. Inhibition of TOR Signaling by Rapamycin in Chlamydomonas

A. Rapamycin Action and the CrFKBP12 Protein

The macrolide antibiotic rapamycin is produced by the bacterium *Streptomyces hygroscopicus* and was originally identified as a potent antifungal agent with immunosuppressive activity and anticancer properties [21, 22]. The mechanism of action of rapamycin is unique and both receptor and functional target were identified in *S. cerevisiae* [1]. Rapamycin first binds the FKBP12 immunophilin and this complex specifically inhibits TOR. However, the molecular mechanism by which rapamycin negatively acts on TOR is still unclear.

Although rapamycin potently inhibits cell growth in most eukaryotes, the vegetative growth of Arabidopsis and other plants such as *Oryza sativa*, *Nicotiana tabacum*, or *Brassica napus* is not sensitive to rapamycin [13], likely due to the inability of plant FKBP12 to bind this drug [23]. Unlike plants, growth of the unicellular green alga *Chlamydomonas reinhardtii* is sensitive to rapamycin, indicating the presence of a rapamycin-sensitive TOR signaling cascade in this photosynthetic organism [12]. Treatment of Chlamydomonas cells with rapamycin causes a significant increase in cell size [12, 19] (Figure 12.3), which results in a cell cycle arrest (J. L. Crespo, unpublished data). Similar to other eukaryotes, the negative effects of rapamycin on Chlamydomonas cell growth are mediated by FKBP12. The *FKBP12* gene is highly conserved in the Chlamydomonas genome, and an evolutionary analysis revealed that Chlamydomonas FKBP12 (CrFKBP12) does not group with plant homologs and it is closer to higher eukaryotes than other unicellular organisms like fungi. Yeast complementation assays have demonstrated that CrFKBP12 can functionally replace yeast FKBP12 in mediating rapamycin action and demonstrated that unlike plant FKBP12, CrFKBP12 binds rapamycin *in vivo* [12].

Compared to other eukaryotes, the concentration of rapamycin required to inhibit Chlamydomonas cell growth is higher, likely due to a minor affinity of CrFKBP12 to rapamycin. A detailed study of the CrFKBP12 protein has demonstrated that this protein exhibits an affinity to rapamycin lower than its yeast or mammalian homologues [12]. Crystal structure of the human FKBP12-rapamycin complex revealed that rapamycin binds FKBP12 through hydrophobic contacts and hydrogen bonds with key residues in a hydrophobic pocket of the protein [24]. CrFKBP12 has a conserved drug-binding pocket. Except for Gln at position 53, all residues that are predicted to establish hydrogen bonds with rapamycin (Asp37, Glu54, Ile56, and Tyr82) are conserved in CrFKBP12. The absence of this Gln residue in the drug-binding pocket might cause reduced affinity of CrFKBP12 to rapamycin. This hypothesis was confirmed by the analysis of two different CrFKBP12 mutants with more similarity to yeast and human FKBP12s in the region adjacent to Gln53 [12]. Yeast complementation studies demonstrated that mutations introduced in CrFKBP12 significantly increased the affinity to rapamycin and strongly suggested that the Gln residue that is not conserved in CrFKBP12 plays an important role in rapamycin binding. Like in yeasts, expression of these CrFKBP12 mutants in Chlamydomonas cells resulted in increased sensitivity to rapamycin, demonstrating that the reduced sensitivity to this drug of Chlamydomonas is due to a reduced affinity of CrFKBP12 to rapamycin.

A Chlamydomonas mutant that exhibited complete resistance to rapamycin has been identified in a screen for mutants resistant to this drug. A detailed analysis of this mutant, named *rap2*, revealed a DNA reorganization in the *CrFKBP12* locus. As a consequence of this genomic reorganization, the expression of *CrFKBP12* gene and the CrFKBP12 protein cannot be detected in *rap2* mutant cells [12]. Expression of CrFKBP12 in *rap2* cells fully restored the rapamycin sensitivity of the *rap2* mutant, which demonstrated that rapamycin resistance was due to the absence of the CrFKBP12 protein. The characterization of the *rap2* mutant also demonstrated that the CrFKBP12 immunophilin is not required for cell viability, since growth of *rap2* mutant and wild-type cells is similar under optimal or stress conditions [12].

B. Interaction of the Rapamycin-CrFKBP12 Complex with CrTOR

Inhibition of Chlamydomonas cell growth by rapamycin strongly suggested that the rapamycin-CrFKBP12 complex must interact with CrTOR, since rapamycin action requires formation of a FKBP12-rapamycin-TOR ternary complex. The analysis of the amino acid sequence of the FRB domain of CrTOR shows that critical residues that establish hydrophobic interactions with FKBP12-rapamycin are conserved. Pull-down assays performed with the FRB domain of CrTOR and the FKBP12 protein from total extracts of Chlamydomonas have demonstrated that rapamycin-CrFKBP12 targets CrTOR [12]. The FRB domain of CrTOR was able to interact with FKBP12 in the presence of rapamycin, whereas no binding was detected in the absence of the drug. Therefore, unlike its plant homologue [13], CrFKBP12 mediates rapamycin action and interacts with the FRB domain of the Chlamydomonas TOR kinase in the presence of rapamycin.

IV. TOR Complexes

A. Conserved TORC Components in Nonphotosynthetic Eukaryotes

The TOR kinase resides in two distinct multiprotein complexes, termed TORC1 and TORC2, which modulate two major signaling branches in eukaryotic cells. TOR complexes were initially identified in yeasts [9, 10] and subsequently in mammals [25–29]. More recently, TOR complexes have been reported in other eukaryotes [30–32]. TORC1 and TORC2 differ in protein composition, although some proteins are common in both complexes. In addition to the TOR kinase, the LST8 protein is present in most TORC1 and TORC2 complexes described so far [9, 10, 28, 31, 32]. Together with TOR and LST8, TORC1 is composed of the KOG1/raptor protein, whereas the core of TORC2 also contains AVO1/hSIN1 and AVO3/rictor. The similar composition of TORCs in widely divergent kingdoms such as metazoans and fungi suggests that these complexes are broadly conserved through evolution. In most eukaryotes, TORC1 mediates the rapamycin-sensitive signaling branch that regulates translation, ribosome biogenesis, autophagy, and metabolic pathways, whereas TORC2-controlled processes are insensitive to rapamycin [2, 11]. However, a recent study of TOR signaling in the protozoa *Trypanosoma brucei* revealed that despite protein composition and functions of TORC1 and TORC2 being conserved in this primitive eukaryote, rapamycin acts on TORC2 and has no effect on TORC1 [30]. Whether this different mode of action of rapamycin is exclusive of trypanosomes or is conserved in other organisms remains to be investigated.

B. TOR Complexes in Plants and Algae

The current knowledge about control of cell growth by TOR in photosynthetic organisms is limited compared to other eukaryotes. The TOR kinase has been identified in the model photosynthetic systems *Arabidopsis thaliana* and *Chlamydomonas reinhardtii* [12, 13]. The AtTOR and CrTOR proteins contain domains highly conserved in TOR kinases, including the FKBP12-rapamycin-binding (FRB) domain, the C-terminal kinase domain, N-terminal HEAT repeats, and the FAT and FATC domains characteristic of phosphatidylinositol 3-kinase related kinases. A comparative study of TOR proteins from different eukaryotes indicates that the closest homologues to CrTOR are found in plants and green algae. A single *TOR* gene seems to be conserved in the unicellular red alga *Cyanidioschyzon merolae*, which is considered to be one of the most primitive alga that diverged near the root of the red lineage and occupies a basal evolutionary position within eukaryotes [33]. Interestingly, the *C. merolae* TOR kinase seems to be phylogenetically distant from other photosynthetic organisms (Figure 12.1).

1. TORC1

The presence of CrTOR in high-molecular-mass complexes has been demonstrated in Chlamydomonas, which suggests the conservation of TOR complexes in photosynthetic organisms [16]. Supporting this model, components of TORC1 are conserved in plants and algae. Two genes are coding for KOG1/raptor homologues in the Arabidopsis genome, *AtRaptor1* and *AtRaptor2* [14, 15, 17, 34], while a single *KOG1/raptor* gene appears to be present in the nuclear genome of Chlamydomonas (J. L. Crespo, unpublished observation). The AtRaptor1 and AtRaptor2 proteins from Arabidopsis are 76% identical and, like other raptor homologues, contain HEAT repeats, WD40 motifs, and the raptor N-terminal conserved/caspase motif. Direct interaction of AtRaptor1 with the heat repeats of AtTOR has been demonstrated, indicating that at least TORC1 might be conserved in plants [17]. Besides KOG1/raptor and TOR proteins, LST8 homologues have been identified in photosynthetic eukaryotes. Chlamydomonas contains a functional *LST8* gene, *CrLST8*, whose product has high identity to yeast and mammalian LST8 proteins [16]. CrLST8 forms part of a rapamycin-sensitive TOR complex, since it can be copurified with CrTOR and CrFKBP12 in the presence of rapamycin [16]. Yeast and mammalian LST8s bind to the kinase domain of TOR, and this interaction is required for full catalytic activity of TOR [28, 35]. CrLST8 is able to bind to purified kinase domain of CrTOR, suggesting that CrLST8 may perform a similar function in Chlamydomonas. LST8 functions must be conserved in photosynthetic organisms, since yeast complementation assays have demonstrated that CrLST8 is able to functionally and structurally replace endogenous yeast LST8 [16]. Physical interaction of CrLST8 with CrTOR and possibly with other proteins may play important roles in regulating the CrTOR pathway, since CrLST8 consists entirely of seven WD-40 domains that mediate protein–protein interactions. Three critical residues for CrLST8 function have been identified in this protein [16]. Mutations at positions Asp106, Thr228, or Ser230 of CrLST8 result in the complete inactivation of the protein and a failure to interact with yeast TOR2. These findings directly link cell growth to the interaction of LST8 and TOR proteins. In plants, two putative LST8 homologues can be identified in the Arabidopsis genome (*AtLST8.1*, At3g18140; and *AtLST8.2*, At2g22040), while the rice genome appears to contain a single *LST8* gene (OsLST8, Os03g47780). Whether plant LST8 proteins are functionally conserved or may establish direct interactions with TOR is at present unknown.

A

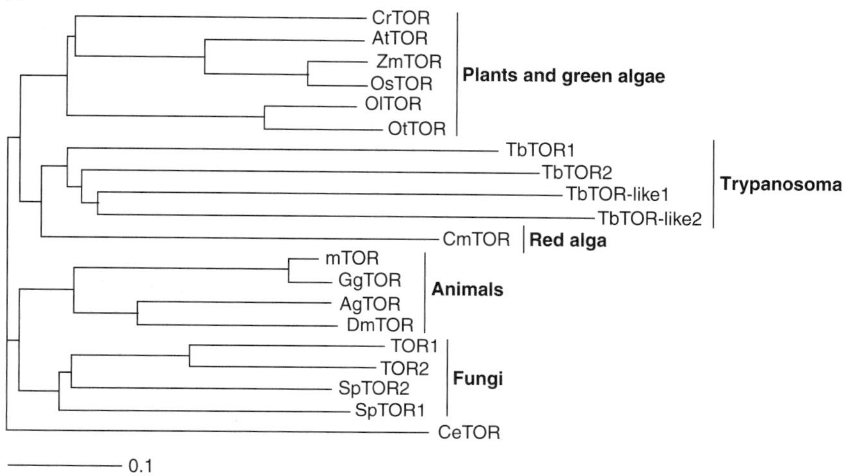

B

S. cerevisiae	*H. sapiens*	*A. thaliana*	*C. reinhardtii*	*O. tauri*	*O. lucimarinus*	*C. merolae*
TORC1						
TOR1/2	mTOR	AtTOR	CrTOR	OtTOR	OlTOR	CmTOR
KOG1	Raptor	AtRaptor1A AtRaptor1B	CrKOG1	OtKOG1	OlKOG1	CmKOG1
LST8	mLST8	AtLST8.1 AtLST8.2	CrLST8	OtLST8	OlLST8	CmLST8
TCO89	–	–	–	–	–	–
TORC2						
TOR2	mTOR	AtTOR	CrTOR	OtTOR	OlTOR	CmTOR
AVO1	hSIN1	–	–	–	–	–
AVO2	–	–	–	–	–	–
AVO3	Rictor	–	–	–	–	–
BIT61	–	–	–	–	–	–

Fig. 12.1. Phylogenetic analysis of TOR and components of TORC1 and TORC2 in various eukaryotes. (A) Neighbor-Joining tree of TOR proteins from representative organisms. CrTOR, *Chlamydomonas reinhardtii*; AtTOR, *Arabidopsis thaliana*; ZmTOR, *Zea mays*; OsTOR, *Oryza sativa*; OlTOR, *Ostreococcus lucimarinus*; OtTOR, *Ostreococcus tauri*; TbTOR, *Trypanosoma brucei*; CmTOR, *Cyanidioschyzon merolae*; GgTOR, *Gallus gallus*; AgTOR, *Anopheles gambiae*; DmTOR, *Drosophila melanogaster*; SpTOR, *Schyzosaccharomyces pombe*; CeTOR, *Caenorhabditis elegans*. (B) Evolutionarily conserved protein components of TORC1 and TORC2 from various eukaryotes. *O. tauri* genome web: http://genome.jgi-psf.org/Ostta4/Ostta4.home.html; *O. lucimarinus* genome web: http://genome.jgi-psf.org/Ost9901_3/Ost9901_3.home.html; *C. merolae* genome web: http://merolae.biol.s.u-tokyo.ac.jp/. -, no homologue found.

Genome sequencing of other green alga species, such as *Ostreococcus tauri* and *Ostreococcus lucimarinus* [36], or the red alga *Cyanidioschyzon merolae* [37], the smallest free-living eukaryotes known to date, has revealed that the core TORC1 proteins TOR, LST8, and KOG1/raptor are widely conserved in these unicellular organisms (Figure 12.1).

2. *TORC2*

Unlike TORC1 components, no obvious homologues exist for the TORC2-specific proteins AVO1/hSIN1 and AVO3/rictor in plants and algae, raising the question of whether TORC2 is structurally conserved in photosynthetic organisms. Regardless of the absence of TORC2 components, it is likely that plants and algae functionally maintain a TORC2 complex, although the proteins that constitute this putative complex may substantially differ from their yeast and mammalian counterparts.

C. Cellular Localization of TOR Complexes

Biochemical fractionation and immunofluorescence studies of Chlamydomonas cells have revealed that CrTOR and CrLST8 are peripherally associated to internal membranes, suggesting that TOR complexes may localize on these membranous sites [16]. Mammalian and yeast TORs have been detected in different cellular compartments, including plasma, vacuolar, and ER membranes, as well as in the nucleus, the mitochondria and the Golgi apparatus, and despite significant discrepancies, these studies indicate that TOR complexes are primarily membrane associated [10, 38–45]. In Chlamydomonas cells, CrTOR and CrLST8 have a common localization at the peribasal body region, which is enriched in membranes [16]. However, CrTOR is also detected in discrete punctae that are abundant in regions adjacent to the plasma membrane, while CrLST8 is more abundant around the nucleus (Figure 12.2). This cellular distribution of CrTOR and CrLST8 is in part coincident with the staining pattern of the ER marker BiP, which also accumulates in the peribasal body region, indicating that TOR complexes may associate with membranes from the ER system in Chlamydomonas, although additional localization at other cellular compartments might be possible [16]. Like in other evolutionarily distant eukaryotes [30, 38, 40, 41], the association of TORC proteins to ER membranes in Chlamydomonas emphasizes the important role that this cellular compartment may play in TOR signaling.

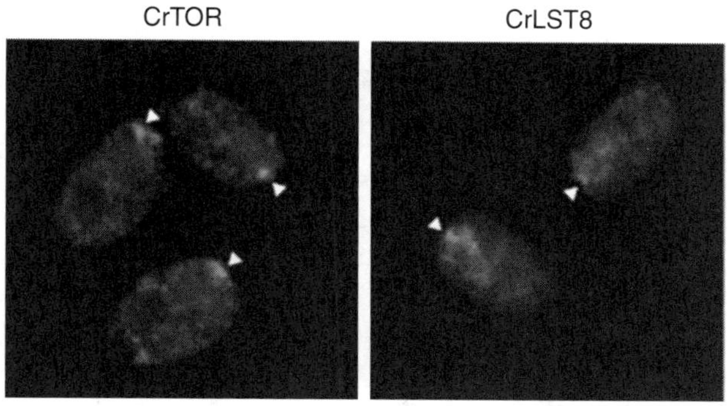

FIG. 12.2. Cellular localization of CrTOR and CrLST8 proteins in Chlamydomonas. CrTOR localizes to discrete spots that are more abundant in the proximity of the plasma membrane and the basal bodies. CrLST8 is concentrated around the nucleus and, like CrTOR, the basal body. CrTOR and CrLST8 are shown in green, and DNA in blue. Arrowheads indicate the apical end of cells. See Ref. [16] for details about CrTOR and CrLST8 localization.

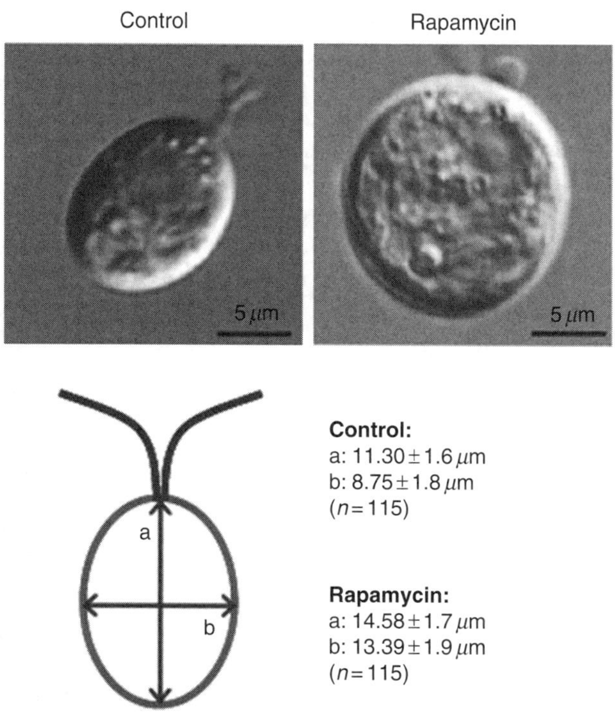

FIG. 12.3. Effect of TOR inhibition by rapamycin on Chlamydomonas cell size. Images of Chlamydomonas cells showing an increase in cell size caused by the rapamycin-mediated inhibition of TOR signaling. Cells were grown in rich (acetate) medium and were treated or not with 500 nM rapamycin for 8 h. See Refs. [12, 19] for further details about rapamycin effects on Chlamydomonas.

V. Control of Autophagy by TOR

A. TORC1 Inhibits Autophagy

Autophagy is a catabolic membrane-trafficking process whereby cells recycle cytosolic proteins and organelles. During this degradative process, a large number of cytosolic components are nonselectively enclosed within a double-membrane structure (autophagosome) and delivered to the vacuole for degradation to recycle-needed nutrients or degrade toxic components [46, 47]. Autophagy is conserved among yeast, animal, and plant cells, indicating that many of the structural and regulatory components of this cellular process must be conserved. Accordingly, genomic approaches have revealed that a significant number of autophagy-related genes (*ATG*) that participate in the activation and regulation of this process are evolutionarily conserved in photosynthetic and nonphotosynthetic eukaryotes [48, 49].

Autophagy is triggered when the cell perceives a reduction in the availability of nutrients. TORC1 has been identified as an essential component in the molecular mechanism that transmits this starvation signal to the autophagic machinery [50]. TORC1 inhibits autophagy by negatively regulating the association between the Ser/Thr kinase ATG1 with ATG13, a regulatory subunit of ATG1 [51]. Under nutrient-rich conditions, TORC1 promotes phosphorylation of ATG1 and ATG13, and when phosphorylated, these proteins cannot interact and consequently, ATG1 kinase remains inactive. TORC1 inhibition causes rapid dephosphorylation of ATG1 and ATG13, which results in ATG1 activation and autophagy induction. The molecular mechanism of autophagy regulation by TORC1 signaling appears to be conserved from yeasts to animals [52–54].

Homologues to yeast *ATG* genes have been identified in plants such as Arabidopsis or maize [49]. These homologues include autophagy regulatory components such as ATG1, ATG13, and the TOR kinase. However, at present it is unknown whether TOR controls autophagy in plants or ATG1 and ATG13 plant homologues are essential for autophagy.

B. Monitoring Autophagy: The ATG8 Protein

The development of specific markers or assays of autophagy has significantly increased our knowledge about the process of autophagy. The ATG8 protein is one of the most effective markers used to trace autophagy in different organisms. The highly conserved ATG8 protein associates with the autophagosome membrane and remains associated with the mature autophagosome until this specialized vesicle fuses with the vacuole/lysosome [55]. A conserved C-terminal Gly residue of ATG8 covalently binds to phosphatidylethanolamine, and this modified form tightly associates with autophagosomes. In general, activation of autophagy in the cell can be monitored through the detection of lipidated ATG8 forms or by a modification in the cellular localization of this protein. In plants, the use of ATG8 as an autophagy marker has revealed that this process is active at a basal level under normal growth conditions and is induced upon nutrient limitation or in response to oxidative stress or pathogens [56–61]. However, the presence of multiple ATG8 isoforms in plant genomes reflects a greater complexity of autophagy in these organisms compared to yeast and animal cells [61–63].

C. TOR Inhibits Autophagy in Chlamydomonas

A detailed study of Chlamydomonas cells treated with rapamycin led to the initial hypothesis that TOR controls autophagy in photosynthetic eukaryotes. Treatment of Chlamydomonas cells with rapamycin resulted in a pronounced increase of vacuole size that resembled autophagic-like processes [12]. This rapamycin-induced vacuolization is mediated by formation of a rapamycin-CrFKBP12 complex, since no vacuolization was detected in a Chlamydomonas mutant lacking CrFKBP12 upon rapamycin treatment. An analysis of the Chlamydomonas genome revealed that putative homologues of yeast autophagy genes, including the *CrATG8* gene, are highly conserved in this green alga [49]. Together, these findings strongly suggested that autophagy is conserved in Chlamydomonas cells and TOR might play an important role in the control of this process. The recent establishment of CrATG8 as a specific autophagy marker in Chlamydomonas has demonstrated that indeed this catabolic process occurs in green alga and is inhibited by a rapamycin-sensitive TOR signaling pathway [19, 64]. Treatment of Chlamydomonas cells with rapamycin results in increased abundance of CrATG8 and the detection of modified forms of this protein. CrATG8 localization was also altered in response to rapamycin treatment (Figure 12.4). Under optimal growth conditions, the CrATG8 signal is restricted to discrete punctate structures that might correspond to the phagophore assembly site (PAS) described in yeast [65]. However, autophagy induction by rapamycin treatment drastically modified the cellular localization of CrATG8 and larger spots distributed throughout the cytoplasm. Based on these findings, TOR may play an important role in signaling nutritional inputs to proteins involved in autophagy activation and autophagosome formation. Although autophagy and the TOR kinase are conserved in plants, a role of this signaling pathway in autophagy regulation has not been demonstrated in higher photosynthetic eukaryotes. It remains to be determined whether, as reported in other systems, inhibition of TOR by rapamycin triggers autophagy through the activation of the ATG1 kinase, which is conserved in Chlamydomonas.

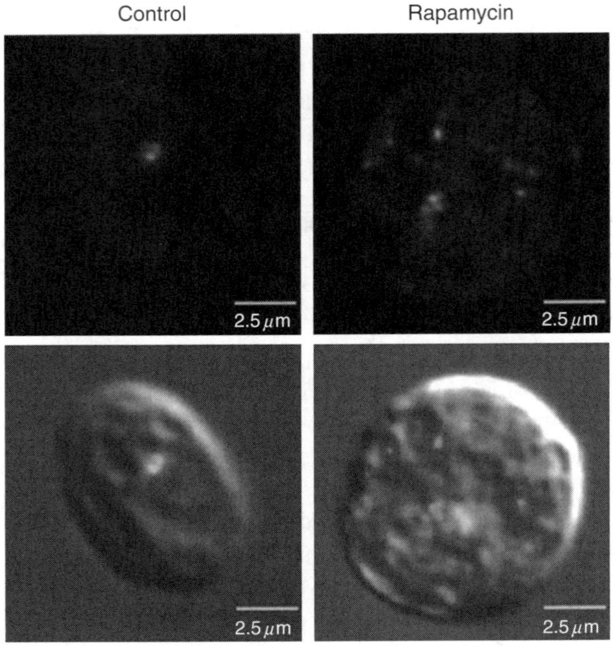

FIG. 12.4. Activation of autophagy by rapamycin in Chlamydomonas. The cellular distribution of the CrATG8 protein can be used to monitor autophagy in Chlamydomonas. In exponentially growing cells, CrATG8 localizes to punctate structures, preferentially in a single spot. Rapamycin treatment results in a pronounced detection of CrATG8 in more punctate structures and an increase in cell size. Images show immunofluorescence localization of CrATG8 in untreated and rapamycin-treated cells. See Ref. [19] for details.

D. INDUCTION OF AUTOPHAGY BY STRESS IN CHLAMYDOMONAS

Autophagy is induced when cells are subjected to nutrient limitation or under certain stress conditions. The study of the CrATG8 protein in Chlamydomonas has revealed that autophagy is induced by nutrient starvation [19, 64]. Nitrogen or carbon limitation resulted in a significant increase in CrATG8 protein abundance as well as the detection of modified forms of CrATG8, both landmarks of autophagy activation. Moreover, the cellular localization of CrATG8 was altered upon nutrient starvation, and similar to rapamycin treatment, CrATG8 signal was detected in numerous punctate structures all over the cell. In addition to nutrient depletion, autophagy is induced in Chlamydomonas when cells enter into the stationary growth phase or in response to the induction of oxidative stress or the accumulation of unfolded proteins in the ER (ER stress) [19, 64]. All these adverse conditions resulted in increased CrATG8 abundance, the detection of modified CrATG8, and a drastic change in the cellular localization of this protein. Interestingly, ER stress causes a massive accumulation of CrATG8 at this cellular compartment, indicating that autophagy is important for the mobilization and degradation of unfolded proteins in photosynthetic organisms. Given the central role of the TOR pathway in perceiving and transmitting nutrient and stress signals to the growth machinery in eukaryotes, it is likely that the CrTOR kinase may participate in the control of autophagy in response to stress in Chlamydomonas.

VI. Perspectives

The study of TOR in photosynthetic organisms is in its infancy. Despite recent advances in plant and alga TOR research, little is known about the molecular mechanisms by which this signaling pathway promotes photosynthetic cell growth. Efforts should concentrate in the identification of components of this signaling cascade that may act upstream or downstream of TOR. TORC1 is conserved in plants and algae, and some TORC1-controlled processes have already been demonstrated to be regulated by TOR in these organisms [17–19]. It remains to be investigated whether other TORC1-conserved readouts such as the control of ribosome biogenesis or the transcriptional regulation of central metabolic pathways mediate TORC1 signaling in photosynthetic eukaryotes. Unlike TORC1, TORC2 components are absent in plants and algae. However, it is at present unknown whether this signaling cascade is functionally conserved in these organisms. Loss-of-function studies of the *TOR* gene in plants or algae may give an answer to this intriguing question. On the other hand, the identification of upstream regulators of TOR will provide valuable information about nutritional or stress inputs to this kinase. Carbon assimilation is critical for cell growth in photosynthetic organisms, and, given the central role that TOR plays in nutrient signaling, a model where TOR integrates signals from carbon metabolism to promote cell growth is feasible. Given the relevance of photosynthetic organisms in global biomass production, a better knowledge of the TOR pathway in cell growth control in plants and algae may potentially redound to benefits in industrial sectors like agriculture or biofuel production.

Acknowledgments

This work was supported by the Spanish Ministry of Science and Innovation (grant no. BFU2009-07368 to J.L.C.)

References

1. Heitman, J., Movva, N.R., and Hall, M.N. (1991). Targets for cell cycle arrest by the immunosuppressant rapamycin in yeast. *Science* 253:905–909.
2. Laplante, M., and Sabatini, D.M. (2009). mTOR signaling at a glance. *J Cell Sci* 122:3589–3594.
3. Plas, D.R., and Thomas, G. (2009). Tubers and tumors: rapamycin therapy for benign and malignant tumors. *Curr Opin Cell Biol* 21:230–236.
4. Polak, P., and Hall, M.N. (2009). mTOR and the control of whole body metabolism. *Curr Opin Cell Biol* 21:209–218.
5. Soulard, A., Cohen, A., and Hall, M.N. (2009). TOR signaling in invertebrates. *Curr Opin Cell Biol* 21:825–836.
6. Crespo, J.L., and Hall, M.N. (2002). Elucidating TOR signaling and rapamycin action: lessons from *Saccharomyces cerevisiae. Microbiol Mol Biol Rev* 66:579–591, table of contents.
7. Helliwell, S.B., Wagner, P., Kunz, J., Deuter-Reinhard, M., Henriquez, R., and Hall, M.N. (1994). TOR1 and TOR2 are structurally and functionally similar but not identical phosphatidylinositol kinase homologues in yeast. *Mol Biol Cell* 5:105–118.
8. Kunz, J., Henriquez, R., Schneider, U., Deuter-Reinhard, M., Movva, N.R., and Hall, M.N. (1993). Target of rapamycin in yeast, TOR2, is an essential phosphatidylinositol kinase homolog required for G1 progression. *Cell* 73:585–596.
9. Loewith, R., *et al.* (2002). Two TOR complexes, only one of which is rapamycin sensitive, have distinct roles in cell growth control. *Mol Cell* 10:457–468.
10. Wedaman, K.P., Reinke, A., Anderson, S., Yates, J., 3rd, McCaffery, J.M., and Powers, T. (2003). Tor kinases are in distinct membrane-associated protein complexes in Saccharomyces cerevisiae. *Mol Biol Cell* 14:1204–1220.
11. Wullschleger, S., Loewith, R., and Hall, M.N. (2006). TOR signaling in growth and metabolism. *Cell* 124:471–484.
12. Crespo, J.L., Diaz-Troya, S., and Florencio, F.J. (2005). Inhibition of target of rapamycin signaling by rapamycin in the unicellular green alga Chlamydomonas reinhardtii. *Plant Physiol* 139:1736–1749.
13. Menand, B., *et al.* (2002). Expression and disruption of the Arabidopsis TOR (target of rapamycin) gene. *Proc Natl Acad Sci USA* 99:6422–6427.
14. Anderson, G.H., Veit, B., and Hanson, M.R. (2005). The Arabidopsis AtRaptor genes are essential for post-embryonic plant growth. *BMC Biol* 3:12.
15. Deprost, D., Truong, H.N., Robaglia, C., and Meyer, C. (2005). An Arabidopsis homolog of RAPTOR/KOG1 is essential for early embryo development. *Biochem Biophys Res Commun* 326:844–850.
16. Diaz-Troya, S., Florencio, F.J., and Crespo, J.L. (2008). Target of rapamycin and LST8 proteins associate with membranes from the endoplasmic reticulum in the unicellular green alga Chlamydomonas reinhardtii. *Eukaryot Cell* 7:212–222.
17. Mahfouz, M.M., Kim, S., Delauney, A.J., and Verma, D.P. (2006). Arabidopsis TARGET OF RAPAMYCIN interacts with RAPTOR, which regulates the activity of S6 kinase in response to osmotic stress signals. *Plant Cell* 18:477–490.
18. Deprost, D., *et al.* (2007). The Arabidopsis TOR kinase links plant growth, yield, stress resistance and mRNA translation. *EMBO Rep* 8:864–870.
19. Perez-Perez, M.E., Florencio, F.J., and Crespo, J.L. (2010). Inhibition of TOR signaling and stress activate autophagy in *Chlamydomonas reinhardtii. Plant Physiol* 152:1874–1888.

20. Sormani, R., *et al.* (2007). Saccharomyces cerevisiae FKBP12 binds Arabidopsis thaliana TOR and its expression in plants leads to rapamycin susceptibility. *BMC Plant Biol* 7:26.
21. Thomson, A.W., Turnquist, H.R., and Raimondi, G. (2009). Immunoregulatory functions of mTOR inhibition. *Nat Rev Immunol* 9:324–337.
22. Vezina, C., Kudelski, A., and Sehgal, S.N. (1975). Rapamycin (AY-22, 989), a new antifungal antibiotic. *I. Taxonomy of the producing streptomycete and isolation of the active principle. J Antibiot (Tokyo)* 28:721–726.
23. Xu, Q., Liang, S., Kudla, J., and Luan, S. (1998). Molecular characterization of a plant FKBP12 that does not mediate action of FK506 and rapamycin. *Plant J* 15:511–519.
24. Choi, J., Chen, J., Schreiber, S.L., and Clardy, J. (1996). Structure of the FKBP12-rapamycin complex interacting with the binding domain of human FRAP. *Science* 273:239–242.
25. Hara, K., *et al.* (2002). Raptor, a binding partner of target of rapamycin (TOR), mediates TOR action. *Cell* 110:177–189.
26. Jacinto, E., *et al.* (2004). Mammalian TOR complex 2 controls the actin cytoskeleton and is rapamycin insensitive. *Nat Cell Biol* 6:1122–1128.
27. Kim, D.H., *et al.* (2002). mTOR interacts with raptor to form a nutrient-sensitive complex that signals to the cell growth machinery. *Cell* 110:163–175.
28. Kim, D.H., *et al.* (2003). GbetaL, a positive regulator of the rapamycin-sensitive pathway required for the nutrient-sensitive interaction between raptor and mTOR. *Mol Cell* 11:895–904.
29. Sarbassov, D.D., *et al.* (2004). Rictor, a novel binding partner of mTOR, defines a rapamycin-insensitive and raptor-independent pathway that regulates the cytoskeleton. *Curr Biol* 14:1296–1302.
30. Barquilla, A., Crespo, J.L., and Navarro, M. (2008). Rapamycin inhibits trypanosome cell growth by preventing TOR complex 2 formation. *Proc Natl Acad Sci USA* 105:14579–14584.
31. Hayashi, T., *et al.* (2007). Rapamycin sensitivity of the Schizosaccharomyces pombe tor2 mutant and organization of two highly phosphorylated TOR complexes by specific and common subunits. *Genes Cells* 12:1357–1370.
32. Matsuo, T., Otsubo, Y., Urano, J., Tamanoi, F., and Yamamoto, M. (2007). Loss of the TOR kinase Tor2 mimics nitrogen starvation and activates the sexual development pathway in fission yeast. *Mol Cell Biol* 27:3154–3164.
33. Nozaki, H., *et al.* (2003). The phylogenetic position of red algae revealed by multiple nuclear genes from mitochondria-containing eukaryotes and an alternative hypothesis on the origin of plastids. *J Mol Evol* 56:485–497.
34. Anderson, G.H., and Hanson, M.R. (2005). The Arabidopsis Mei2 homologue AML1 binds AtRaptor1B, the plant homologue of a major regulator of eukaryotic cell growth. *BMC Plant Biol* 5:2.
35. Wullschleger, S., Loewith, R., Oppliger, W., and Hall, M.N. (2005). Molecular organization of target of rapamycin complex 2. *J Biol Chem* 280:30697–30704.
36. Derelle, E., *et al.* (2006). Genome analysis of the smallest free-living eukaryote Ostreococcus tauri unveils many unique features. *Proc Natl Acad Sci USA* 103:11647–11652.
37. Nozaki, H., *et al.* (2007). A 100%-complete sequence reveals unusually simple genomic features in the hot-spring red alga Cyanidioschyzon merolae. *BMC Biol* 5:28.
38. Aronova, S., Wedaman, K., Anderson, S., Yates, J., 3rd, and Powers, T. (2007). Probing the membrane environment of the TOR kinases reveals functional interactions between TORC1, actin, and membrane trafficking in Saccharomyces cerevisiae. *Mol Biol Cell* 18:2779–2794.

39. Berchtold, D., and Walther, T.C. (2009). TORC2 plasma membrane localization is essential for cell viability and restricted to a distinct domain. *Mol Biol Cell* 20:1565–1575.
40. Drenan, R.M., Liu, X., Bertram, P.G., and Zheng, X.F. (2004). FKBP12-rapamycin-associated protein or mammalian target of rapamycin (FRAP/mTOR) localization in the endoplasmic reticulum and the Golgi apparatus. *J Biol Chem* 279:772–778.
41. Flinn, R.J., Yan, Y., Goswami, S., Parker, P.J., and Backer, J.M. (2010). The late endosome is essential for mTORC1 signaling. *Mol Biol Cell* 21:833–841.
42. Kim, J.E., and Chen, J. (2000). Cytoplasmic-nuclear shuttling of FKBP12-rapamycin-associated protein is involved in rapamycin-sensitive signaling and translation initiation. *Proc Natl Acad Sci USA* 97:14340–14345.
43. Kunz, J., Schneider, U., Howald, I., Schmidt, A., and Hall, M.N. (2000). HEAT repeats mediate plasma membrane localization of Tor2p in yeast. *J Biol Chem* 275:37011–37020.
44. Li, H., Tsang, C.K., Watkins, M., Bertram, P.G., and Zheng, X.F. (2006). Nutrient regulates Tor1 nuclear localization and association with rDNA promoter. *Nature* 442:1058–1061.
45. Sturgill, T.W., Cohen, A., Diefenbacher, M., Trautwein, M., Martin, D.E., and Hall, M.N. (2008). TOR1 and TOR2 have distinct locations in live cells. *Eukaryot Cell* 7:1819–1830.
46. He, C., and Klionsky, D.J. (2009). Regulation mechanisms and signaling pathways of autophagy. *Annu Rev Genet* 43:67–93.
47. Nakatogawa, H., Suzuki, K., Kamada, Y., and Ohsumi, Y. (2009). Dynamics and diversity in autophagy mechanisms: lessons from yeast. *Nat Rev Mol Cell Biol* 10:458–467.
48. Bassham, D.C. (2009). Function and regulation of macroautophagy in plants. *Biochim Biophys Acta* 1793:1397–1403.
49. Diaz-Troya, S., Perez-Perez, M.E., Florencio, F.J., and Crespo, J.L. (2008). The role of TOR in autophagy regulation from yeast to plants and mammals. *Autophagy* 4:851–865.
50. Noda, T., and Ohsumi, Y. (1998). Tor, a phosphatidylinositol kinase homologue, controls autophagy in yeast. *J Biol Chem* 273:3963–3966.
51. Kamada, Y., Funakoshi, T., Shintani, T., Nagano, K., Ohsumi, M., and Ohsumi, Y. (2000). Tor-mediated induction of autophagy via an Apg1 protein kinase complex. *J Cell Biol* 150:1507–1513.
52. Chang, Y.Y., and Neufeld, T.P. (2009). An Atg1/Atg13 complex with multiple roles in TOR-mediated autophagy regulation. *Mol Biol Cell* 20:2004–2014.
53. Hosokawa, N., *et al.* (2009). Nutrient-dependent mTORC1 association with the ULK1-Atg13-FIP200 complex required for autophagy. *Mol Biol Cell* 20:1981–1991.
54. Jung, C.H., *et al.* (2009). ULK-Atg13-FIP200 complexes mediate mTOR signaling to the autophagy machinery. *Mol Biol Cell* 20:1992–2003.
55. Geng, J., and Klionsky, D.J. (2008). The Atg8 and Atg12 ubiquitin-like conjugation systems in macroautophagy. 'Protein modifications: beyond the usual suspects' review series. *EMBO Rep* 9:859–864.
56. Doelling, J.H., Walker, J.M., Friedman, E.M., Thompson, A.R., and Vierstra, R.D. (2002). The APG8/12-activating enzyme APG7 is required for proper nutrient recycling and senescence in Arabidopsis thaliana. *J Biol Chem* 277:33105–33114.
57. Hanaoka, H., *et al.* (2002). Leaf senescence and starvation-induced chlorosis are accelerated by the disruption of an Arabidopsis autophagy gene. *Plant Physiol* 129:1181–1193.
58. Liu, Y., Schiff, M., Czymmek, K., Talloczy, Z., Levine, B., and Dinesh-Kumar, S.P. (2005). Autophagy regulates programmed cell death during the plant innate immune response. *Cell* 121:567–577.
59. Xiong, Y., Contento, A.L., and Bassham, D.C. (2005). AtATG18a is required for the formation of autophagosomes during nutrient stress and senescence in Arabidopsis thaliana. *Plant J* 42:535–546.

60. Xiong, Y., Contento, A.L., and Bassham, D.C. (2007). Disruption of autophagy results in constitutive oxidative stress in Arabidopsis. *Autophagy* 3:257–258.
61. Yoshimoto, K., *et al.* (2004). Processing of ATG8s, ubiquitin-like proteins, and their deconjugation by ATG4s are essential for plant autophagy. *Plant Cell* 16:2967–2983.
62. Chung, T., Suttangkakul, A., and Vierstra, R.D. (2009). The ATG autophagic conjugation system in maize: ATG transcripts and abundance of the ATG8-lipid adduct are regulated by development and nutrient availability. *Plant Physiol* 149:220–234.
63. Ketelaar, T., Voss, C., Dimmock, S.A., Thumm, M., and Hussey, P.J. (2004). Arabidopsis homologues of the autophagy protein Atg8 are a novel family of microtubule binding proteins. *FEBS Lett* 567:302–306.
64. Perez-Perez, M.E., and Crespo, J.L. (2010). Autophagy in the model alga *Chlamydomonas reinhardtii*. *Autophagy* 6:562–563.
65. Xie, Z., and Klionsky, D.J. (2007). Autophagosome formation: core machinery and adaptations. *Nat Cell Biol* 9:1102–1109.

13

mTORC1 and mTORC2 in Energy Homeostasis

MARION CORNU • MICHAEL N. HALL

Biozentrum, University of Basel
Basel, Switzerland

I. Abstract

Energy homeostasis is the regulation of energy uptake, storage, and expenditure such that an organism maintains a balanced metabolic state. It requires communication between the so-called metabolic organs, including the central nervous system (CNS), the pancreas, adipose tissue, skeletal muscle, and the liver. For example, nutrients stimulate the pancreas and adipose tissue to secrete hormones such as insulin and leptin, respectively. The hypothalamus in the CNS in turn integrates hormonal and nutrient signals to control energy uptake (appetite), storage, and expenditure. As a nutrient, hormone and energy sensor active in all metabolic organs, mTOR plays a central role in mediating energy homeostasis. Here, we review mTORC1 and mTORC2 in metabolic tissues and their role in the regulation of appetite, glucose homeostasis, and energy balance. The deregulation of mTOR signaling can lead to energy imbalance and metabolic disorders such as obesity and type 2 diabetes.

II. Introduction

The target of rapamycin (TOR) is an evolutionarily conserved serine–threonine kinase that regulates cell growth and metabolism [1]. TOR exists in two structurally and functionally distinct complexes, TORC1 and TORC2.

THE ENZYMES, Vol. XXVIII

ISSN NO: 1874-6047
DOI: 10.1016/S1874-6047(10)28013-7

In complex with the immunophilin FKBP12, rapamycin binds and inhibits TORC1. In mammals, mTORC1 is activated in response to nutrients (amino acids and possibly other nutrients), growth factors such as insulin and IGF-1, and cellular energy status (ATP). The best characterized substrates of mTORC1 are eIF4E binding protein (4E-BP) and S6 kinase (S6K) through which mTORC1 controls protein synthesis. mTORC2 is activated by growth factors, by an unknown mechanism, and possibly by other unknown inputs. The best characterized substrates of mTORC2 are the AGC family kinases Akt (also known as PKB) and SGK1. Here, we review the roles of mTORC1 and mTORC2 in whole body energy homeostasis. In particular, we describe studies in which *TSC1*, *TSC2*, *raptor*, or *rictor* are knocked out specifically in metabolic tissues, the hypothalamus, pancreatic β-cells, adipose tissue, muscle, and the liver. The tumor suppressor proteins TSC1 and TSC2 form a complex that inhibits mTORC1 by acting as a GAP for the small GTPase and mTORC1 activator Rheb. Thus, ablation of *TSC1* or *TSC2* results in hyperactive mTORC1 signaling [2, 3]. raptor is an essential and specific subunit of mTORC1, and its removal results in loss of mTORC1 activity. Likewise, rictor is an essential and specific subunit of mTORC2, and its removal results in loss of mTORC2 activity. Full body ablation of TSC1 [2], TSC2 [4, 5], or any component of mTORC1 or mTORC2 [6–11] is embryonically lethal. mTOR signaling was assumed early on to have a particularly important role in metabolic tissues, and thus in energy homeostasis, because it is activated by nutrients, insulin, and energy.

Amino acids, especially leucine, activate mTORC1 via the Rag GTPases [12, 13]. Amino acids induce GTP loading of the Rag proteins. GTP-loaded Rag then binds raptor and promotes localization of mTORC1 to a compartment that contains Rheb [12]. The Ste20-related kinase MAP4K3 [14] and the class III PI3K mVps34 [15, 16] have also been implicated in the activation of mTORC1 by amino acids. While the mechanism by which MAP4K3 activates mTORC1 is unknown, it has been proposed that mVps34 activates mTORC1 signaling through amino acids-induced extracellular calcium influx activating calmodulin. Calmodulin binds and activates mVps34, leading to the production of phosphatidylinositol-3-phosphate and, by an unknown mechanism, mTORC1 activation [17].

III. mTORC1 in the Hypothalamus

The hypothalamus is important in the control of energy homeostasis. In particular, the hypothalamic arcuate nucleus (ARC) is a key region for sensing nutrients and hormones, such as insulin or fat-derived leptin, and thereby relaying information on energy status to other parts of the hypothalamus and body. Leptin and insulin bind receptors highly expressed in the ARC. The ARC contains at least two distinct populations of neurons involved in mediating the effect of these hormones on whole body storage or utilization of energy. The orexigenic neurons synthesize two peptides, neuropeptide Y (NPY) and agouti-related protein (AgRP), whose expression is increased in fasted or leptin-deficient mice [18, 19] and which promote food intake. The anorexigenic neurons synthesize the pro-opimelanocortin (POMC) peptide that is processed into α-melanocyte-stimulating hormone (α-MSH) and which reduces food intake. In addition to hormones, the ARC neurons also sense nutrients, such as glucose, amino acids, and fatty acids, which inhibit NPY–AgRP neurons and activate POMC neurons to signal satiety.

mTORC1 and its downstream target S6K are widely expressed throughout the central nervous system (CNS) including AgRP/NPY and POMC neurons. In the ARC, but not in the paraventricular nucleus (PVN) of the hypothalamus or in extra-hypothalamic areas, mTORC1 signaling is inhibited in fasted rats and is activated upon refeeding or leptin injection [20]. Furthermore, a leucine-supplemented diet inhibits NPY expression and stimulates POMC expression [21]. These observations suggest that, in response to nutrients and hormones, mTORC1 signaling in neurons of the ARC inhibits NPY and stimulates POMC to reduce food intake. However, *TSC1* ablation and thus hyperactive mTORC1 signaling in multiple regions of the mouse hypothalamus, including both anorexigenic and orexigenic neurons, enhanced NPY expression and reduced POMC expression and caused hyperphagia [22]. Surprisingly and contrary to the findings of others, this suggests that mTORC1 stimulates NPY and inhibits POMC to increase food intake. To examine the role of chronic mTORC1 activation in a hypothalamic neuronal subtype, Mori *et al.* [22] also generated mice harboring a disruption of *TSC1* specifically in POMC neurons. In these mice, injected leptin failed to decrease food intake and body weight as compared to control mice. Thus, mTORC1 signaling in POMC neurons inhibits the anorexigenic function of these neurons [22], although POMC mRNA levels were unaffected by mTORC1 hyperactivation specifically in POMC neurons. In summary, hypothalamic mTORC1 appears to have an important role in controlling appetite. However, the apparent discrepancy of whether hypothalamic mTORC1 signaling is anorexigenic or orexigenic remains to be resolved. The fact that mTORC1 is a nutrient sensor suggests that it is anorexigenic. Knockout of *raptor* specifically in AgRP/NPY and POMC neurons should be informative in clarifying whether hypothalamic mTORC1 is anorexigenic or orexigenic. It would also be of interest to determine the roles of hypothalamic Rag GTPases, mVPS34, MAPK4K3 in mTORC1-mediated regulation of appetite by amino acids.

The hypothalamus expresses lipogenic enzymes such as fatty acid synthase (FAS) and acetyl CoA carboxylase (ACC) [23]. Studies using pharmacological or genetic inhibitors have suggested that FAS and ACC of the CNS also have a role in the regulation of food intake and body weight [24–28]. Intracerebral administration of the natural FAS inhibitor cerulenin or the synthetic inhibitor C75 induces dose-dependent anorexia and weight loss in rodents, possibly because of accumulation of the FAS substrate malonyl-CoA [27–29]. Cerulenin or C75 also induced S6K1 and S6 phosphorylation in the mediobasal hypothalamus [29], suggesting that these treatments activate mTORC1. Furthermore, rapamycin treatment or S6K knockout blocked the anorexic effect of cerulenin or C75, indicating that mTORC1 signaling is required for the anorexic effect [29]. The mechanism by which cerulenin and C75 activate mTORC1 signaling is not known but may involve inhibition of AMP-dependent protein kinase (AMPK). AMPK negatively regulates mTORC1 in response to low cellular energy (see below). Finally, the effects of cerulenin and C75 also suggest that hypothalamic mTORC1 is anorexigenic.

AMPK is a serine–threonine protein kinase activated upon fasting or glucose deprivation and is inhibited upon refeeding or administration of glucose, insulin, or leptin. Thus, AMPK and mTORC1 are regulated in opposite ways and have antagonistic functions in regulating energy homeostasis. AMPK inhibits mTORC1 signaling by phosphorylating and activating the TSC1–TSC2 complex [30]] and by directly phosphorylating raptor in mTORC1 [31]. In the hypothalamus, AMPK promotes food intake [32]. The precise mechanism involved in AMPK-mediated regulation of food intake is incompletely understood but may involve inhibition of ACC. ACC converts acetyl-CoA to malonyl-CoA [33]. In the hypothalamus, AMPK leads to a reduction in malonyl-CoA levels which in turn leads to an increase in food intake [32, 34]. Hypothalamic AMPK could also increase food uptake via inhibition of anorexigenic mTORC1, but such a link remains to be demonstrated.

The circadian clock coordinates physiological and behavioral processes, including food intake, according to daily light–dark cycles. Recently, mTORC1 activation has been shown to mediate photic (light) entrainment of the circadian clock. In mammals, the master circardian clock is located in the suprachiasmatic nuclei (SCN) of the anterior hypothalamus. Photic input, relayed from the retina to the SCN through the retinohypothalamic tract, induces release of glutamate and pituitary adenylate cyclase-activating peptide (PACAP) in the SCN. Glutamate and PACAP then bind their postsynaptic receptors on SCN neurons activating the ERK MAPK signaling pathway. Light-mediated ERK activation stimulates S6K phosphorylation in an mTORC1-dependent manner [35]. Moreover, mTORC1 is reported to regulate photic entrainment via expression of the circadian clock proteins PERIOD1 and PERIOD2 [36]. The aforementioned observations underscore the multifaceted complexity and importance of hypothalamic mTORC1 in the regulation of energy homeostasis.

IV. mTORC1 in Pancreatic β-Cells

The number of β-cells of the endocrine pancreas and their capacity to secrete insulin are modulated in response to metabolic demand. β-Cell mass is regulated by a dynamic balance of neogenesis, growth, proliferation, and apoptosis. A reduction in β-cell mass or function and a consequential failure of the endocrine pancreas to maintain adequate insulin secretion underlie the pathogenesis of both type 1 and type 2 diabetes. Evidence for the involvement of mTORC1 in the regulation of β-cell mass and function comes from *in vivo* studies using genetically modified mice. Mice deficient for S6K1 [37] or for its direct substrate S6 [38] display decreased β-cell mass and hypoinsulinemia, suggesting that mTORC1-S6K1 signaling stimulates β-cell growth and function. To better understand the role of mTORC1 in the regulation of β-cell mass and insulin secretion, several investigators generated mice with hyperactive mTORC1 signaling specifically in β-cells. These included β-cell-specific *TSC1* [39] or *TSC2* knockout mice [40, 41] and transgenic mice overexpressing Rheb in β-cells [42]. Hyperactive mTORC1 signaling increased islet mass and enhanced insulin secretion, at least in young mice [39–42]. These effects were reversed by rapamycin treatment [40, 41, 43]. However, disruption of *TSC2* in β-cells increased islet mass by increasing β-cell size and number [40], whereas *TSC1* disruption or Rheb overexpression increased only β-cell size [39, 42]. Despite the mild discrepancy, these studies indicate that mTORC1 controls β-cell growth and function.

LKB1 is a serine–threonine kinase and a potent tumor suppressor whose inactivation has been described in Peutz–Jeghers syndrome (PJS), characterized by gastrointestinal hamartomatous polyps. LKB1 phosphorylates and activates AMPK and other kinases. In β-cells, AMPK activation has been implicated in the control of β-cell survival and has been shown to antagonize glucose-stimulated insulin secretion [44–48]. Three groups recently reported the effect of knocking out *LKB1* in β-cells [49–51]. They demonstrated that deletion of *LKB1* leads to increased insulin production and enhanced β-cell size. LKB1 controls β-cell size through mTORC1 signaling, as assessed by reduced AMPK phosphorylation and increased S6K phosphorylation [49–51]. Thus, LKB1–AMPK signaling negatively controls β-cell size by inhibiting mTORC1.

To examine the effect of mTORC1 inhibition on whole body metabolism *Psammomys obesus*, a gerbil used as a model for diet-induced type 2 diabetes, was treated with rapamycin [52]. Rapamycin treatment exacerbated the insulin resistance in these animals, as indicated by increased blood glucose levels. In *P. obesus*, rapamycin also induced β-cell apoptosis and reduced glucose-stimulated insulin secretion and synthesis [52], suggesting that mTORC1 signaling is beneficial in β-cells. However, in a different context, mTORC1 appears to have a detrimental effect on β-cells. Insulin receptor substrate is a defining component in the insulin/IGF signaling pathway in both β-cells and cells of peripheral organs. IRS2, the most abundant and functionally important IRS family member in β-cells, is upstream of mTORC1 and is required for β-cell growth, proliferation, and survival [53–55]. Inappropriate activation of the mTORC1 pathway in β-cells by chronic exposure to glucose or IGF-1 activates a negative feedback loop, from mTORC1-S6K to IRS2, which ultimately leads to proteasomal degradation of IRS2 and apoptosis. These effects in β-cells are prevented by treatment with rapamycin or the proteasomal inhibitor lactacystin [56], suggesting that inhibition of the negative feedback loop from mTORC1 to IRS2 by rapamcyin treatment may have a therapeutic effect. Thus, although mTORC1 is essential for β-cell growth, deregulated mTORC1 activity can contribute to β-cell pathophysiology.

V. mTORC1 and mTORC2 in Adipose Tissue

White adipose tissue is a fat storage depot and an endocrine organ that secretes hormones and cytokines such as leptin, adiponectin, and TNF-α to control energy homeostasis, lipid metabolism, and fertility [57–59]. Excess adipose tissue contributes to the development of many metabolic disorders, including obesity, insulin resistance, type 2 diabetes, cardiovascular diseases, and cancer. The fat body in *Drosophila*, a tissue equivalent to adipose tissue in mammals, regulates full body growth in response to nutrients. The downregulation of TOR signaling specifically in the fat body causes a global growth defect [60]. Given the role of mTOR as a nutrient sensor and its involvement in metabolic disorders [1, 61, 62], the role of mTORC1 and mTORC2 in adipose tissue was assessed *in vitro* and *in vivo*. Several studies demonstrated that long-term rapamycin treatment inhibits differentiation of mouse and human preadipocytes in culture [63–69]. In addition, knockdown of raptor in 3T3-L1 preadipocytes or loss of one *raptor* allele (*raptor*$^{+/-}$) in mouse embryonic fibroblasts (MEFs) reduced adipogenesis [69], suggesting that mTORC1 is required for the differentiation and maintenance of adipocytes. Other *in vitro* studies showed that rapamycin-mediated mTORC1 inhibition prevents expression of the adipogenic transcription factors peroxisome proliferator-activated receptor-γ (PPAR-γ) and CCAAT/enhancer binding protein-α (C/EBP-α) [64, 66, 67, 69]. Consistent with these observations, *TSC2* deletion-mediated mTORC1 hyperactivation in cultured cells stimulated adipogenesis through the increased expression of PPAR-γ [70], suggesting that mTORC1 controls adipogenesis through PPARγ. However, *in vivo* studies with knockout mice revealed a different result. Mice lacking raptor specifically in adipose tissue (*raptor*$^{ad-/-}$) are lean, due to a reduction in the size and the number of adipocytes [69]. Surprisingly, PPAR-γ and C/EBP-α expression were unaffected in epididymal white adipose tissue of the *raptor*$^{ad-/-}$ mice, suggesting that in animals there must be an mTORC1-independent mechanism to maintain PPAR-γ and C/EBP-α. The leanness of *raptor*$^{ad-/-}$ mice was due to enhanced energy expenditure, likely as a result of UCP1-mediated mitochondrial uncoupling and enhanced mitochondrial respiration [69]. These observations are in agreement with the phenotype of full-body S6K1 knockout mouse. S6K1 null mice are also lean and protected against diet-induced obesity, due to increased energy expenditure [71]. Conversely, knockout of 4E-BP1 and 4E-BP2, mTORC1 negative effectors, results in reduced energy expenditure and increased sensitivity to diet-induced obesity [72]. Together, these findings strongly suggest that mTORC1 controls adipose metabolism and thereby energy homeostasis.

Mice lacking mTORC2, specifically in adipose tissue (*rictor*$^{ad-/-}$), have an unexpected phenotype. They display enhanced body size because of an increase in lean tissue mass, with little to no difference in the total amount of fat mass [73, 74]. Furthermore, the *rictor*$^{ad-/-}$ mice are hyperinsulinemic and exhibit elevated levels of circulating IGF-1. These studies suggest that mTORC2 in adipose tissue controls whole body growth and glucose metabolism by inhibiting insulin production by the pancreas and IGF-1 secretion by the liver [73]. The physiological relevance of this negative regulation of the pancreas and liver by adipose mTORC2 may be to maintain insulin and IGF-1 homeostasis. mTORC2 is itself activated by insulin and IGF-1. Hence, excessive levels of either hormone would activate mTORC2 signaling in adipose tissue and this, in turn, would downregulate insulin and IGF-1 production by the pancreas and liver to restore appropriate levels of the two hormones. The observation that adipose-specific ablation of the IGF-1 receptor confers a phenotype similar to that of mTORC2 knockout is consistent with such a negative feedback loop [75]. However, the mechanism by which mTORC2 in adipose tissue signals to the pancreas and the liver remains to be determined.

VI. mTORC1 and mTORC2 in Muscle

Muscle is the major consumer of glucose in response to insulin stimulation and thus plays a major role in energy homeostasis. As demonstrated in adult rodents, mTORC1 positively regulates skeletal muscle fiber size and prevents muscle atrophy [76]. Additionally, overexpression of a constitutively active form of Akt promotes muscle hypertrophy via both stimulation of growth and inhibition of atrophy pathways [77, 78]. Rapamycin treatment blocks Akt-mediated muscle hypertrophy [78]. Furthermore, mice deficient for S6K1 display atrophy of skeletal muscle [79]. To assess the function of mTORC1 in muscle *in vivo*, Bentzinger *et al.* generated muscle-specific *raptor* knockout mice. The raptor-deficient muscle became progressively dystrophic and exhibited decreased oxidative capacity and increased glycogen content [80]. These effects were due to downregulation of genes such as *PGC-1α* involved in mitochondrial biogenesis and to hyperactivation of Akt [80]. Reduced PGC-1α levels likely contributed to the dystrophic phenotype of the knockout mice since muscle-specific *PGC-1α* knockout mice display a myopathic phenotype [81]. The hyperactivation of Akt was due to loss of the negative feedback loop from mTORC1 to IRS. Finally, the muscle-specific *raptor* knockout mice were slightly glucose intolerant despite Akt hyperactivation in muscle. The glucose intolerance could be due to the enhanced levels of glycogen and reduced oxidative capacity [80].

Rapamycin treatment or knockdown of mTOR or raptor in muscle cells *in vitro* reduces mitochondrial gene expression and oxygen consumption, whereas $TSC2^{-/-}$ MEFs show increased expression of mitochondrial genes [82]. These effects are through regulation of PGC-1α and the transcription factor ying yang 1 (YY1) [82, 83]. Surprisingly, muscle of mice deficient for S6K1 and S6K2 display increased mitochondrial biogenesis and enhanced PGC-1α expression [84]. Thus, mitochondria are a major downstream target of mTORC1 signaling. However, the effect of mTORC1 on mitochondria can vary depending on the tissue. mTORC1 appears to stimulate mitochondrial function in muscle whereas, as described earlier, it inhibits mitochondrial activity in adipose tissue.

The knockout of *rictor* specifically in muscle results in little to no phenotype [80, 85]. These mice are slightly glucose intolerant and display decreased Akt phosphorylation at Ser473 and reduced insulin-mediated phosphorylation of the Akt substrate AS160. AS160 mediates insulin-stimulated translocation of the glucose transporter GLUT4 to the plasma membrane. Thus, the slight glucose intolerance is likely due to a defect in insulin-stimulated delivery of GLUT4 to the plasma membrane in muscle [85]. Finally, muscle deficient for both raptor and rictor show a marked increase in Akt phosphorylation at Thr308 and Ser473, as observed in muscle lacking raptor alone [80]. These results suggest that a kinase other than mTORC2 is able to phosphorylate Akt at Ser473 when mTORC1 is absent.

VII. mTORC1 in the Liver

The incidence of nonalcoholic fatty liver disease (NAFLD) is increasing rapidly in industrialized countries. NAFLD encompasses a wide spectrum of diseases, ranging from hepatic steatosis (also called fatty liver), with benign prognosis, to a potentially progressive form, nonalcoholic steatohepatitis (NASH), which can lead to liver fibrosis and cirrhosis. Cirrhosis may progress to hepatocellular carcinoma (HCC). The hallmark of NAFLD is excessive fat accumulation, mainly triglycerides, in hepatocytes and is associated with features of the metabolic syndrome, including obesity, insulin resistance, and type 2 diabetes. Accumulation of fatty acids is due to *de novo* lipogenesis from excess carbohydrates which are esterified with glycerol-3-phosphate to generate triglycerides [86]. In response to nutritional cues, such as a high-carbohydrate diet, insulin activates key enzymes of the glycolytic and lipogenic pathways through the transcriptional factor sterol regulatory element-binding protein (SREBP)-1c. Insulin stimulates both expression and proteolytic activation of SREBP-1c. SREBPs are basic helix-loop-helix-leucine zipper (bHLH-LZ) transcription factors synthesized as inactive precursors bound to the endoplasmic reticulum (ER) membrane. The ER-bound SREBP precursor is proteolytically cleaved to release an active, amino-terminal domain termed nuclear SREBP (nSREBP). nSREBP translocates into the nucleus where it binds and activates target genes [87]. In mammals, three SREBP isoforms have been described, SREBP-1a, SREBP-1c, and SREBP-2. SREBP-1a and SREBP-1c are encoded by a single gene and produced by alternative splicing. SREBP-1a promotes transcription of genes involved in the synthesis of cholesterol, fatty acids, and triglyceride, whereas SREBP-1c preferentially activates genes involved in fatty acid synthesis [88]. SREBP-2 is encoded by a separate gene and promotes cholesterol synthesis [89, 90]. SREBP-1c and SREBP-2 are expressed mainly in the liver, adipose tissue, and brain [91], whereas SREBP-1a is expressed mainly in cell lines and tissues with high capacity for cell proliferation such as intestine and spleen [91]. Mice overexpressing a constitutively active form of SREBP-1c (nSREBP-1c) develop hepatic steatosis due to activation of lipogenic genes [88, 92]. Conversely, liver of mice specifically deficient for SREBP-1c show a reduction in mRNAs encoding enzymes of fatty acid and triglyceride synthesis [93]. Furthermore, SREBP-1c$^{-/-}$ *ob/ob* mice are protected against hepatic steatosis as compared to *ob/ob* mice [94].

The observation that the insulin-Akt pathway controls SREBP-1c and expression of lipogenic genes prompted investigation of a role for mTORC1 in liver lipid metabolism. Independent studies using primary hepatocytes or hepatic cell lines demonstrated that rapamycin-mediated mTORC1 inhibition decreases expression of acetyl-CoA carboxylase (ACC1), glycerol-3-phosphate acyltransferase (GPAT) [95], and stearoyl-CoA desaturase 1 (SCD1) [96]. Furthermore, using immortalized human retinal pigment epithelial cells (RPE) expressing a conditionally activated version of Akt1 kinase (myrAkt-ER), Porstmann *et al.* demonstrated that rapamycin treatment or raptor knockdown blocks expression of FAS and ATP-citrate lyase (ACLY) in response to Akt activation and prevents Akt-dependent lipogenesis [97]. Rapamycin blocked Akt-induced nuclear accumulation of SREBP-1, lipid synthesis, and phosphoglyceride accumulation [97]. These studies indicate that mTORC1 controls lipogenic genes in primary hepatocytes and other cells. The mechanism by which mTORC1 controls SREBP-1 proteolytic cleavage is still unclear and should be assessed in primary hepatocytes. Interestingly, silencing of SREBP-1 also restricts cell growth suggesting that the PI3K-Akt-mTORC1 pathway regulates cell growth by regulating protein and lipid synthesis [97].

Insulin reduces hepatic glucose production by inhibiting FoxO1, a transcription factor of key gluconeogenic genes such as phosphoenolpyruvate carboxykinase (PEPCK). Insulin activates SREBP-1c and inhibits PEPCK expression by signaling through the conventional IRS-PI3K-Akt pathway. However, in insulin-resistant cells, insulin loses its ability to suppress gluconeogenesis but is still able to stimulate hepatic lipid synthesis. This suggests that there is a bifurcation in the insulin signaling pathway downstream of Akt. To investigate this possibility, Li *et al.* [98] examined insulin signaling in freshly isolated primary hepatocytes. They found that rapamycin blocks insulin-stimulated expression of SREBP-1c but has no effect on insulin-mediated inhibition of PEPCK. Similar effects were observed *in vivo*, in livers of fasted rodents after intraperitoneal injection of rapamycin. Thus, mTORC1 controls SREBP-1c but not PEPCK, and the bifurcation in the insulin signaling pathway is upstream of mTORC1. Inhibition of S6K did not affect insulin-stimulated SREBP-1c expression, suggesting that mTORC1 controls SREBP-1c expression via an effector other than S6K [98].

Metformin is a widely prescribed antidiabetic drug that increases insulin sensitivity and lowers blood glucose and blood lipid content by activating AMPK in the liver [99, 100]. In hepatocytes, metformin inhibits glucose production and lipogenic gene expression. These effects are suppressed by treating cells with an AMPK inhibitor or by expression of kinase-inactive

AMPKα [101, 102]. AMPK inhibits mTORC1 signaling and thereby prevents the negative feedback loop that downregulates insulin signaling. In agreement with the effects of metformin, overexpression of a dominant negative form of raptor in the liver of insulin-resistant mice improved glucose tolerance and restored insulin sensitivity by reducing the negative feedback loop [103]. These findings suggest that inappropriate mTORC1 activity in the liver contributes to diabetes and that metformin, at least in part, targets mTORC1 as part of its therapeutic effect.

VIII. Conclusion

Many studies have examined the roles of mTORC1 and mTORC2 in metabolic tissues and have shown that the two mTOR complexes are important players in energy homeostasis. However, these studies are largely in their infancy, as much remains to be elucidated. For example, how mTOR signaling in subregions of the hypothalamus controls appetite and how this is integrated with mTOR signaling in other metabolic tissues are particularly fascinating questions. Ongoing studies should not only shed more light on the role of the mTORCs in energy homeostasis but should also lead the development of antiobesity and antidiabetes drugs.

Acknowledgments

We acknowledge support from the Swiss National Science Foundation, the Swiss Cancer League, the Louis-Jeantet Foundation, and the Canton of Basel.

References

1. Wullschleger, S., Loewith, R., and Hall, M.N. (2006). TOR signaling in growth and metabolism. *Cell* 124:471–484.
2. Kwiatkowski, D.J., *et al.* (2002). A mouse model of TSC1 reveals sex-dependent lethality from liver hemangiomas, and up-regulation of p70S6 kinase activity in Tsc1 null cells. *Hum Mol Genet* 11:525–534.
3. El-Hashemite, N., Zhang, H., Henske, E.P., and Kwiatkowski, D.J. (2003). Mutation in TSC2 and activation of mammalian target of rapamycin signalling pathway in *renal angiomyolipoma*. *Lancet* 361:1348–1349.
4. Kobayashi, T., Minowa, O., Kuno, J., Mitani, H., Hino, O., and Noda, T. (1999). Renal carcinogenesis, hepatic hemangiomatosis, and embryonic lethality caused by a germ-line Tsc2 mutation in mice. *Cancer Res* 59:1206–1211.
5. Onda, H., Lueck, A., Marks, P.W., Warren, H.B., and Kwiatkowski, D.J. (1999). Tsc2(+/−) mice develop tumors in multiple sites that express gelsolin and are influenced by genetic background. *J Clin Invest* 104:687–695.

6. Gangloff, Y.G., *et al.* (2004). Disruption of the mouse mTOR gene leads to early postimplantation lethality and prohibits embryonic stem cell development. *Mol Cell Biol* 24:9508–9516.
7. Murakami, M., *et al.* (2004). mTOR is essential for growth and proliferation in early mouse embryos and embryonic stem cells. *Mol Cell Biol* 24:6710–6718.
8. Guertin, D.A., *et al.* (2006). Ablation in mice of the mTORC components raptor, rictor, or mLST8 reveals that mTORC2 is required for signaling to Akt-FOXO and PKCalpha, but not S6K1. *Dev Cell* 11:859–871.
9. Jacinto, E., *et al.* (2006). SIN1/MIP1 maintains rictor-mTOR complex integrity and regulates Akt phosphorylation and substrate specificity. *Cell* 127:125–137.
10. Shiota, C., Woo, J.T., Lindner, J., Shelton, K.D., and Magnuson, M.A. (2006). Multiallelic disruption of the rictor gene in mice reveals that mTOR complex 2 is essential for fetal growth and viability. *Dev Cell* 11:583–589.
11. Yang, Q., Inoki, K., Ikenoue, T., and Guan, K.L. (2006). Identification of Sin1 as an essential TORC2 component required for complex formation and kinase activity. *Genes Dev* 20:2820–2832.
12. Sancak, Y., *et al.* (2008). The Rag GTPases bind raptor and mediate amino acid signaling to mTORC1. *Science* 320:1496–1501.
13. Kim, E., Goraksha-Hicks, P., Li, L., Neufeld, T.P., and Guan, K.L. (2008). Regulation of TORC1 by Rag GTPases in nutrient response. *Nat Cell Biol* 10:935–945.
14. Findlay, G.M., Yan, L., Procter, J., Mieulet, V., and Lamb, R.F. (2007). A MAP4 kinase related to Ste20 is a nutrient-sensitive regulator of mTOR signalling. *Biochem J* 403:13–20.
15. Nobukuni, T., *et al.* (2005). Amino acids mediate mTOR/raptor signaling through activation of class 3 phosphatidylinositol 3OH-kinase. *Proc Natl Acad Sci USA* 102:14238–14243.
16. Byfield, M.P., Murray, J.T., and Backer, J.M. (2005). hVps34 is a nutrient-regulated lipid kinase required for activation of p70 S6 kinase. *J Biol Chem* 280:33076–33082.
17. Gulati, P., *et al.* (2008). Amino acids activate mTOR complex 1 via Ca2+/CaM signaling to hVps34. *Cell Metab* 7:456–465.
18. Mizuno, T.M., Kleopoulos, S.P., Bergen, H.T., Roberts, J.L., Priest, C.A., and Mobbs, C.V. (1998). Hypothalamic pro-opiomelanocortin mRNA is reduced by fasting and [corrected] in ob/ob and db/db mice, but is stimulated by leptin. *Diabetes* 47:294–297.
19. Shutter, J.R., Graham, M., Kinsey, A.C., Scully, S., Luthy, R., and Stark, K.L. (1997). Hypothalamic expression of ART, a novel gene related to agouti, is up-regulated in obese and diabetic mutant mice. *Genes Dev* 11:593–602.
20. Cota, D., *et al.* (2006). Hypothalamic mTOR signaling regulates food intake. *Science* 312:927–930.
21. Ropelle, E.R., *et al.* (2008). A central role for neuronal AMP-activated protein kinase (AMPK) and mammalian target of rapamycin (mTOR) in high-protein diet-induced weight loss. *Diabetes* 57:594–605.
22. Mori, H., *et al.* (2009). Critical role for hypothalamic mTOR activity in energy balance. *Cell Metab* 9:362–374.
23. Pocai, A., *et al.* (2006). Restoration of hypothalamic lipid sensing normalizes energy and glucose homeostasis in overfed rats. *J Clin Invest* 116:1081–1091.
24. Hu, Z., Dai, Y., Prentki, M., Chohnan, S., and Lane, M.D. (2005). A role for hypothalamic malonyl-CoA in the control of food intake. *J Biol Chem* 280:39681–39683.
25. Clegg, D.J., Wortman, M.D., Benoit, S.C., McOsker, C.C., and Seeley, R.J. (2002). Comparison of central and peripheral administration of C75 on food intake, body weight, and conditioned taste aversion. *Diabetes* 51:3196–3201.

26. Wortman, M.D., Clegg, D.J., D'Alessio, D., Woods, S.C., and Seeley, R.J. (2003). C75 inhibits food intake by increasing CNS glucose metabolism. *Nat Med* 9:483–485.
27. Lam, T.K., Schwartz, G.J., and Rossetti, L. (2005). Hypothalamic sensing of fatty acids. *Nat Neurosci* 8:579–584.
28. Loftus, T.M., *et al.* (2000). Reduced food intake and body weight in mice treated with fatty acid synthase inhibitors. *Science* 288:2379–2381.
29. Proulx, K., Cota, D., Woods, S.C., and Seeley, R.J. (2008). Fatty acid synthase inhibitors modulate energy balance via mammalian target of rapamycin complex 1 signaling in the central nervous system. *Diabetes* 57:3231–3238.
30. Inoki, K., Zhu, T., and Guan, K.L. (2003). TSC2 mediates cellular energy response to control cell growth and survival. *Cell* 115:577–590.
31. Gwinn, D.M., *et al.* (2008). AMPK phosphorylation of raptor mediates a metabolic checkpoint. *Mol Cell* 30:214–226.
32. Minokoshi, Y., *et al.* (2004). AMP-kinase regulates food intake by responding to hormonal and nutrient signals in the hypothalamus. *Nature* 428:569–574.
33. Ruderman, N.B., *et al.* (2003). AMPK as a metabolic switch in rat muscle, liver and adipose tissue after exercise. *Acta Physiol Scand* 178:435–442.
34. Gao, S., *et al.* (2007). Leptin activates hypothalamic acetyl-CoA carboxylase to inhibit food intake. *Proc Natl Acad Sci USA* 104:17358–17363.
35. Cao, R., Lee, B., Cho, H.Y., Saklayen, S., and Obrietan, K. (2008). Photic regulation of the mTOR signaling pathway in the suprachiasmatic circadian clock. *Mol Cell Neurosci* 38:312–324.
36. Cao R, Li A, Cho HY, Lee B, Obrietan K. Mammalian target of rapamycin signaling modulates photic entrainment of the suprachiasmatic circadian clock. *J Neurosci* 30:6302-6314.
37. Pende, M., *et al.* (2000). Hypoinsulinaemia, glucose intolerance and diminished beta-cell size in S6K1-deficient mice. *Nature* 408:994–997.
38. Ruvinsky, I., *et al.* (2005). Ribosomal protein S6 phosphorylation is a determinant of cell size and glucose homeostasis. *Genes Dev* 19:2199–2211.
39. Mori, H., *et al.* (2009). Critical roles for the TSC-mTOR pathway in {beta}-cell function. *Am J Physiol Endocrinol Metab* 297:E1013–E1022.
40. Rachdi, L., *et al.* (2008). Disruption of Tsc2 in pancreatic beta cells induces beta cell mass expansion and improved glucose tolerance in a TORC1-dependent manner. *Proc Natl Acad Sci USA* 105:9250–9255.
41. Shigeyama, Y., *et al.* (2008). Biphasic response of pancreatic beta-cell mass to ablation of tuberous sclerosis complex 2 in mice. *Mol Cell Biol* 28:2971–2979.
42. Hamada, S., *et al.* (2009). Upregulation of the mammalian target of rapamycin complex 1 pathway by Ras homolog enriched in brain in pancreatic beta-cells leads to increased beta-cell mass and prevention of hyperglycemia. *Diabetes* 58:1321–1332.
43. Bussiere, C.T., Lakey, J.R., Shapiro, A.M., and Korbutt, G.S. (2006). The impact of the mTOR inhibitor sirolimus on the proliferation and function of pancreatic islets and ductal cells. *Diabetologia* 49:2341–2349.
44. da Silva, Xavier G., Leclerc, I., Varadi, A., Tsuboi, T., Moule, S.K., and Rutter, G.A. (2003). Role for AMP-activated protein kinase in glucose-stimulated insulin secretion and preproinsulin gene expression. *Biochem J* 371:761–774.
45. Richards, S.K., Parton, L.E., Leclerc, I., Rutter, G.A., and Smith, R.M. (2005). Overexpression of AMP-activated protein kinase impairs pancreatic {beta}-cell function in vivo. *J Endocrinol* 187:225–235.
46. Leclerc, I., *et al.* (2004). Metformin, but not leptin, regulates AMP-activated protein kinase in pancreatic islets: impact on glucose-stimulated insulin secretion. *Am J Physiol Endocrinol Metab* 286:E1023–E1031.

47. Kefas, B.A., *et al.* (2003). AICA-riboside induces apoptosis of pancreatic beta cells through stimulation of AMP-activated protein kinase. *Diabetologia* 46:250–254.
48. Riboulet-Chavey, A., Diraison, F., Siew, L.K., Wong, F.S., and Rutter, G.A. (2008). Inhibition of AMP-activated protein kinase protects pancreatic beta-cells from cytokine-mediated apoptosis and CD8+ T-cell-induced cytotoxicity. *Diabetes* 57:415–423.
49. Fu, A., *et al.* (2009). Loss of Lkb1 in adult beta cells increases beta cell mass and enhances glucose tolerance in mice. *Cell Metab* 10:285–295.
50. Granot, Z., *et al.* (2009). LKB1 regulates pancreatic beta cell size, polarity, and function. *Cell Metab* 10:296–308.
51. Sun G, *et al.* (2010). LKB1 deletion with the RIP.Cre-transgene modifies pancreatic {beta}-cell morphology and enhances insulin secretion in vivo. *Am J Physiol Endocrinol Metab*. 298:E1261–E1273
52. Fraenkel, M., *et al.* (2008). mTOR inhibition by rapamycin prevents beta-cell adaptation to hyperglycemia and exacerbates the metabolic state in type 2 diabetes. *Diabetes* 57:945–957.
53. Hennige, A.M., *et al.* (2003). Upregulation of insulin receptor substrate-2 in pancreatic beta cells prevents diabetes. *J Clin Invest* 112:1521–1532.
54. Withers, D.J., Burks, D.J., Towery, H.H., Altamuro, S.L., Flint, C.L., and White, M.F. (1999). Irs-2 coordinates Igf-1 receptor-mediated beta-cell development and peripheral insulin signalling. *Nat Genet* 23:32–40.
55. Lingohr, M.K., Dickson, L.M., McCuaig, J.F., Hugl, S.R., Twardzik, D.R., and Rhodes, C.J. (2002). Activation of IRS-2-mediated signal transduction by IGF-1, but not TGF-alpha or EGF, augments pancreatic beta-cell proliferation. *Diabetes* 51:966–976.
56. Briaud, I., Dickson, L.M., Lingohr, M.K., McCuaig, J.F., Lawrence, J.C., and Rhodes, C.J. (2005). Insulin receptor substrate-2 proteasomal degradation mediated by a mammalian target of rapamycin (mTOR)-induced negative feedback down-regulates protein kinase B-mediated signaling pathway in beta-cells. *J Biol Chem* 280:2282–2293.
57. Gimeno, R.E., and Klaman, L.D. (2005). Adipose tissue as an active endocrine organ: recent advances. *Curr Opin Pharmacol* 5:122–128.
58. Shi, Y., and Burn, P. (2004). Lipid metabolic enzymes: emerging drug targets for the treatment of obesity. *Nat Rev Drug Discov* 3:695–710.
59. Kershaw, E.E., and Flier, J.S. (2004). Adipose tissue as an endocrine organ. *J Clin Endocrinol Metab* 89:2548–2556.
60. Colombani, J., Raisin, S., Pantalacci, S., Radimerski, T., Montagne, J., and Leopold, P. (2003). A nutrient sensor mechanism controls Drosophila growth. *Cell* 114:739–749.
61. Guertin, D.A., and Sabatini, D.M. (2007). Defining the role of mTOR in cancer. *Cancer Cell* 12:9–22.
62. Polak, P., and Hall, M.N. (2009). mTOR and the control of whole body metabolism. *Curr Opin Cell Biol* 21:209–218.
63. Bell, A., Grunder, L., and Sorisky, A. (2000). Rapamycin inhibits human adipocyte differentiation in primary culture. *Obes Res* 8:249–254.
64. Cho, H.J., Park, J., Lee, H.W., Lee, Y.S., and Kim, J.B. (2004). Regulation of adipocyte differentiation and insulin action with rapamycin. *Biochem Biophys Res Commun* 321:942–948.
65. El-Chaar, D., Gagnon, A., and Sorisky, A. (2004). Inhibition of insulin signaling and adipogenesis by rapamycin: effect on phosphorylation of p70 S6 kinase vs eIF4E-BP1. *Int J Obes Relat Metab Disord* 28:191–198.
66. Gagnon, A., Lau, S., and Sorisky, A. (2001). Rapamycin-sensitive phase of 3 T3-L1 preadipocyte differentiation after clonal expansion. *J Cell Physiol* 189:14–22.

67. Kim, J.E., and Chen, J. (2004). regulation of peroxisome proliferator-activated receptor-gamma activity by mammalian target of rapamycin and amino acids in adipogenesis. *Diabetes* 53:2748–2756.
68. Yeh, W.C., Bierer, B.E., and McKnight, S.L. (1995). Rapamycin inhibits clonal expansion and adipogenic differentiation of 3 T3-L1 cells. *Proc Natl Acad Sci USA* 92:11086–11090.
69. Polak, P., Cybulski, N., Feige, J.N., Auwerx, J., Ruegg, M.A., and Hall, M.N. (2008). Adipose-specific knockout of raptor results in lean mice with enhanced mitochondrial respiration. *Cell Metab* 8:399–410.
70. Zhang, H.H., *et al.* (2009). Insulin stimulates adipogenesis through the Akt-TSC2-mTORC1 pathway. *PLoS One* 4:e6189.
71. Um, S.H., *et al.* (2004). Absence of S6K1 protects against age- and diet-induced obesity while enhancing insulin sensitivity. *Nature* 431:200–205.
72. Le Bacquer, O., *et al.* (2007). Elevated sensitivity to diet-induced obesity and insulin resistance in mice lacking 4E-BP1 and 4E-BP2. *J Clin Invest* 117:387–396.
73. Cybulski, N., Polak, P., Auwerx, J., Ruegg, M.A., and Hall, M.N. (2009). mTOR complex 2 in adipose tissue negatively controls whole-body growth. *Proc Natl Acad Sci USA* 106:9902–9907.
74. Kumar A, *et al.* (2010). Fat cell-specific ablation of rictor in mice impairs insulin-regulated fat cell and whole body glucose and lipid metabolism. Diabetes. 59:1397–1406
75. Kloting, N., *et al.* (2008). Autocrine IGF-1 action in adipocytes controls systemic IGF-1 concentrations and growth. *Diabetes* 57:2074–2082.
76. Bodine, S.C., *et al.* (2001). Akt/mTOR pathway is a crucial regulator of skeletal muscle hypertrophy and can prevent muscle atrophy in vivo. *Nat Cell Biol* 3:1014–1019.
77. Pallafacchina, G., Calabria, E., Serrano, A.L., Kalhovde, J.M., and Schiaffino, S. (2002). A protein kinase B-dependent and rapamycin-sensitive pathway controls skeletal muscle growth but not fiber type specification. *Proc Natl Acad Sci USA* 99:9213–9218.
78. Izumiya, Y., *et al.* (2008). Fast/Glycolytic muscle fiber growth reduces fat mass and improves metabolic parameters in obese mice. *Cell Metab* 7:159–172.
79. Ohanna, M., *et al.* (2005). Atrophy of S6K1(−/−) skeletal muscle cells reveals distinct mTOR effectors for cell cycle and size control. *Nat Cell Biol* 7:286–294.
80. Bentzinger, C.F., *et al.* (2008). Skeletal muscle-specific ablation of raptor, but not of rictor, causes metabolic changes and results in muscle dystrophy. *Cell Metab* 8:411–424.
81. Handschin, C., *et al.* (2007). Skeletal muscle fiber-type switching, exercise intolerance, and myopathy in PGC-1alpha muscle-specific knock-out animals. *J Biol Chem* 282:30014–30021.
82. Cunningham, J.T., Rodgers, J.T., Arlow, D.H., Vazquez, F., Mootha, V.K., and Puigserver, P. (2007). mTOR controls mitochondrial oxidative function through a YY1-PGC-1alpha transcriptional complex. *Nature* 450:736–740.
83. Schieke, S.M., *et al.* (2006). The mammalian target of rapamycin (mTOR) pathway regulates mitochondrial oxygen consumption and oxidative capacity. *J Biol Chem* 281:27643–27652.
84. Aguilar, V., *et al.* (2007). S6 kinase deletion suppresses muscle growth adaptations to nutrient availability by activating AMP kinase. *Cell Metab* 5:476–487.
85. Kumar, A., Harris, T.E., Keller, S.R., Choi, K.M., Magnuson, M.A., and Lawrence, J.C., Jr. (2008). Muscle-specific deletion of rictor impairs insulin-stimulated glucose transport and enhances Basal glycogen synthase activity. *Mol Cell Biol* 28:61–70.
86. Postic, C., and Girard, J. (2008). Contribution of de novo fatty acid synthesis to hepatic steatosis and insulin resistance: lessons from genetically engineered mice. *J Clin Invest* 118:829–838.

87. Eberle, D., Hegarty, B., Bossard, P., Ferre, P., and Foufelle, F. (2004). SREBP transcription factors: master regulators of lipid homeostasis. *Biochimie* 86:839–848.
88. Shimano, H., Horton, J.D., Shimomura, I., Hammer, R.E., Brown, M.S., and Goldstein, J.L. (1997). Isoform 1c of sterol regulatory element binding protein is less active than isoform 1a in livers of transgenic mice and in cultured cells. *J Clin Invest* 99:846–854.
89. Horton, J.D., *et al.* (2003). Combined analysis of oligonucleotide microarray data from transgenic and knockout mice identifies direct SREBP target genes. *Proc Natl Acad Sci USA* 100:12027–12032.
90. Horton, J.D., Shimomura, I., Brown, M.S., Hammer, R.E., Goldstein, J.L., and Shimano, H. (1998). Activation of cholesterol synthesis in preference to fatty acid synthesis in liver and adipose tissue of transgenic mice overproducing sterol regulatory element-binding protein-2. *J Clin Invest* 101:2331–2339.
91. Shimomura, I., Shimano, H., Horton, J.D., Goldstein, J.L., and Brown, M.S. (1997). Differential expression of exons 1a and 1c in mRNAs for sterol regulatory element binding protein-1 in human and mouse organs and cultured cells. *J Clin Invest* 99:838–845.
92. Becard, D., Hainault, I., Azzout-Marniche, D., Bertry-Coussot, L., Ferre, P., and Foufelle, F. (2001). Adenovirus-mediated overexpression of sterol regulatory element binding protein-1c mimics insulin effects on hepatic gene expression and glucose homeostasis in diabetic mice. *Diabetes* 50:2425–2430.
93. Liang, G., Yang, J., Horton, J.D., Hammer, R.E., Goldstein, J.L., and Brown, M.S. (2002). Diminished hepatic response to fasting/refeeding and liver X receptor agonists in mice with selective deficiency of sterol regulatory element-binding protein-1c. *J Biol Chem* 277:9520–9528.
94. Yahagi, N., *et al.* (2002). Absence of sterol regulatory element-binding protein-1 (SREBP-1) ameliorates fatty livers but not obesity or insulin resistance in Lep(ob)/Lep(ob) mice. *J Biol Chem* 277:19353–19357.
95. Brown, N.F., Stefanovic-Racic, M., Sipula, I.J., and Perdomo, G. (2007). The mammalian target of rapamycin regulates lipid metabolism in primary cultures of rat hepatocytes. *Metabolism* 56:1500–1507.
96. Mauvoisin, D., Rocque, G., Arfa, O., Radenne, A., Boissier, P., and Mounier, C. (2007). Role of the PI3-kinase/mTor pathway in the regulation of the stearoyl CoA desaturase (SCD1) gene expression by insulin in liver. *J Cell Commun Signal* 1:113–125.
97. Porstmann, T., *et al.* (2008). SREBP activity is regulated by mTORC1 and contributes to Akt-dependent cell growth. *Cell Metab* 8:224–236.
98. Li S, Brown MS, Goldstein JL, Bifurcation of insulin signaling pathway in rat liver: mTORC1 required for stimulation of lipogenesis, but not inhibition of gluconeogenesis. *Proc Natl Acad Sci USA* **107**: 3441-3446.
99. Shaw, R.J., *et al.* (2005). The kinase LKB1 mediates glucose homeostasis in liver and therapeutic effects of metformin. *Science* 310:1642–1646.
100. Tzatsos, A., and Kandror, K.V. (2006). Nutrients suppress phosphatidylinositol 3-kinase/Akt signaling via raptor-dependent mTOR-mediated insulin receptor substrate 1 phosphorylation. *Mol Cell Biol* 26:63–76.
101. Zhou, G., *et al.* (2001). Role of AMP-activated protein kinase in mechanism of metformin action. *J Clin Invest* 108:1167–1174.
102. Zang, M., *et al.* (2004). AMP-activated protein kinase is required for the lipid-lowering effect of metformin in insulin-resistant human HepG2 cells. *J Biol Chem* 279:47898–47905.
103. Koketsu, Y., *et al.* (2008). Hepatic overexpression of a dominant negative form of raptor enhances Akt phosphorylation and restores insulin sensitivity in K/KAy mice. *Am J Physiol Endocrinol Metab* 294:E719–E725.

14

TOR Signaling and Aging

MALENE HANSEN[a] • PANKAJ KAPAHI[b]

[a]*Sanford-Burnham Medical Research Institute*
La Jolla, California, USA

[b]*Buck Institute for Age Research*
Novato, California, USA

I. Abstract

Aging remains an interesting, yet incompletely understood biological phenomenon. Intriguingly, aging has been found to be subject to profound genetic influence, as evidenced from extensive studies in model organisms from yeast to mice. Valuable insights from these organisms have shown that many conserved genes impact life span, including the nutrient sensor and kinase Target of Rapamycin (TOR). TOR is emerging as a key regulator of life span and health span, and the mechanism(s) by which TOR operates to affect organismal aging are currently under intense investigation. TOR regulates several important biological processes that could modulate aging. Two such TOR-regulated processes, protein translation and autophagy, a cellular recycling process, have received particular attention and have been shown to affect life span in a conserved fashion. Other processes such as metabolism and stress responses are also emerging to play critical roles. Here, we review current knowledge on how these TOR-regulated processes affect life span in common model organisms, such as yeast, worms, flies, and mice. We also discuss how the nutrient-sensing TOR pathway appears to be critically important for mediating the longevity effects of dietary restriction, a potent environmental method of life span extension by nutrient limitation observed in a multitude of different model organisms.

THE ENZYMES, Vol. XXVIII

ISSN NO: 1874-6047
DOI: 10.1016/S1874-6047(10)28014-9

II. Introduction

The TOR protein was originally identified in the yeast *Saccharomyces cerevisiae* as the target of rapamycin [1], a growth-inhibitory and immunosuppressing compound originally identified in a soil bacterium on the Easter Island Rapa Nui in 1975 [2]. Subsequently, decades of research have shown that TOR is present in all eukaryotes, operating in a highly conserved, nutrient-sensitive signaling pathway to regulate multiple biological processes, including growth and metabolism [3] (see also other chapters in this book).

TOR exists in at least two growth-regulatory complexes with different functions, TOR complex I (TORC1, which is sensitive to rapamycin) and TORC2 (which is insensitive to rapamycin) [4]. While TORC2 controls spatial aspects of growth within the cell, TORC1 integrates mitogen and nutrient signals to control cell proliferation and cell size [3]. As a consequence, TORC1 has been implicated in a large number of human diseases, including diabetes, obesity, heart disease, and cancer, that are characterized by aberrant cell growth and proliferation [5, 6]. Interestingly, a common risk factor for these diseases is aging, and growing evidence has directly implicated the TORC1 pathway in determining life span in multiple model systems. Specifically, inhibition of TORC1 (hereafter referred to as TOR), for example by rapamycin treatment, extends life span in yeast, worms, flies, and mice [7, 8]. In further support of this longevity role, TOR is emerging as a robust mediator of the beneficial effects of dietary restriction, a method known to increase life span and protect against age-related diseases by reducing nutrient intake without malnutrition in many species [9]. Interestingly, TOR activity is downregulated in multiple model organisms in response to dietary restriction, consistent with TOR's role as a nutrient sensor. TOR integrates a variety of upstream signals, which aside from nutrients like amino acids, also include growth factors, energy status, and various stressors [3]. According to the type of input, TOR responds by regulating a number of outputs, including protein translation, autophagy, metabolism, and stress responses (Figure 14.1). Below, we discuss how these processes represent possible conserved mechanisms by which TOR, as well as dietary restriction, modulate aging.

III. TOR and Aging in *S. cerevisiae*

Life span in the unicellular organisms, *S. cerevisiae* is measured in two different ways, either as replicative or chronological life span. Replicative life span measures the number of times a mother cell can produce

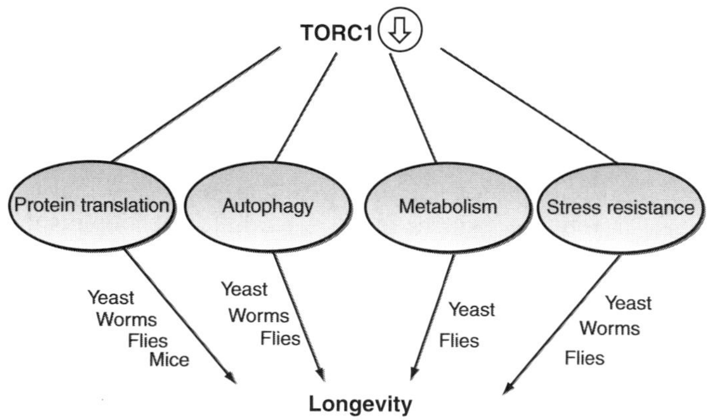

FIG. 14.1. Overview of TORC1-regulated processes with conserved effects on aging. Inhibition of TORC1 extends the life span of multiple model organisms by affecting several biological processes, including protein translation, the cellular recycling pathway autophagy, metabolism, and stress responses. TORC1 regulates these processes in response to a variety of upstream inputs, including growth factors, energy, and stress. See text for more details.

a daughter cell (~30–40 divisions) [10]. Chronological life span is measured by the length of time cells can survive in a quiescent state while retaining the ability to reenter the cell cycle upon stimulation by appropriate cues (median survival ~6-8 days) [11]. This latter type of aging has been proposed to represent features of postmitotic, senescent cells.

S. cerevisiae has been successfully used to investigate the role of TOR in aging, and *TOR1* can affect both replicative and chronological aging in yeast (a second *TOR* gene exists in yeast, *TOR2*, but this gene is essential). Deletion of *TOR1* [12–14], or inhibition of TORC1 by rapamycin treatment [14, 15] extends the chronological and replicative life span of yeast by ~15–100%, with the most potent effects on chronological life span.

TOR is also likely to mediate, at least in part, the effects of dietary restriction in *S. cerevisiae*. In yeast, dietary restriction can be imposed by a reduction in glucose levels, and this treatment does not further extend the long replicative life span of a *TOR1* deletion mutant [13]. As discussed in detail in the following section, changes in protein translation, autophagy, metabolism, and stress resistance have been implicated to exert TOR's effects on yeast life span.

A. Protein Translation

TOR is known to be a key regulator of protein translation, and inhibition of TOR has been extensively studied in multiple organisms, including yeast, and shown to decrease ribosome biogenesis as well as the process of protein translation itself. For example, when TOR activity is diminished, phosphorylation of a key substrate, ribosomal-protein S6 kinase (S6K) is reduced, which, in turn, leads to the dephosphorylation and inactivation of translation elongation factor 2 (eEF2) kinase, thereby inhibiting global translation [16]. TOR inhibition similarly reduces the phosphorylation of translation-initiation factor 4 binding proteins (eIF4-BPs) [17, 18]. As a consequence, eIF4-BPs bind to eIF4E, which impairs the recruitment of the 40S ribosomal subunit to the cap structure present at the 5′-end of eukaryotic mRNAs [19]. The mRNA cap-binding protein eIF4E is a component of the eIF4E complex, which also contains the scaffold protein eIF4G and the RNA helicase eIF4A [20]. Moreover, TOR regulates ribosomal biogenesis by multiple mechanisms, including the transcription of mRNAs encoding ribosomal proteins [3].

Several of these conserved components of the translational machinery have been linked to aging. For example, Kaeberlein and coworkers conducted a screen in *S. cerevisiae* to identify novel longevity genes and found that gene inactivation not only of *TOR1*, but also of the S6K homolog *SCH9* as well as many genes encoding ribosomal subunits extended yeast replicative life span [13]. Interestingly, mostly ribosomal proteins from the large 60S subunit were identified (albeit a few components from the small subunit have also been identified [21]). The mechanism by which depletion of 60S ribosomal subunits appears to regulate aging involves the nutrient-responsive transcription factor, Gcn4 [22]. Upon starvation, the global rate of protein synthesis decreases, but production of Gcn4 protein increases by an enhancement in ribosomal loading onto *GCN4* mRNA [23]. *GCN4* is required for deletion of *TOR1*, *SCH9*, and ribosomal subunits to extend life span [22], suggesting that Gcn4 functions as an effector for the life span effects mediated by reduced activity of the TOR pathway. Gcn4 is also likely to be an effector of dietary restriction in yeast, as Gcn4 levels increase under glucose restriction [22], and *GCN4* is partially required for dietary restriction to extend yeast life span [22]. Moreover, glucose restriction does not further extend the replicative life span of yeast deficient in *SCH9* [13, 24, 25], as is the case for *TOR1* inactivation [13]. Taken together, these experiments suggest a critical role for Tor1 and protein translation, via the transcription factor Gcn4, in the longevity response to dietary restriction in *S. cerevisiae* (See also Table 14.1).

TABLE 14.1

COMPONENTS WITH ROLES IN TOR-MEDIATED LONGEVITY RESPONSES IN VARIOUS SPECIES

Model organism	Proteins involved	DR role?[a]	References
Yeast	Tor1/TOR	Yes	[13, 22]
	Shc9/S6K	Yes	[13]
	Gnc4, ribosomal subunits	Yes	[22]
	Autophagy	Yes	[26, 27]
	Mitochondrial ETC genes	Yes	[12]
	Msn2/4		[15, 28]
	Rim15	Yes	[28]
	Gis1	Yes	[28]
	Pnc1	Yes	[15]
	Glycerol synthesis	Yes	[29]
Worms	LET-363/TOR	Yes	[30, 31]
	DAF-15/Raptor		[32–34]
	RHEB-1	Yes	[35]
	RSKS-1/S6K	Yes	[31, 36]
	Translation factors		[31, 37–40]
	Ribosomal proteins		[31, 37, 38]
	Autophagy	Yes	[34, 41, 83, 83a, 84]
	PHA-4/FOXA	Yes	[42, 73]
	DAF-16/FOXO	Yes	[32, 35, 43]
	HIF-1, EGL-9, VHL-1, IRE-1	Yes	[33, 45, 46, 88, 89]
	TOR-interacting proteins		[48]
Flies	dTOR, dTSC1/2, dS6K	Yes	[49, 50]
	d4E-BP, mitochondrial ETC	Yes	[51]
Mice	TOR		[52]
	S6K		[53]

[a]DR, dietary restriction. This column notes whether the longevity effects associated with protein are implicated in the longevity response to dietary restriction.

B. AUTOPHAGY

TOR was first identified as a pivotal regulator of the process of macroautophagy (hereafter referred to as autophagy) in yeast [54]. Autophagy is a process in which portions of the cytoplasm, including mitochondria and other organelles, are degraded under stress-full conditions, for example, nutrient limitation, allowing cellular macromolecules to be catabolized and recycled. During autophagy, large double-membrane vesicles, called autophagosomes, are generated and degraded in lysosomes, together with their contents. The breakdown products are subsequently recycled to the cytoplasm [55]. The regulation of the autophagy process has been studied extensively in yeast, where autophagy is controlled by the *ATG* genes, many of which have functional orthologs in other organisms [55–57]. For example, TOR inhibits the protein kinase Atg1, which mediates an early activation step in the autophagic process [58]. In response to Atg1 activity, the Vps34 complex, which contains the Class III phosphatidylinositol-3-kinase Vps34 as well as Atg6/Vps30, the ortholog of the mammalian gene

Beclin1, stimulates and nucleates the formation of autophagosomes [56, 59, 60]. Other Atg proteins are important for vesicle expansion, for example, the autophagosomal membrane-localized protein Atg8/LC3, and finally retrieval, including the transmembrane protein Atg9 [55].

Importantly, several autophagy genes, including *ATG1*, are required for rapamycin to extend the chronological life span in yeast, suggesting a role for autophagy in TOR-mediated longevity [26]. Consistent with a critical role for autophagy in yeast aging, autophagy genes have also been found to be required for normal chronological life span in yeast [27, 61]. So far, a role for autophagy genes in yeast replicative aging has not been reported. In response to starvation, autophagy is known to degrade various components in the cell, including mitochondria and long-lived proteins [55]. A future challenge will be to investigate what type of cargo is being turned over in the cells with low TOR levels, or when subjected to nutrient limitation.

C. Metabolism

TOR also plays a role in metabolism, including energy sensing via mitochondrial signaling in yeast. TOR has been associated with mitochondria [62], and TOR activity is regulated in a redox-sensitive manner [5], suggesting TOR as a possible link between nutrient/energy sensing and mitochondrial activity. Interestingly, the connection between TOR and mitochondria seems to also play a role in life span. For example, inactivation of TOR has been shown to extend the chronological life span of *S. cerevisiae*, primarily by increasing mitochondrial respiration via increased translation specifically of mitochondria-encoded electron transport complex (ETC) subunits [12]. While the exact mechanism for how TOR signals to mitochondria is currently unclear, it is possible that life span extension can result from an increase in mitochondrial oxygen consumption that might limit intracellular oxygen and thereby reduce damage by reactive oxygen species [63].

D. Stress Responses

Life span extension by TOR inhibition in *S. cerevisiae* is likely to be mediated by regulators of various stress responses. For example, the two stress-responsive transcription factors, *MSN2* and *MSN4*, are required for the life span extension induced by rapamycin [15]. TOR inhibition causes a translocation of these transcription factors to the nucleus [14, 15], where they increase the expression of the nicotinamidase gene *PCN1*, an inducer of another longevity gene, the nicotinamide adenine dinucleotide (+)-dependent deacetylase *SIR2*. Moreover, the serine/threonine kinase Rim15 positively regulates Msn2/4 and the stress-responsive transcription factor Gis1, which are required for the extension of the chronological life span caused by *TOR1* deletion or by dietary restriction [28]. Consistent with these observations, inactivation of TOR signaling improves resistance to heat and oxidative stress [14]. Interestingly, this stress resistance might be caused, at least in part, by switching metabolism to enhance glycerol biosynthesis [29]. Deletion of glycerol biosynthesis genes upregulated in long-lived *TOR1* and *SCH9* mutants is sufficient to reverse the chronological life span extension and to reduce the stress resistance observed in *SCH9* mutants [29], suggesting the interesting possibility that glycerol production enhances stress resistance and longevity in yeast.

As mentioned earlier, life span extension by TOR mutation and dietary restriction is also dependent on the transcription factor *GCN4*, which aside from responding to nutrient levels, is also required for activating genes in response to endoplasmic reticulum (ER) stress [64]. It remains unknown whether *TOR1* mutant yeast is resistant to ER stress and which *GCN4*-regulated genes are responsible for TOR-mediated lifespan effects.

IV. TOR and Aging in *C. elegans*

The microscopic soil nematode, *Caenorhabditis elegans*, has proven especially useful in elucidating the role of multiple longevity genes with potential function in the TOR pathway. With a short life span of ~2–3 weeks, physiological complexity with multiple tissues (despite its small size ~1 mm in length), and genetic tractability, (for example, feeding RNAi allows for inactivation of genes specifically in the adult animal) the worm has been extremely useful to identify and characterize the role of longevity genes tied to TOR function.

Partial inhibition of the single gene *tor* (*let-363*) in *C. elegans* has been shown to extend life span. For example, RNAi inhibition of *tor* specifically during adulthood extends the life span of wild-type adult animals ~20–100% [30, 31]. As the adult *C. elegans* is postmitotic, this result points to a specific role for TOR, which is required for growth during development [65], in adult longevity. Similarly to the results with *tor*, heterozygous animals carrying a mutation in the TORC1 regulatory-associated protein *daf-15*/Raptor also live long [32]. Importantly, the *daf-15*/Raptor gene provides an interaction point with the well-characterized longevity pathway, i.e. the insulin/IGF-1 signaling pathway [66]. Life span extensions induced by mutations in components of the conserved *daf-2*/insulin/IGF-1/PI3K/AKT pathway require the FOXO transcription factor DAF-16 to extend life span [66]. The longevity phenotype of *daf-15*/raptor is suppressed by *daf-16*/FOXO and *daf-15*/raptor is a transcriptional target of *daf-16*/FOXO [32]. As TOR RNAi does not further extend the life span of *daf-2*/insulin/IGF-1-like receptor mutants [30, 31], these observations support links between the insulin/IGF-1 signaling pathway and TOR signaling in aging, as observed for other functions of TOR, such as growth [67, 68]. Interestingly, another interacting partner of TOR, Rictor, seems to have opposite effects on longevity, as inactivation of this gene shortens *C. elegans* life span [69]. Lastly, inactivation of a predicted activator of TOR, the small GTPase *rheb-1*, has been reported to extend the life span of worms [35]. Taken together, the data from *C. elegans* strongly support a role for the TOR pathway in aging.

As in yeast, the longevity effects of dietary restriction in *C. elegans* also appear to be, at least in part, mediated by TOR. In worms, dietary restriction is typically induced either by dilution of the food source *E. coli* or by genetic mutation, for instance of the acetylcholine-receptor *eat-2*, which renders the animals incapable of pumping in normal amounts of food [43, 70, 71]. Alternatively, *C. elegans* can enjoy life span extension by a regimen of intermittent fasting [35]. Several lines of evidence suggest a role for TOR in the longevity effects induced by dietary restriction. First, just like dietary restriction, *tor* inhibition extends the life span of *C. elegans* when food intake is reduced during adulthood only [30, 72]. Second, both *tor* and the TOR-activating GTPase *rheb-1* are involved in the longevity extension induced by intermittent fasting in *C. elegans* [35]. Third, *tor* inhibition does not further extend the life span of dietary-restricted *eat-2* mutants [31]. Fourth, genetic analysis supports an overlap in mechanisms between TOR and dietary restriction. The life span extension of worms with reduced TOR levels, just like dietary-restricted worms [73], requires the FOXA transcription factor PHA-4 to extend the life span as discussed in detail below [42, 73]. Taken together, these data strongly support a conserved role for TOR in mediating the beneficial effects residing from dietary restriction on life span.

A. Protein Translation

Multiple studies have shown a direct role for protein translation in the longevity of *C. elegans*. Inactivation of multiple components of the protein translation machinery, for example ribosomal proteins, or regulators of the process, for example S6K and multiple translation-initiation factors, including eIF2b, eIF4G, eIF4E, and eIF4H, can significantly extend the life span of *C. elegans* [31, 36–40, 74, 75]. Consistent with their proposed function, inhibition of these genes reduces overall protein translation rates [31, 36]. Interestingly, reduction of regulators of translation, including of TOR itself, has been observed to make these animals more resistant to different stresses, including heat and starvation, suggesting a possible mechanism for how these animals live longer [31, 36, 39, 75]. However, a more systematic analysis of the types of stress resistance induced by different protein translation mutants is needed to firmly establish this point. Interestingly, the genetic requirements appear to differ between the different ways of inhibiting protein synthesis. For example, reduction of eIFs, but not inhibition of TOR, ribosomal proteins, or S6K, requires the DAF-16/FOXO transcription factor to extend life span [31, 74]. It is interesting to note that while *C. elegans* does not appear to have a 4E-BP homolog, 4E-BP is a transcriptional target of FOXO in flies and in mammals [47, 76, 77]. Consistent with the specificity in regard to DAF-16/FOXO, inactivation of S6K, just like TOR, but not of eIFs, instead requires another forkhead transcription factor PHA-4/FOXA to promote longevity [42]. The mammalian orthologs of PHA-4, FoxA4, FoxA2, and FoxA3, all play key roles in development and act later in life to regulate glucagon production and glucose homeostasis in response to fasting. In *C. elegans*, PHA-4/FOXA plays an essential role in the development of the foregut [78], and the lethality of *pha-4*/FOXA mutants can be suppressed by TOR and S6K inactivation [42], suggesting a critical interaction between TOR signaling and PHA-4/FOXA. Taken together, these observations suggest that at least two different ways of extending longevity by inhibiting global protein translation exist in *C. elegans*. Further experiments will be required to fully characterize and compare these different pathways in more detail.

The mechanism by which S6K extends lifespan has been investigated in some detail. Inactivation of the regulatory subunit of the AMP-activated kinase AMPK, called AAK-2 in worms, rescues the long lifespan S6K mutants as well as the reduced size and fecundity of these animals [53]. Moreover, AAK-2 is hyperphosphorylated in these animals [53], suggesting that S6K can modulate longevity by post-translational mechanisms. AAK-2 is required for at least some forms of dietary restriction to extend lifespan in *C. elegans* [79], possibly by regulating TOR. When energy levels drop, AMP levels increase, causing AMPK to get activated. This activation is known to lead to the inhibition of TOR activity [80–82]. Hence, these observations suggest the existence of a positive feedback-signaling loop between AMPK, TOR, S6K, and possibly dietary restriction–a complex signaling network that awaits much further investigation.

B. Autophagy

A direct link between the cellular recycling process autophagy and organismal aging was first made in *C. elegans* (e.g., in long-lived *daf-2*/insulin/IGF-1 receptor mutants) [83]. As in yeast, TOR regulates autophagy in *C. elegans*, since animals with reduced levels of TOR or *daf-15*/Raptor heterozygotes have increased levels of autophagy [34]. Consistent with this, rapamycin also induces autophagy in *C. elegans* (C. Kumsta and

M. Hansen, unpublished observations). This induction in autophagy is critical for the life span extension observed in animals with reduced TOR activity. For example, inactivation during adulthood of genes with autophagy functions, for example, *vps-34* and *bec-1*, are required for *daf-15*/Raptor heterozygotes to live long [34], and *tor* RNAi fails to extend the life span of short-lived autophagy mutants, including *atg-1*/*unc-51* and *atg-18* mutants [41]. Consistent with an overlap in mechanisms between TOR and dietary restriction, dietary-restricted animals have increased autophagy [34, 41] and require autophagy genes during adulthood to live long [34, 41, 84]. Moreover, the induction of autophagy in dietary-restricted *eat-2* mutants requires the transcription factor PHA-4/FOXA [34], suggesting that autophagy is transcriptionally regulated in response to nutrients. Taken together, these data strongly suggest that TOR, as well as dietary restriction, relies on the upregulation of the autophagy process to extend *C. elegans* life span.

C. Metabolism

As observed in other organisms [3], TOR regulates fat metabolism in *C. elegans*. Both reduction of TOR and inactivation of the TOR interactor Rictor increase fat levels in *C. elegans* [30, 69, 85]. While lipolysis has recently been linked to longevity in worms [86], it is unclear whether the fat phenotype observed in animals with reduced TOR activity is directly linked to longevity, yet inactivation of Rictor decreases *C. elegans* life span [69].

D. Stress Responses

As mentioned earlier, reduction of TOR increases the stress resistance of *C. elegans*. TOR regulates stress modulators in mammalian cells, including the hypoxia-responsive transcription factor HIF-1 [87]. HIF-1 was recently linked to *C. elegans* longevity in several studies [33, 45, 46, 88, 89], which showed HIF-1 to function in a complex context-dependent manner, either promoting or inhibiting longevity depending on the conditions. For example, inhibition of *hif-1* increases longevity at 25 °C [33], and *hif-1* inhibition does not further extend the life span of *daf-15*/Raptor mutants [33], suggesting that the mechanism by which TOR and HIF-1 extend life span might be overlapping. Interestingly, *hif-1* appears to function through an ER stress pathway, as *hif-1* mutants require an ER transmembrane protein that senses misfolded proteins in the ER lumen, IRE-1, to extend life span [33]. Interestingly, the response to longevity by *hif-1* is nutrient-dependent, and genetic epistasis analysis suggests that *hif-1* functions downstream of S6K to mediate life span extension [33]. Taken together, these data suggest that dietary restriction and TOR inhibition can extend life span via an ER stress- and hypoxia-related mechanism in *C. elegans*, still to be fully investigated.

V. TOR and Aging in *Drosophila*

The fruit fly *Drosophila* is also a frequently used model organism to study the mechanism by which conserved longevity genes such as TOR affect organismal aging. *Drosophila* has a mean life span of ~3 months, yet flies have increased physiological complexity, including a simple heart and a complex brain.

As is the case in yeast and worms, dTOR reduction extends the mean life span in *Drosophila*. Specifically, inactivation of *dTOR* or activation of the upstream negative regulators tuberous sclerosis complex gene 1 (*dTSC1*), and *dTSC2*, extends the mean life span of flies by ~20–30% [49, 50]. Similarly, reduction of dTOR pathway activity protects against age-related, organ-specific declines in the fly, as these conditions improve heart function with age [49, 90]. Consistent with these observations, rapamycin can also promote a ~10% increase in mean life span in *Drosophila* [88]. Interestingly, upregulation of *dTSC2* specifically in the fat body, the major site for fat storage in flies, is sufficient for life span-extending effects in *Drosophila* [49, 50]. This observation suggests a key role for fat-storing tissues, as also observed in long-lived flies overexpressing dFOXO [44, 91], and in long-lived FIRKO (fat-specific insulin receptor knockout) mice [92]. While large amounts of data show that signaling through the TOR pathway can act both in parallel to but also interact with the insulin/IGF-1 pathway in flies [67, 68], only few experiments have directly explored a possible link in life span. For example, rapamycin causes a life span increase of weak insulin/IGF-1 pathway mutants, suggesting that additional mechanisms are involved [88]. Clearly, more experiments, including genetic epistasis analysis, are needed to further clarify the interaction between the dTOR and the insulin/IGF-1 signaling pathways in regard to aging.

Dietary restriction in flies, either by restriction of total food [93–95] or by reduction of the amount of yeast in the food [50, 96], potently extends life span, and dTOR and dietary restriction may act by an overlapping mechanism. For example, the long life span of mutants overexpressing dTSC1 and dTSC2 cannot be further extended by dietary restriction [68]. However, a combination of rapamycin and dietary restriction, maximized for life span extension, actually can cause some additional longevity effects when compared to each single treatment [88], indicating that the effects of rapamycin, and by extension TOR, are not completely identical to the mechanisms by which dietary restriction extends the life span. Indeed, it is possible that rapamycin has additional TOR-independent targets that have yet to be identified.

A. Protein Translation

Consistent with a role for protein translation in fly aging, several downstream effectors of dTOR affect longevity. For example, dominant-negative S6K fly mutants have an extended life span [68]. Likewise, flies overexpressing an activated form of the translational repressor d4E-BP live long [51], while animals with reduced d4E-BP levels are short-lived [97]. Interestingly, d4E-BP mediates differential translation of specific targets with an extended 5′-UTR, including nuclear-encoded mitochondrial proteins [51], similar to the situation observed in yeast with reduced TOR activity. Importantly, the longevity effects of these differentially expressed targets are likely to reside from overlapping mechanisms with the longevity response to dietary restriction [51]. Consistent with these observations, d4E-BP is required for maximal life span extension upon dietary restriction in flies [51].

B. Autophagy

dTOR regulates autophagy in adult flies in response to starvation [98], and dTOR mutants have increased autophagy [98]. Importantly, overexpression of the upstream autophagy-specific protein kinase Atg1 is sufficient to potently induce autophagy, yet this high increase in autophagy results in apoptotic cell death, and adult animals have reduced viability [99]. In contrast, localized overexpression of the autophagic vesicle marker LC3/Atg8 using a specific promoter for expression in neurons is sufficient to extend the life span of the fly by more than 50% [100]. Consistent with results from other organisms, autophagy gene *Atg5* is required for rapamycin to extend the life span in flies [88], and fly mutants with reduced levels of autophagy components are short-lived [41, 101]. These experiments suggest that increases in autophagy could be beneficial to the organism, yet dose- and tissue-specific induction of autophagy are likely to be critical to obtain such systemic effects.

C. Metabolism

Similar to the link observed in yeast, translational regulation of mitochondrial proteins, including ETC components, has also been associated with the longevity benefits of dTOR [51]. In a study to find differentially translated mRNA targets induced under dietary restriction, nuclear-encoded mitochondrial genes were found to be highly expressed in long-lived flies subjected to dietary restriction, and these genes were required for dietary restriction to extend life span [51]. Consistent with this, mitochondrial processes, including mitochondrial protein density, as well as cytochrome C oxidase and complex II activity, are increased in a d4E-BP-dependent fashion in response to dietary restriction [51]. It is unclear whether dTOR reduction *per se* would lead to exactly the same changes as the ones dietary restriction induces in flies, yet future experiments should help clarify this aspect. Taken together, these observations support the idea that a deficiency of mitochondrial function with age, observed in multiple species, may be responsible for organismal aging [102]. On the other hand, the increased translation of ETC genes as a response to dietary restriction may serve as a protective mechanism to increase mitochondrial efficiency and maintain the function of the ETC to ensure sustained ATP production, which is otherwise known to decline with age [102].

dTOR also regulates fat metabolism in flies as *dTOR* mutation has been found to reduce fat levels, and this decrease in fat levels correlates with an increase in the expression of the lipase Brummer [84]. In contrast, rapamycin treatments appears to increase fat storage in *Drosophila* [88]. Taken together, the regulation of fat metabolism by dTOR is likely to be complex and awaits further investigation, including if this regulation is linked to longevity.

D. Stress Responses

dTOR inhibition in adult flies by rapamycin increases stress resistance to both starvation and oxidative stress [88], and d4E-BP is required for the latter [97]. However, partial reduction of dTOR, despite resulting in life span extension, is not sufficient to induce these beneficial effects on the organism [49]. Clearly, the unexpected differences between these results need further examination. Interestingly, a recent report has suggested that sestrins, which are conserved proteins that accumulate in cells exposed to stress, act as feedback inhibitors of dTOR to prevent age-related pathologies, including lipid accumulation and decline in heart function [103]. dTOR signaling was proposed to induce sestrins by a mechanism that relies on increasing reactive oxygen species, potentially as a result of reduced levels of autophagy [103]. It remains to be investigated whether overexpression of sestrins is sufficient to alleviate age-linked pathologies or longevity.

VI. TOR and Aging in Mammals

While invertebrate model organisms have been extremely valuable in elucidating TOR function in longevity, the mouse, with a mean life span of ~2 years, represents a very practical model for longevity experiments to confirm conservation in higher mammals. A key example is that inhibition of TOR was recently found to extend the life span of rodents, similarly to what has been described earlier for yeast, worm, and flies. Specifically, it was found that feeding rapamycin to three different cohorts of (i.e., genetically heterogenous) mice significantly extended both female and male life span by ~10–15% [52]. Importantly, some of the mice were fed rapamycin starting in late adulthood (~600 days, or ~50 months, of age), and these mice still enjoyed a life span extension. Similarly, another study demonstrated that rapamycin treatment initiated at mid-age (~22–24 months) increased life span and abrogated age-related loss in immune function and boosted hematopoiesis [104]. Taken together, these studies suggest that inhibiting the TOR pathway not only extends the life span in mammals, but might also work to prevent age-related diseases when initiated late in the animal's life.

A. Protein Translation

Consistent with a role for TOR in life span in mammals, *S6K1* mutant mice were recently reported to be long-lived [53]. Importantly, the ~20% increase in life span observed in *S6K1*$^{-/-}$ animals is accompanied by an increased resistance to a variety of age-related pathologies, including bone, immune, and motor dysfunction, as well as better insulin sensitivity compared to wild-type animals. *S6K1*$^{-/-}$ mice were also smaller and less fat than normal animals, consistent with a previous report that *S6K1*$^{-/-}$ mice are resistant to a high-fat diet and have improved insulin sensitivity [105]. Interestingly, these physiological effects were only prominent in females, similar to the gender effects observed in long-lived insulin/IGF-1 signaling mutants, including mice without the insulin receptor substrate *Irs1* [106, 107]. As S6K is known to regulate IRS, it is possible that some of the effects of S6K are mediated via the insulin/IGF-1 signaling pathway. Moreover, long-lived Ames dwarf mice with mutations in a gene called *Prop-1* that disrupts pituitary gland development and as a consequence have reduced levels of circulating growth hormone and insulin-like growth factor 1, show decreased protein synthesis possibly via inhibition of TOR signaling [108]. Consistent with links between *S6K* and *Irs1*, a high degree of overlap in transcriptional changes in the liver was observed between genes regulated in *S6K1*$^{-/-}$ mutants [53] and in *Irs1*$^{-/-}$ mice [106]. Moreover, similar overlaps were observed in mice subjected to dietary restriction [109]. Taken together, these observations are consistent with the existence of common mechanisms underlying the longevity effects of S6K, IRS, and dietary restriction on aging, and suggest that therapeutic interventions of S6K1 could mimic dietary restriction and provide protection against age-related diseases.

B. Metabolism

Aside from S6K1's function on protein translation, S6K1 also has anabolic effects on the cellular energy respondent AMPK [110]. Specifically, *S6K1*$^{-/-}$ mice have increased activity of AMPK [110], as also seen in *C. elegans* S6K mutants. Consistent with this, the AMPK activator aminoimidazole carboxamide ribonucleotide (AICAR) more potently induces AMPK activity in hepatocytes from *S6K1*$^{-/-}$ mutants than from wild-type animals [53]. Moreover, a strong overlap exists between gene categories associated with lipid metabolism in the muscle of *S6K*$^{-/-}$ mutants compared to wild-type animals [53]. It will be interesting to investigate the role of metabolism in the extended life span observed in S6K$^{-/-}$ mice in more detail in the future.

VII. Conclusion and Future Perspectives

TOR is emerging as a key modulator of life span in multiple organisms, suggesting a conserved longevity role possibly also in humans. Likewise, TOR is a strong candidate regulator of, at least in part, the beneficial effects of dietary restriction. TOR is likely to mediate longevity effects by regulating several key biological processes. These processes include protein translation, autophagy, metabolism, and stress responses. The precise mechanism by which TOR regulates these processes, and how they affect aging is not yet clear. For example, does reducing protein translation by inhibiting different regulators of the process, for example, S6K and eIFs, regulate life span by different mechanisms, and which downstream effectors do they engage to do this? Likewise, it is unknown how the cellular recycling process of autophagy would contribute to organismal aging in terms of which types of material this process is degrading in long-lived TOR mutants, as well as in dietary-restricted animals. Other outstanding questions relate to the role of fat metabolism in animals with reduced TOR activity, as well as the exact physiological role of increasing stress responses in animals. Another major challenge will be to elucidate the tissue-specific requirements for TOR and TOR's effectors, as well as to better understand the crosstalk of TOR with other longevity pathways, including the insulin/IGF-1 signaling pathway. Moreover, TOR is likely to utilize other effectors than the ones discussed here to regulate longevity, and the exact upstream signals that TOR responds to in order to modulate organismal aging have not yet been identified.

As evidenced by the important findings summarized here, such endeavors might be well worth initiating in the simpler model organisms. Acquired new insight should not only provide a better molecular understanding of aging, but is also likely to shed light on the link between diet and various age-related disorders such as diabetes, neurodegeneration, and cancer. Such knowledge might ultimately facilitate development of treatments for these serious diseases.

Acknowledgments

The authors wish to apologize for references that were left out in the interest of space, and to thank Drs. Rolf Bodmer and Caroline Kumsta for reading the manuscript. This work was supported by grants from the Ellison Medical Research Foundation (MH and PK), American Federation for Aging Research (MH and PK), the Hillblom Foundation (PK), a Nathan Shock Startup award (PK), and the NIH National Institute on Aging (PK).

References

1. Brown, E.J., *et al.* (1994). A mammalian protein targeted by G1-arresting rapamycin-receptor complex. *Nature* 369:756–758.
2. Vezina, C., Kudelski, A., and Sehgal, S.N. (1975). Rapamycin (AY-22, 989), a new antifungal antibiotic. I. Taxonomy of the producing streptomycete and isolation of the active principle. *J Antibiot (Tokyo)* 28:721–726.
3. Wullschleger, S., Loewith, R., and Hall, M.N. (2006). TOR signaling in growth and metabolism. *Cell* 124:471–484.
4. Martin, D.E., and Hall, M.N. (2005). The expanding TOR signaling network. *Curr Opin Cell Biol* 17:158–166.
5. Sarbassov, D.D., Ali, S.M., and Sabatini, D.M. (2005). Growing roles for the mTOR pathway. *Curr Opin Cell Biol* 17:596–603.
6. Inoki, K., Corradetti, M.N., and Guan, K.L. (2005). Dysregulation of the TSC–mTOR pathway in human disease. *Nat Genet* 37:19–24.
7. Kapahi, P., *et al.* (2010). With TOR, less is more: a key role for the conserved nutrient-sensing TOR pathway in aging. *Cell Metab* 11:453–465.
8. Evans, D.S., Kapahi, P., Hsueh, W.C., and Kockel, L. (2010). TOR signaling never gets old: aging, longevity and TORC1 activity. *Ageing Res Rev*.
9. Fontana, L., Partridge, L., and Longo, V.D. (2010). Extending healthy life span—from yeast to humans. *Science* 328:321–326.
10. Mortimer, R.K., and Johnston, J.R. (1959). Life span of individual yeast cells. *Nature* 183:1751–1752.
11. Fabrizio, P., and Longo, V.D. (2003). The chronological life span of *Saccharomyces cerevisiae*. *Aging Cell* 2:73–81.
12. Bonawitz, N.D., Chatenay-Lapointe, M., Pan, Y., and Shadel, G.S. (2007). Reduced TOR signaling extends chronological life span via increased respiration and upregulation of mitochondrial gene expression. *Cell Metab* 5:265–277.
13. Kaeberlein, M., *et al.* (2005). Regulation of yeast replicative life span by TOR and Sch9 in response to nutrients. *Science* 310:1193–1196.
14. Powers, R.W., III, Kaeberlein, M., Caldwell, S.D., Kennedy, B.K., and Fields, S. (2006). Extension of chronological life span in yeast by decreased TOR pathway signaling. *Genes Dev* 20:174–184.

15. Medvedik, O., Lamming, D.W., Kim, K.D., and Sinclair, D.A. (2007). MSN2 and MSN4 link calorie restriction and TOR to sirtuin-mediated lifespan extension in *Saccharomyces cerevisiae*. *PLoS Biol* 5:e261.
16. Wang, X., Li, W., Williams, M., Terada, N., Alessi, D.R., and Proud, C.G. (2001). Regulation of elongation factor 2 kinase by p90(RSK1) and p70 S6 kinase. *EMBO J* 20:4370–4379.
17. Tee, A.R., and Blenis, J. (2005). mTOR, translational control and human disease. *Semin Cell Dev Biol* 16:29–37.
18. Hay, N., and Sonenberg, N. (2004). Upstream and downstream of mTOR. *Genes Dev* 18:1926–1945.
19. Sonenberg, N., and Gingras, A.C. (1998). The mRNA 5′ cap-binding protein eIF4E and control of cell growth. *Curr Opin Cell Biol* 10:268–275.
20. Marintchev, A., and Wagner, G. (2004). Translation initiation: structures, mechanisms and evolution. *Q Rev Biophys* 37:197–284.
21. Chiocchetti, A., *et al.* (2007). Ribosomal proteins Rpl10 and Rps6 are potent regulators of yeast replicative life span. *Exp Gerontol* 42:275–286.
22. Steffen, K.K., *et al.* (2008). Yeast life span extension by depletion of 60s ribosomal subunits is mediated by Gcn4. *Cell* 133:292–302.
23. Hinnebusch, A.G. (2005). Translational regulation of GCN4 and the general amino acid control of yeast. *Annu Rev Microbiol* 59:407–450.
24. Urban, J., *et al.* (2007). Sch9 is a major target of TORC1 in *Saccharomyces cerevisiae*. *Mol Cell* 26:663–674.
25. Fabrizio, P., Pozza, F., Pletcher, S.D., Gendron, C.M., and Longo, V.D. (2001). Regulation of longevity and stress resistance by Sch9 in yeast. *Science* 292:288–290.
26. Alvers, A.L., Wood, M.S., Hu, D., Kaywell, A.C., Dunn, W.A., Jr., and Aris, J.P. (2009). Autophagy is required for extension of yeast chronological life span by rapamycin. *Autophagy* 5:847–849.
27. Alvers, A.L., *et al.* (2009). Autophagy and amino acid homeostasis are required for chronological longevity in *Saccharomyces cerevisiae*. *Aging Cell* 8:353–369.
28. Wei, M., *et al.* (2008). Life span extension by calorie restriction depends on Rim15 and transcription factors downstream of Ras/PKA, Tor, and Sch9. *PLoS Genet* 4:e13.
29. Wei, M., *et al.* (2009). Tor1/Sch9-regulated carbon source substitution is as effective as calorie restriction in life span extension. *PLoS Genet* 5:e1000467.
30. Vellai, T., Takacs-Vellai, K., Zhang, Y., Kovacs, A.L., Orosz, L., and Muller, F. (2003). Genetics: influence of TOR kinase on lifespan in *C. elegans*. *Nature* 426:620.
31. Hansen, M., Taubert, S., Crawford, D., Libina, N., Lee, S.J., and Kenyon, C. (2007). Lifespan extension by conditions that inhibit translation in *Caenorhabditis elegans*. *Aging Cell* 6:95–110.
32. Jia, K., Chen, D., and Riddle, D.L. (2004). The TOR pathway interacts with the insulin signaling pathway to regulate *C. elegans* larval development, metabolism and life span. *Development* 131:3897–3906.
33. Chen, D., Thomas, E.L., and Kapahi, P. (2009). HIF-1 modulates dietary restriction-mediated lifespan extension via IRE-1 in *Caenorhabditis elegans*. *PLoS Genet* 5:e1000486.
34. Hansen, M., Chandra, A., Mitic, L.L., Onken, B., Driscoll, M., and Kenyon, C. (2008). A role for autophagy in the extension of lifespan by dietary restriction in *C. elegans*. *PLoS Genet* 4:e24.
35. Honjoh, S., Yamamoto, T., Uno, M., and Nishida, E. (2009). Signalling through RHEB-1 mediates intermittent fasting-induced longevity in *C. elegans*. *Nature* 457:726–730.
36. Pan, K.Z., *et al.* (2007). Inhibition of mRNA translation extends lifespan in *Caenorhabditis elegans*. *Aging Cell* 6:111–119.

37. Chen, D., Pan, K.Z., Palter, J.E., and Kapahi, P. (2007). Longevity determined by developmental arrest genes in *Caenorhabditis elegans*. *Aging Cell* 6:525–533.
38. Curran, S.P., and Ruvkun, G. (2007). Lifespan regulation by evolutionarily conserved genes essential for viability. *PLoS Genet* 3:e56.
39. Tohyama, D., Yamaguchi, A., and Yamashita, T. (2008). Inhibition of a eukaryotic initiation factor (eIF2Bdelta/F11A3.2) during adulthood extends lifespan in *Caenorhabditis elegans*. *FASEB J* 22:4327–4337.
40. T.T. Ching, A.B. Paal, A. Mehta, L. Zhong, A.L. Hsu. (2010). drr-2 encodes an eIF4H that acts downstream of TOR in diet-restriction-induced longevity of *C. elegans*. *Aging Cell* 9:545–557.
41. Toth, M.L., *et al.* (2008). Longevity pathways converge on autophagy genes to regulate life span in *Caenorhabditis elegans*. *Autophagy* 4:330–338.
42. Sheaffer, K.L., Updike, D.L., and Mango, S.E. (2008). The target of rapamycin pathway antagonizes pha-4/FoxA to control development and aging. *Curr Biol* 18:1355–1364.
43. Greer, E.L., and Brunet, A. (2009). Different dietary restriction regimens extend lifespan by both independent and overlapping genetic pathways in *C. elegans*. *Aging Cell* 8:113–127.
44. Hwangbo, D.S., Gershman, B., Tu, M.P., Palmer, M., and Tatar, M. (2004). *Drosophila* dFOXO controls lifespan and regulates insulin signalling in brain and fat body. *Nature* 429:562–566.
45. Müller, R.U., *et al.* (2009). The von Hippel Lindau tumor suppressor limits longevity. *J Am Soc Nephrol* 12:2513–2517.
46. Bellier A., *et al.*, (2009) Hypoxia and the hypoxic response pathway protect against pore-forming toxins in *C. elegans*. *PLoS Pathog* 12:e1000689.
47. Teleman, A.A., *et al.* (2005). 4E-BP functions as a metabolic brake used under stress conditions but not during normal growth. *Genes Dev* 19:1844–1848.
48. Bell, R., *et al.* (2009). A human protein interaction network shows conservation of aging processes between human and invertebrate species. *PLoS Genet* 5:e1000414.
49. Luong, N., *et al.* (2006). Activated FOXO-mediated insulin resistance is blocked by reduction of TOR activity. *Cell Metab* 4:133–142.
50. Kapahi, P., Zid, B.M., Harper, T., Koslover, D., Sapin, V., and Benzer, S. (2004). Regulation of lifespan in *Drosophila* by modulation of genes in the TOR signaling pathway. *Curr Biol* 14:885–890.
51. Zid, B.M., *et al.* (2009). 4E-BP extends lifespan upon dietary restriction by enhancing mitochondrial activity in *Drosophila*. *Cell* 139:149–160.
52. Harrison, D.E., *et al.* (2009). Rapamycin fed late in life extends lifespan in genetically heterogeneous mice. *Nature* 460:392–395.
53. Selman, C., *et al.* (2009). Ribosomal protein S6 kinase 1 signaling regulates mammalian life span. *Science* 326:140–144.
54. Noda, T., and Ohsumi, Y. (1998). Tor, a phosphatidylinositol kinase homologue, controls autophagy in yeast. *J Biol Chem* 273:3963–3966.
55. Levine, B., and Klionsky, D.J. (2004). Development by self-digestion: molecular mechanisms and biological functions of autophagy. *Dev Cell* 6:463–477.
56. Suzuki, K., and Ohsumi, Y. (2007). Molecular machinery of autophagosome formation in yeast, *Saccharomyces cerevisiae*. *FEBS Lett* 581:2156–2161.
57. Melendez, A., and Neufeld, T.P. (2008). The cell biology of autophagy in metazoans: a developing story. *Development* 135:2347–2360.
58. Kamada, Y., Funakoshi, T., Shintani, T., Nagano, K., Ohsumi, M., and Ohsumi, Y. (2000). Tor-mediated induction of autophagy via an Apg1 protein kinase complex. *J Cell Biol* 150:1507–1513.

59. Kihara, A., Kabeya, Y., Ohsumi, Y., and Yoshimori, T. (2001). Beclin-phosphatidylinositol 3-kinase complex functions at the trans-Golgi network. *EMBO Rep* 2:330–335.
60. Kihara, A., Noda, T., Ishihara, N., and Ohsumi, Y. (2001). Two distinct Vps34 phosphatidylinositol 3-kinase complexes function in autophagy and carboxypeptidase Y sorting in *Saccharomyces cerevisiae*. *J Cell Biol* 152:519–530.
61. Matecic, M., *et al.* (2010). A microarray-based genetic screen for yeast chronological aging factors. *PLoS Genet* 6:e1000921.
62. Desai, B.N., Myers, B.R., and Schreiber, S.L. (2002). FKBP12-rapamycin-associated protein associates with mitochondria and senses osmotic stress via mitochondrial dysfunction. *Proc Natl Acad Sci USA* 99:4319–4324.
63. Pan, Y., and Shadel, G.S. (2009). Extension of chronological life span by reduced TOR signaling requires down-regulation of Sch9p and involves increased mitochondrial OXPHOS complex density. *Aging (Albany NY)* 1:131–145.
64. Patil, C.K., Li, H., and Walter, P. (2004). Gcn4p and novel upstream activating sequences regulate targets of the unfolded protein response. *PLoS Biol* 2:E246.
65. Long, X., Spycher, C., Han, Z.S., Rose, A.M., Muller, F., and Avruch, J. (2002). TOR deficiency in *C. elegans* causes developmental arrest and intestinal atrophy by inhibition of mRNA translation. *Curr Biol* 12:1448–1461.
66. Kenyon, C. (2005). The plasticity of aging: insights from long-lived mutants. *Cell* 120:449–460.
67. Oldham, S., and Hafen, E. (2003). Insulin/IGF and target of rapamycin signaling: a TOR de force in growth control. *Trends Cell Biol* 13:79–85.
68. Kapahi, P., and Zid, B. (2004). TOR pathway: linking nutrient sensing to life span. *Sci Aging Knowledge Environ* 2004:PE34.
69. Soukas, A.A., Kane, E.A., Carr, C.E., Melo, J.A., and Ruvkun, G. (2009). Rictor/TORC2 regulates fat metabolism, feeding, growth, and life span in *Caenorhabditis elegans*. *Genes Dev* 23:496–511.
70. Mair, W., and Dillin, A. (2008). Aging and survival: the genetics of life span extension by dietary restriction. *Annu Rev Biochem* 77:727–754.
71. Mair, W., Panowski, S.H., Shaw, R.J., and Dillin, A. (2009). Optimizing dietary restriction for genetic epistasis analysis and gene discovery in *C. elegans*. *PLoS ONE* 4:e4535.
72. Klass, M.R. (1977). Aging in the nematode *Caenorhabditis elegans*: major biological and environmental factors influencing life span. *Mech Ageing Dev* 6:413–429.
73. Panowski, S.H., Wolff, S., Aguilaniu, H., Durieux, J., and Dillin, A. (2007). PHA-4/Foxa mediates diet-restriction-induced longevity of *C. elegans*. *Nature* 447:550–555.
74. Henderson, S.T., Bonafe, M., and Johnson, T.E. (2006). daf-16 protects the nematode *Caenorhabditis elegans* during food deprivation. *J Gerontol A Biol Sci Med Sci* 61:444–460.
75. Syntichaki, P., Troulinaki, K., and Tavernarakis, N. (2007). eIF4E function in somatic cells modulates ageing in *Caenorhabditis elegans*. *Nature* 445:922–926.
76. Southgate, R.J., *et al.* (2007). FOXO1 regulates the expression of 4E-BP1 and inhibits mTOR signaling in mammalian skeletal muscle. *J Biol Chem* 282:21176–21186.
77. Junger, M.A., *et al.* (2003). The Drosophila forkhead transcription factor FOXO mediates the reduction in cell number associated with reduced insulin signaling. *J Biol* 2:20.
78. Mango, S.E. (2009). The molecular basis of organ formation: insights from the *C. elegans* foregut. *Annu Rev Cell Dev Biol* 25:597–628.
79. Greer, E.L., *et al.* (2007). An AMPK–FOXO pathway mediates longevity induced by a novel method of dietary restriction in *C. elegans*. *Curr Biol* 17:1646–1656.
80. Gwinn, D.M., *et al.* (2008). AMPK phosphorylation of raptor mediates a metabolic checkpoint. *Mol Cell* 30:214–226.

81. Inoki, K., and Guan, K.L. (2006). Complexity of the TOR signaling network. *Trends Cell Biol* 16:206–212.
82. Inoki, K., Zhu, T., and Guan, K.L. (2003). TSC2 mediates cellular energy response to control cell growth and survival. *Cell* 115:577–590.
83. Melendez, A., Talloczy, Z., Seaman, M., Eskelinen, E.L., Hall, D.H., and Levine, B. (2003). Autophagy genes are essential for dauer development and life-span extension in *C. elegans*. *Science* 301:1387–1391.
83a. Hars, E.S., Qi, H., Ryazanov, A.G., Jin, S., Cai, L., Hu, C., and Liu, L.F. (2007). Autophagy regulates ageing in *C. elegans*. *Autophagy* 3:93–95.
84. Jia, K., and Levine, B. (2007). Autophagy is required for dietary restriction-mediated life span extension in *C. elegans*. *Autophagy* 3:597–599.
85. Jones, K.T., Greer, E.R., Pearce, D., and Ashrafi, K. (2009). Rictor/TORC2 regulates *Caenorhabditis elegans* fat storage, body size, and development through sgk-1. *PLoS Biol* 7:e60.
86. Wang, M.C., O'Rourke, E.J., and Ruvkun, G. (2008). Fat metabolism links germline stem cells and longevity in *C. elegans*. *Science* 322:957–960.
87. Kaelin, W.G., Jr., and Ratcliffe, P.J. (2008). Oxygen sensing by metazoans: the central role of the HIF hydroxylase pathway. *Mol Cell* 30:393–402.
88. Mehta, R., *et al.* (2009). Proteasomal regulation of the hypoxic response modulates aging in *C. elegans*. *Science* 324:1196–1198.
89. Zhang, Y., *et al.* (2009). The HIF-1 hypoxia-inducible factor modulates lifespan in *C. elegans*. *PLoS One* 4:e6348.
90. Wessells, R., *et al.* (2009). d4eBP acts downstream of both dTOR and dFoxo to modulate cardiac functional aging in *Drosophila*. *Aging Cell* 8:542–552.
91. Giannakou, M.E., Goss, M., Junger, M.A., Hafen, E., Leevers, S.J., and Partridge, L. (2004). Long-lived *Drosophila* with overexpressed dFOXO in adult fat body. *Science* 305:361.
92. Bluher, M., Kahn, B.B., and Kahn, C.R. (2003). Extended longevity in mice lacking the insulin receptor in adipose tissue. *Science* 299:572–574.
93. Clancy, D.J., Gems, D., Hafen, E., Leevers, S.J., and Partridge, L. (2002). Dietary restriction in long-lived dwarf flies. *Science* 296:319.
94. Mair, W., Goymer, P., Pletcher, S.D., and Partridge, L. (2003). Demography of dietary restriction and death in *Drosophila*. *Science* 301:1731–1733.
95. Rogina, B., Helfand, S.L., and Frankel, S. (2002). Longevity regulation by *Drosophila* Rpd3 deacetylase and caloric restriction. *Science* 298:1745.
96. Mair, W., Piper, M.D., and Partridge, L. (2005). Calories do not explain extension of life span by dietary restriction in *Drosophila*. *PLoS Biol* 3:e223.
97. Tettweiler, G., Miron, M., Jenkins, M., Sonenberg, N., and Lasko, P.F. (2005). Starvation and oxidative stress resistance in *Drosophila* are mediated through the eIF4E-binding protein, d4E-BP. *Genes Dev* 19:1840–1843.
98. Scott, R.C., Schuldiner, O., and Neufeld, T.P. (2004). Role and regulation of starvation-induced autophagy in the *Drosophila* fat body. *Dev Cell* 7:167–178.
99. Scott, R.C., Juhasz, G., and Neufeld, T.P. (2007). Direct induction of autophagy by Atg1 inhibits cell growth and induces apoptotic cell death. *Curr Biol* 17:1–11.
100. Simonsen, A., Cumming, R.C., Brech, A., Isakson, P., Schubert, D.R., and Finley, K.D. (2008). Promoting basal levels of autophagy in the nervous system enhances longevity and oxidant resistance in adult *Drosophila*. *Autophagy* 4:176–184.
101. Juhasz, G., Erdi, B., Sass, M., and Neufeld, T.P. (2007). Atg7-dependent autophagy promotes neuronal health, stress tolerance, and longevity but is dispensable for metamorphosis in *Drosophila*. *Genes Dev* 21:3061–3066.

102. McCarroll, S.A., *et al.* (2004). Comparing genomic expression patterns across species identifies shared transcriptional profile in aging. *Nat Genet* 36:197–204.
103. Lee, J.H., *et al.* (2010). Sestrin as a feedback inhibitor of TOR that prevents age-related pathologies. *Science* 327:1223–1228.
104. Chen, C., Liu, Y., and Zheng, P. (2009). mTOR regulation and therapeutic rejuvenation of aging hematopoietic stem cells. *Sci Signal* 2:ra75.
105. Um, S.H., *et al.* (2004). Absence of S6K1 protects against age- and diet-induced obesity while enhancing insulin sensitivity. *Nature* 431:200–205.
106. Selman, C., *et al.* (2008). Evidence for lifespan extension and delayed age-related biomarkers in insulin receptor substrate 1 null mice. *FASEB J* 22:807–818.
107. Holzenberger, M., *et al.* (2003). IGF-1 receptor regulates lifespan and resistance to oxidative stress in mice. *Nature* 421:182–187.
108. Sharp, Z.D., and Bartke, A. (2005). Evidence for down-regulation of phosphoinositide 3-kinase/Akt/mammalian target of rapamycin (PI3K/Akt/mTOR)-dependent translation regulatory signaling pathways in Ames dwarf mice. *J Gerontol A Biol Sci Med Sci* 60:293–300.
109. Dhahbi, J.M., Kim, H.J., Mote, P.L., Beaver, R.J., and Spindler, S.R. (2004). Temporal linkage between the phenotypic and genomic responses to caloric restriction. *Proc Natl Acad Sci USA* 101:5524–5529.
110. Aguilar, V., *et al.* (2007). S6 kinase deletion suppresses muscle growth adaptations to nutrient availability by activating AMP kinase. *Cell Metab* 5:476–487.

15

mTOR Signaling and Human Cancer

NAPHAT CHANTARAVISOOT • FUYUHIKO TAMANOI

Department of Microbiology, Immunology and Molecular Genetics
Jonsson Comprehensive Cancer Center, Molecular Biology Institute
University of California
Los Angeles, California, USA

I. Abstract

A variety of observations point to the idea that dysregulation of the mammalian target of rapamycin (mTOR) signaling is closely associated with human cancer. First, frequent activation of the mTOR signaling is observed in human cancer, as detected by examining tissue samples for the level of phospho-S6K, phospho-S6, or phospho-mTOR. Some of these cases are due to the activation of upstream events such as the activation of Akt. Some genetic disorders associated with benign tumor formation occur when upstream regulators of mTOR are mutated. Second, we have shown that activating mutations in mTOR can be identified in human cancer. This was possible by mining the human cancer genome database. Because the human cancer genome database is rapidly expanding, this raises the possibility that a large number of activating mTOR mutations will be identified in human cancer in the future. Finally, small molecule inhibitors of mTOR have been developed and these inhibitors exhibit efficacy to block tumor growth. Some inhibitors are currently being evaluated in clinical studies.

THE ENZYMES, Vol. XXVIII

ISSN NO: 1874-6047

DOI: 10.1016/S1874-6047(10)28015-0

II. Introduction

mTOR is a 289 kDa protein that belongs to the phosphatidylinositol 3-kinase (PI3K)-related kinase family that includes ATM, ATR, and DNA-dependent protein kinases [1]. These kinases have a unique structure that includes HEAT, FAT, FATC domains as well as a kinase domain that is located in the C-terminal half. The mTOR kinase associates with other proteins to form two different complexes, mTORC1 and mTORC2 [2, 3]. While mTORC1 contains Raptor, mTORC2 contains Rictor. The mTORC1 regulates protein synthesis by phosphorylating S6K as well as 4E-BP1. Phosphorylation of 4E-BP1 relieves negative regulation of eIF-4E. On the other hand, mTORC2 phosphorylates Akt and is reported to be involved in the regulation of actin cytoskeleton. The mTORC1 signaling is a major signaling pathway that controls growth and proliferation in response to growth factors, nutrient, and energy conditions.

The idea that mTOR is a major target of anticancer drug development is supported by the findings that the mTOR signaling is activated in a number of human cancers [4]. We have recently shown that activating mutations of mTOR can be identified in human cancer [5]. This was a breakthrough that was accomplished by mining human cancer genome databases. Although we have identified two mutations in this initial attempt, increasing number of cancer mutations are expected to be identified in the future, as the human cancer genome database is rapidly expanding [6]. The study may reveal important clues regarding the type of cancer that could be associated with mTOR activation. Significance of the mTOR signaling in human cancer can also be gleaned from the study to use rapamycin as an anticancer drug [7, 8]. Clinical ongoing studies are yielding new insight into human cancer and the mTOR signaling. An important advance has been made recently by the development of a new generation of mTOR inhibitors [9]. These compounds are based on ATP analogs that directly interfere with the kinase activity. While rapamycin inhibits mTORC1 preferentially, these new compounds inhibit both mTORC1 and mTORC2. Interesting insights have been gained by the use of these new-generation inhibitors. This chapter is intended to review these developments. Our intention is to highlight the significance of the mTOR signaling in human cancer.

III. Frequent Activation of the mTOR Signaling in Human Cancer

A. Activation of Akt Leading to the Activation of Downstream Events

Activation of the mTOR/S6K signaling is widespread in a number of human cancer cases [10]. First, the mTOR signaling can be activated as a consequence of Akt activation. Akt phosphorylates TSC2, a component of the TSC1/TSC2 complex that acts as a GTPase activating protein (GAP) for Rheb. The phosphorylation inhibits the GAP activity of TSC1/TSC2, thus leading to the activation of Rheb and consequently mTOR. Akt activation is seen in cancer including breast, lung, prostate, colorectal, and kidney [4]. Activation of PI3K because of the activation of receptors such as growth factor receptors leads to the activation of Akt. In addition, loss of Pten function results in the activation of Akt. For example, Pten is mutated in prostate cancer, which then leads to the activation of Akt. Furthermore, activating mutations in the catalytic subunit of PI3K cause Akt activation. These events lead to the activation of downstream events, including mTOR. Endometrial cancer arises from the endometrium or lining of the uterus. Inactivation of TSC2 by loss of expression or phosphorylation occurs frequently in endometrial carcinoma, leading to high-frequency activation of mTOR [11].

B. Hyperphosphorylation of S6K, S6, and mTOR

Experimental evidence pointing to the activation of the mTOR signaling has been obtained by examining hyperphosphorylation of S6K and S6. Since Akt can activate multiple downstream events, detection of hyperphosphorylation of S6K or S6 is more indicative of the mTOR activation. Malignant melanoma has recently been shown to have increased level of mTOR activation as detected by the phosphorylation of S6. Moderate to strong hyperphosphorylation of S6 was observed in 78 out of 107 melanoma samples (73%) in contrast to benign nevi that showed hyperphosphorylation of S6 in only 4% of samples [12], suggesting that the mTOR activation is associated with malignant melanostic lesions. Nearly 40% of adenomas and carcinomas of the colorectum exhibited either partial or whole positive staining for phospho-S6K [13]. Phosphorylation of S6 was detected in 24.5% of hepatocellular carcinoma (HCC) tumor tissue samples [14]. The level of phospho-S6 was associated with elevated Cyclin D1 expression and was correlated with decreased overall survival in patients resected with a clear margin.

Clinical specimens from head and neck squamous cancer (HNSCC) patients were examined by the Head and Neck Cancer Tissue Array Initiative [15]. The results showed that the Akt-mTOR pathway is frequently activated in HNSCC, as a large group of tissue samples displayed active Akt and mTOR. Interestingly, a small group of patients was identified in which the mTOR pathway is activated but not Akt, suggesting that this patient group has mTOR activation that is independent of Akt activation.

Renal clear cell carcinoma (RCC) has been reported to exhibit activation of the mTOR signaling, as moderate to strong phospho-S6 immunoreactivity was found in 17 of 29 clear cell carcinomas (59%). Fourteen of these were also moderately to strongly positive for phospho-mTOR (Ser2448) [16]. Another immunohistochemical study using antibodies against phospho-S6 and phospho-Akt also showed that the mTOR pathway was significantly altered in clear cell renal carcinoma [17]. Phospho-S6 and PTEN showed strong associations with pathological parameters. Immunohistochemical analyses of human gastric adenocarcinoma specimens showed strong expression of phospho-mTOR in 60% of intestinal and 64% of diffuse-type human gastric adenocarcinomas [18]. It was found that RCC specimens exhibit higher levels of phospho-mTOR compared to benign renal parenchyma.

Analysis of 64 gliomas, including high- and low-grade tumors, showed that phospho-mTOR, phospho-S6, and phospho-Akt exhibit increased levels correlated with an increasing grade of malignancy [19]. In particular, with glioblastoma, correlation of phospho-S6 and proliferation was observed.

Extramammary Paget's disease (EMPD), is a rare, slow growing, adenocarcinoma outside of the mammary gland that is usually noninvasive. Of the 32 EMPD tissue samples tested, 29, 27, 26, 29, 26 samples were positive for phospho-Akt, phospho-mTOR, phospho-4E-BP1, phospho-S6K, and phospho-S6, respectively [20].

Myelodysplastic syndromes (MDS) are characterized by ineffective hematopoiesis. They have a high risk of evolution into AML (acute myelogenous leukemia). Analysis of bone marrow mononuclear samples from 20 patients showed that the levels of phospho-mTOR, phospho-S6K, and phospho-4E-BP1 are higher among high-risk MDS patients compared with either low-risk MDS cells or healthy donors [21].

C. Genetic Disorders due to the Activation of the TSC/Rheb/TOR Signaling

A number of genetic disorders have been identified that are due to mutations in the genes involved in the mTOR signaling pathway [22]. Tuberous sclerosis, a genetic disorder associated with the appearance of benign tumors called hamartomas in a number of locations in the body such as kidney and lung, is due to mutations in the *TSC1* or *TSC2* gene [23]. Since TSC1/TSC2 complex functions as a GAP for Rheb, mutations in either of the *TSC* genes result in the loss of GAP function, leading to the activation of Rheb and mTOR. In addition to benign tumors, tuberous sclerosis patients exhibit neurological symptoms such as mental retardation, seizure, and autism.

We have characterized mouse embryonic fibroblast (MEF) obtained from mice with *Tsc1* or *Tsc2* knockout [24]. Constitutive activation of the mTOR signaling as detected by the hyperphosphorylation of S6 was detected in the *Tsc2* knockout MEF. These MEF cells grew even in the absence of serum and did not respond to nutrient starvation. In astrocytes derived from mice with astrocyte-specific conditional knockout of Tsc1, similar activation of the mTOR signaling was observed by examining hyperphosphorylation of S6K and S6 [25]. Interestingly, these astrocyte cells exhibited larger size than the cells from control mice.

Angiomyolipoma belongs to a family of tumors known as perivascular epithelial tumors (PEComas). These tumors have strong association with tuberous sclerosis. Examination of the mTOR signaling in sporadic AMLs and PEComas detected increased levels of phospho-S6K in 15 of 15 non-TSC AMLs accompanied by a reduced level of phospho-Akt [26]. An elevated level of phospho-S6K and reduced phospho-Akt were detected in 14 of 15 cases of extrarenal PEComas. Phospho-S6 immunohistochemistry performed on lymphangioleiomyomatosis-associated and sporadic angiomyolipomas revealed a high incidence of mTOR/S6K/S6 signaling pathway activation [27].

Peutz–Jeghers syndrome, a rare inherited hamartoma characterized by gastrointestinal polyps and mucocutaneous pigmentation, has been reported to be associated with the loss of STK11/LKB1 genes [28]. LKB1 mutation causes decreased AMPK activity, leading to the loss of inhibition of TSC1/TSC2 function. This causes activation of the mTOR signaling.

D. Decreased Expression of Negative Regulators Leading to the Activation of mTOR

Negative regulators of mTOR have been identified. PRAS40 competes with substrate proteins for the binding to mTORC1 [29]. The binding of substrate proteins such as 4E-BP1 is mediated by the TOS motif and this TOS motif is also found in PRAS40. Thus, the binding of PRAS40 leads to inhibition of mTORC1. Phosphorylation of PRAS40 by the insulin signaling pathway results in the decrease of its inhibitory action. The second negative regulator DEPTOR has recently been identified [30]. DEPTOR is a 48 kDa protein identified within the mTOR immunoprecipitates. DEPTOR depletion in cells activates mTORC1 and mTORC2 signaling. Interestingly, the level of DEPTOR mRNA was significantly lower in most cancer types, including melanoma, thyroid, cervical, and prostate cancer. On the other hand, DEPTOR mRNA level was found to be high in a subset of multiple myeloma that is characterized with translocations involving the genes for cyclin D1 or D3 or for the c-MAF and MAFB transcription factors.

IV. Identification of mTOR Mutations in Human Cancer

We have recently identified activating mutations in human cancer [5]. This was the first time that activating mutations were identified in human cancer. While only two mutations have been identified so far, our work may mark the beginning of a search to identify mTOR mutations. It is worth pointing out that our study represents the culmination of a series of researches we have conducted regarding TOR. We first carried out random mutagenesis studies to obtain a proof of concept that a single amino acid change can confer constitutive activation of mTOR. We then moved to the analysis of human cancer genome database to identify constitutive active mutations.

A. Random Mutagenesis of Fission Yeast Tor Provided Proof of Evidence that a Single Amino acid Change can Confer Constitutive Activation of mTOR

Despite the significance of the mTOR signaling pathway in human cancer, no activating mutations of mTOR were reported until our discovery. This was rather surprising and the situation contrasts with other cancer-related genes such as Ras and Raf [31]. In the case of Ras, more than 90% of samples from pancreatic cancer contained activating mutations. A variety of mutations in the tumor suppressor such as p53 have been reported [32]. We thought that the failure to identify mTOR mutations could be due to low incidence. Thus, successful identification is expected to require extensive analysis. Before embarking on the pursuit, we first decided to obtain proof of concept that mTOR can actually be activated by a single amino acid change.

To gain insight into the question whether mTOR can be activated by a single amino acid change, we decided to take advantage of a yeast model system that is amenable to extensive screen. We first needed to establish that the TSC/Rheb/TOR signaling is conserved in fission yeast by characterizing the counterparts of TSC, Rheb, TOR, S6K, and S6 in fission yeast [33].

In fact, we have recently shown that all the TSC/Rheb/TOR/S6K/S6 signaling components are conserved in fission yeast. Like the mammalian system, fission yeast has two TOR complexes, TORC1 and TORC2. TORC1 is the one that signals phosphorylation of S6K and S6, while TORC2 is involved in amino acid uptake. There are two *Tor* genes in fission yeast, *Tor1* and *Tor2*. TORC1 contains Tor2 as well as Mip1, a homolog of Raptor. *Tor2* is an essential gene. We identified temperature-sensitive mutants and showed that the inhibition of Tor2 leads to growth arrest with small rounded cells arrested in the G1 phase of the cell cycle [34]. These cells are very similar to the cells exposed to nitrogen starvation. The inhibition of Tor2 led to hyper-mating phenotype. In addition, genes involved in mating were upregulated when Tor2 was inhibited, as revealed by microarray analysis. These results led us to suggest that Tor2 is a dual regulator positively regulating growth and negatively regulating sexual differentiation [34].

We have carried out extensive mutagenesis of *Tor2* to identify hyperactive mutants [35]. This was accomplished by mutagenizing the *Tor2* gene and then looking for activating mutations by examining decreased mating phenotypes. In addition, the ability of the *Tor2* mutant cells to bypass the requirement for Rheb was also examined. In fact, the initial mutant was discovered by mutagenizing yeast cells and then looking for mutants that can bypass the requirement for the presence of Rheb to grow. We found that the mutant had a single amino acid change in Tor2. These studies led to the identification of 22 single amino acid changes, any one of which can cause hyperactivation of Tor2. Interestingly, these mutations were clustered into a small number of regions that are located at the FAT domain and the kinase domain.

Representative mutations of *Tor2* were introduced into *mTOR* by PCR mutagenesis, and the mutant mTOR was expressed in mammalian cells to examine whether a single amino acid change can confer constitutive activation of mTOR. HEK293 cells expressing the mutant mTOR were exposed to nutrient starvation, and phosphorylation of S6K, S6, or 4E-BP1 was examined. Constitutive activation of the mTOR signaling was observed. This study also led to an important observation concerning dominant effects of the mutation [35]. It is known that mTOR is dimerized, raising the possibility that mTORC1 containing both the mutant mTOR and the wild-type mTOR exhibits constitutive active phenotypes. To examine this point, we coexpressed the mutant mTOR tagged with AU1 and the wild-type mTOR tagged with FLAG, and the cells were nutrientstarved. Anti-AU1 antibody was used to pull down a dimer that contains the mutant mTOR. *In vitro* kinase assay of the immunoprecipitate showed that the dimer that contains the mutant mTOR is constitutively active.

B. Mining the Human Cancer Genome Database

We have recently reported, for the first time, that single amino acid changes conferring constitutive activation of mTOR can be identified in human cancer [5]. This was possible by mining the human cancer genome databases. We first examined the Sanger Center Cancer Genome Database, COSMIC, for any changes in mTOR. The database contained information from 750 tumor samples, including the following tumor types: breast, central nervous system, hematopoietic and lymphoid tissue, kidney, large intestine, lung, ovary, pancreas, pleura, skin, stomach, testis, upper aerodigestive tract, and urinary tract. Ten mutations in the *mTOR* gene were identified. Two mutations did not cause amino acid changes and one mutation caused truncation. A mutation M135T was identified twice. Six different amino acid changes identified are shown in Figure 15.1.

Four mutations, A8S, S2215Y, P2476Y, and R22505P, have been examined by introducing each mutation into mTOR and examining whether expression of the mutant mTOR causes constitutive activation of mTOR in the absence of nutrients. This was evaluated by transfecting HEK293T cells with mTOR mutant or with the wild-type mTOR followed by serum and amino acid starvation. Lysates of these cells were used to examine activation of the mTORC1 and mTORC2 signaling. Activation of the mTOR signaling as detected by the phosphorylation of S6K was observed

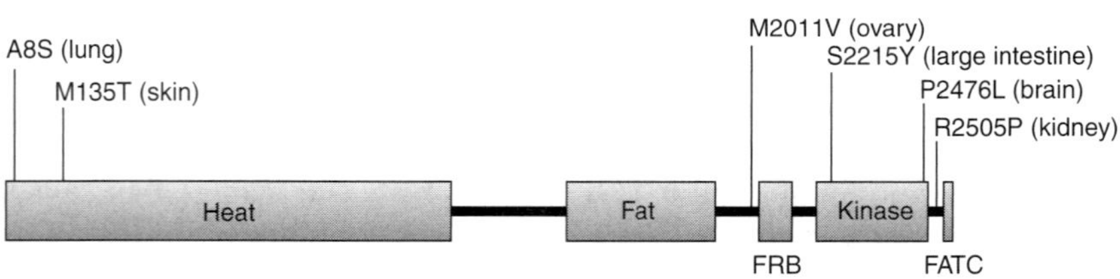

Mutation	Tissue (cancer type)	COSMIC ID
A8S	Lung (large cell carcinoma)	753390
M135T	Skin (malignant melanoma)	753546
M2011V	Ovary (mucinous carcinoma)	753344
S2215Y	Large intestine (adenocarcinoma)	753426
P2476L	Brain (glioma)	753417
R2505P	Kidney (clear cell renal cell carcinoma)	948155

Fig. 15.1. Single amino acid changes of mTOR identified by mining human cancer genome database. Four of these mutations were introduced into mTOR separately, and their ability to confer constitutive activation of mTOR was examined. This identified two mutations, S2215Y and R2505P, as constitutively active mutations of mTOR. Reproduced from Ref. [5].

even in the absence of nutrients when S2215Y or R2505P mutant was expressed. With the S2215Y mutant, phosphorylation of S6 and 4E-BP1 was shown. On the other hand, the mTORC2 signaling as detected by the phosphorylation of Akt was not affected by the expression of mTOR mutants. We then prepared mTORC1 from the nutrient-starved cells expressing mutant mTOR by precipitating with anti-Raptor antibody, and its activity was examined using 4E-BP1. *In vitro* kinase activity of the S2215Y mutant was elevated compared to the wild-type control. This result is important, as it shows that the mTORC1 machinery is in an activated state. On the other hand, mTORC2 kinase activity appears to remain unaffected even when the mutant mTOR is expressed. This was assessed by isolating mTORC2 by using anti-Rictor antibody and incubating with Akt. No significant change was observed compared to mTORC2 isolated from cells expressing the wild-type mTOR. Mutant mTORC1 exhibits increased kinase activity.

We have shown that the expression of cancer mTOR mutations causes constitutive activation of mTOR so that cells are unresponsive to nutrient starvation [5]. Thus, HEK293T cells expressing the S2215Y mutant exhibit a higher percentage of S-phase cells and lower percentage of G0/G1 phase cells compared with the wild-type control. Another consequence of nutrient starvation is the decrease in cell size that can be detected by the forward scatter analysis. Cells expressing the S2215Y mutant do not decrease their cell size upon nutrient starvation. We have also found that the cells expressing the S2215Y mutant exhibit an altered response to 1-butanol, an inhibitor of PLaseD that is implicated in mTOR activation.

Thus, the approach to mine the human cancer database appears productive. There are other databases that can be mined. This includes Cancer Genome Anatomy Project of the National Cancer Institute (NIH) that compiles information on SNPs. We have examined this database and identified eight different base changes within the coding region of mTOR (FRAP1). However, most of these changes did not result in amino acid changes. Other resources include Gene Expression Omnibus and Short Read Archive. Mining of these databases combined with experiments to examine constitutive activation of the mTOR signaling should identify additional mTOR mutations in human cancer.

V. Inhibitors of the mTOR Signaling

Because the mTOR signaling pathway is activated in a wide range of human cancers, this signaling pathway has been a focus of anticancer therapy. A variety of mTOR inhibitors have been developed and they are used to probe the idea that the inhibition of mTOR is effective for inhibiting tumor formation.

Rapamycin was originally identified as an antifungal antibiotic secreted by *Streptomyces hygroscopicus* in 1975 and later, it has been widely used as an immunosuppressant, and more recently in various cancer treatments according to its property to inhibit the mTOR signaling pathway [8, 36]. Rapamycin binds FKBP12 (FK506 binding protein) and this complex binds the FRB domain of mTOR. Rapamycin and its analogs (rapalogs) are being evaluated in clinical trials for cancer therapy.

Recent studies demonstrating the presence of two mTOR complexes, mTORC1 and mTORC2, led to the realization that rapamycin is more selective against mTORC1. This may explain the observations that rapamycin has limited inhibition of mTOR. First, mTORC2 is relatively insensitive to rapamycin [37]. In addition, rapamycin-resistant functions of mTORC1 have been suggested, based on the observation that rapamycin is able to inhibit phosphorylation of S65 residue of 4E-BP1 but is unable to inhibit phosphorylation of other three residues, T37/46, and T70 [38]. These observations led to the development of a new generation of mTOR inhibitors. Many of these are ATP-competitive kinase inhibitors with the ability to potently inhibit all mTOR-related functions [9].

In this chapter, we will discuss the new-generation mTOR inhibitors. For rapamycin, please refer to a review by Abraham et al., (Chemistry and Pharmacology of Rapamycin and Its Derivatives, Chapter 17 of The Enzymes Volume 27).

A. A New Class of mTOR Inhibitors that is ATP-Competitive

A number of groups have developed small molecules that can target ATP-binding sites of mTOR, resulting in significant inhibition of both mTORC1 and mTORC2 complexes. The number of new inhibitors is increasing rapidly, and the effects of these inhibitors are examined *in vitro* and *in vivo*. Figure 15.2 shows these new inhibitors that target the kinase domain of mTOR, as well as rapamycin and rapamycin analogs that target the FRB domain.

In 2008, potent ATP-competitive inhibitors of mTOR were found during an attempt to perform kinome profiling for dual inhibitors of tyrosine and phosphoinositide kinases. PP121, PP242, [39], and PP30 are pyrazolopyrimidines that can target direct upstream and downstream effectors of mTORC1 and mTORC2. They were named mTOR kinase domain inhibitors (TORKinibs) [40]. The TORKinibs, particularly PP242, that exhibits distinctive selectivity against mTOR, was the first novel ATP competitor whose effects were described. The efficiency of PP242 is high with IC_{50} of

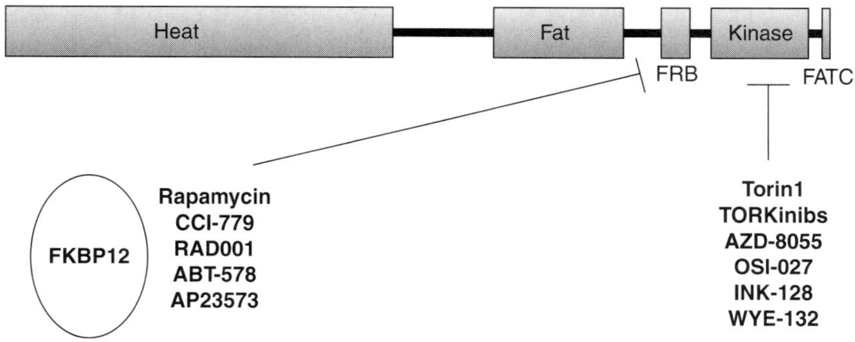

FIG. 15.2. Two classes of small molecule inhibitors against mTOR. Rapamycin and its analogs bind FKBP12 and the resulting complex binds the FRB domain of mTOR. ATP-competitive inhibitors bind directly to the ATP-binding pocket within the kinase domain of mTOR.

8 nM. PP242 can shut down phosphorylation of p70S6K, and phosphorylation of Akt on S473 completely, and partially on T308 of pAkt. S6K and Akt are targets of mTORC1 and mTORC2, respectively. PP242 can also effectively inhibit phosphorylation of 4E-BP1, another direct downstream target of mTORC1 on T37/46 and S65, although it does not affect T70 that is believed to be targeted by other kinases.

OSI-027 and orally available OXA-01 have been developed by OSI pharmaceutical Inc [41]. They are adenine-mimetic compounds that exhibit mTOR inhibition with IC_{50} value of 4 nM. Complete inhibition of phosphorylation of S6K, T37/46 residues of 4E-BP1, and Akt on S473 in a dose-dependent manner was shown. Whether they could inhibit phosphorylation of Akt on T308 has not been reported.

Reports of another family of ATP-competitive inhibitors classified as morpholine-based mTORC1/2 inhibitors based on the presence of a morpholino pyrazolopyrimidine core structure have been released since 2009. Members of this group are Pfizer's (Wyeth) series of inhibitors, WAY-600, WYE-687, WYE-354, [42] and WYE-125132 [43]. WAY-600, WYE-687, and WYE-354 were developed from the initial molecule WAY-001 [42]. All three compounds were reported to fully suppress phosphorylation of S6K, 4E-BP1 on T37/46, as well as S473 residue of Akt, but not on T308. Interestingly, these molecules were capable of inhibiting phosphorylation of 4E-BP1 on T70. This was the first demonstration of a complete loss of T70. Furthermore, WYE-132, the most recently developed morpholino pyrazolopyrimidine, was shown to inhibit all targets similar to WYE-354 [43]. WYE-132 could partially decrease the level of Akt phosphorylation on T308, making it more potent than its relatives [43].

The small molecule Ku-0063794 from AstraZeneca promoted a highly selective inhibition of phosphorylated targets of mTORC1/2, including a complete suppression of phosphorylation of S6K, and p4E-BP1 on T37/46, but the phosphorylation of T70 was not tested [44]. In addition, Ku-0063794 is able to inhibit phosphorylation of Akt on S473 and surprisingly shut down phosphorylation of T308 residue more efficiently than other inhibitors. In addition, this compound was reported to possess much better specificity than PP242.

AZD8055 is another highly selective ATP-competitive mTOR kinase inhibitor that is structurally related to WAY/WYE series and Ku-0063794 [45]. As a result of its excellent selectivity, AZD8055 exhibited potent inhibition against phosphorylation of S6K and on 4E-BP1 at T37/46; however, the suppression of phosphorylation of T70 was not investigated. In terms of inhibiting mTORC2, this inhibitor can shut down phosphorylation of S473 of Akt, while suppression of the phosphorylation of T308 requires higher concentration.

Torin1 was shown to efficiently disable mTORC1 kinase activity at low concentration [46]. Full inhibition of the phosphorylation of pS6K, p4E-BP1 (T37/46) and modest inhibition of the phosphorylation of T70 of p4E-BP1 were demonstrated. Moreover, while the potent shut down of phosphorylation of Akt on S473 was reported, Torin1 was unable to inhibit the phosphorylation of T308 residue of Akt.

INK128 (Intellikine Inc) was also proposed to have excellent potency in suppressing both complexes of mTOR [47]. The data presented showed that INK-128 could inhibit phosphorylation of Akt on S473, 4E-BP1 on S65, and S6 on S235/236. However, effects of the phosphorylation on T37/46, T70 (4E-BP1), as well as on T308 (Akt), have not been reported.

AR-mTOR-26 (BIOPHARMA) is another inhibitor that was developed recently [49]. It was shown to have exceptional efficiency to inhibit mTORC1/2 compared to those of other compounds that were also synthesized by BIOPHARMA. AR-mTOR-26 can totally inhibit phosphorylation of S6 (S235/236), 4E-BP1 (T37/46), and Akt (S473) and substantially decrease the level of phosphorylation of T308 of Akt.

B. *In Vivo* Experiments

The ability of the ATP-competitive mTOR inhibitors to induce suppression of tumor growth or apoptosis of malignancies was compared to that of rapamycin and rapalogs. Some ATP-competitive inhibitors have been evaluated using various cancer models, and experiments with several types of xenografts were performed to examine their antitumor and/or anticancer effects.

PP242 was recently shown to induce cell death in acute leukemia models shown to be insensitive to rapamycin [48]. Moreover, it was shown to have no toxicity against normal cells *in vivo*. AZD8055, an effective orally available drug, was reported to have a good *in vivo* antitumor activity [45]. Significant inhibition of tumor in mice bearing a wide range of human tumor xenografts, such as breast, lung, colon, and glioma, was demonstrated. These effects were associated with dose-dependent pharmacodynamic effects on phospho-S6, phospho-4EBP1, and phospho-Akt. WYE-132 is another potent inhibitor whose oral treatments to multiple mouse models were reported [43]. Its preclinical antitumor activity in single-agent treatments against breast, glioma, lung, and renal tumors was substantial. OSI-027 was determined to exhibit greater antitumor activity in several preclinical xenograft models compared to rapamycin [50]. For instance, OSI-027 induced efficient tumor growth inhibition in mice bearing colon cancer (GEO and Colo205 models), models that were known to be insensitive to rapamycin. Oral administration of INK128 resulted in significant inhibition of angiogenesis in *in vivo* renal cell carcinoma models [51]. AR-mTOR-26, a new efficacious mTOR selective inhibitor, showed substantial dose-dependent effects on inhibiting tumor growth investigated in prostate and lung xenograft models [49].

In addition, these ATP-competitive mTOR inhibitors were tested in combination with known chemotherapeutic agents and other potent compounds such as PI3K inhibitors. Several studies have shown that the combination therapies not only provide much stronger antitumor activity, but also lead to complete regression of tumors.

VI. Future Prospects

Our identification of activating mTOR mutations in human cancer raises the possibility that there is a wide range of activating mTOR mutations that could be identified in human cancer. Because the effort to sequence human cancer genome is advancing rapidly, we can expect that a significant number of mTOR activating mutations will be identified in the future. It will be interesting to see whether higher percentages of mTOR mutation can be found in certain types of cancer. Within a short period of time, many orally bioavailable ATP-competitive inhibitors of mTOR have been developed and they are being investigated in preclinical animal studies. At present, INK128, OSI-027, and AZD8055 have progressed into clinical trials, while WYE-132 is undergoing preclinical studies. Knowledge obtained from these studies should provide insight into the significance of the mTOR signaling in human cancer.

Acknowledgment

This work is supported by the NIH grant CA41996. We would like to thank Dr. Shripad Bhagwat for the information on mTOR inhibitors.

References

1. Abraham, R.T. (2004). PI 3-indase related kinases: 'big' players in stress-induced signaling pathways. *DNA Repair (Amst)* 3:883–887.
2. Polak, P., and Hall, M.N. (2009). mTOR and the control of whole body metabolism. *Curr Opin Cell Biol* 21:209–218.
3. Laplante, M., and Sabatini, D.M. (2009). mTOR signaling at a glance. *J Cell Sci* 122:3589–3594.
4. Memmott, R.M., and Dennis, P.A. (2009). Akt-dependent and independent mechanisms of mTOR regulation in cancer. *Cell Signal* 21:656–664.
5. T. Sato, A. Nakashima, L. Guo, K. Coffman, F. Tamanoi, Single amino-acid changes that confer constitutive activation of mTOR are discovered in human cancer, Oncogene 29 2746–2752.
6. Stratton, M.R., Campbell, P.J., and Futreal, P.A. (2009). The cancer genome. *Nature* 458:719–724.
7. Johnson, B.E., Jackman, D., and Janne, P.A. (2007). Rationale for a phase I trial of erlotinib and the mammalian target of rapamycin inhibitor everolimus (RAD001) for patients with relapsed non small cell lung cancer. *Clin Cancer Res* 13:s468–s4631.
8. Wang, X., Yue, P., Kim, Y.A., Fu, H., Khuri, F.R., and Sun, S.Y. (2008). Enhancing mammalian target of rapamycin (mTOR)-targeted cancer therapy by preventing mTOR/raptor inhibition-initiated, mTOR/rictor-independent Akt activation. *Cancer Res* 68:7409–7418.
9. S.V. Bhagwat, A.P. Crew, Novel inhibitors of mTORC1 and mTORC2, Curr Opin Investig Drugs 11 638–645.
10. Yuan, T.L., and Cantley, L.C. (2008). PI3K pathway alterations in cancer: variations on a theme. *Oncogene* 27:5497–5510.
11. Lu, K.H., *et al.* (2009). Loss of tuberous sclerosis complex-2 function and activation of mammalian target of rapamycin signaling in endometrial carcinoma. *Clin Cancer Res* 14:2543–2550.
12. Karbowiniczek, M., Spittle, C.S., Morrison, T., Wu, H., and Hesnke, E.P. (2008). mTOR is activated in the majority of malignant malanomas. *J Invest Dermatol* 128:980–987.
13. Nozawa, H., Watanabe, T., and Nagawa, H. (2007). Phosphorylation of ribosomal p70 S6 kinase and rapamycin sensitivity in human colorectal cancer. *Cancer Lett* 251:105–113.
14. Baba, H.A., *et al.* (2009). Phosphorylation of p70S6 kinase predicts overall survival in patients with clear margin-resected hepatocellular carcinoma. *Liver Int* 29:399–405.
15. Molinolo, A.A., *et al.* (2007). Dissecting the Akt/mammalian target of rapamycin signaling network: emerging results from the head and neck cancer tissues array initiative. *Clin Cancer Res* 13:4964–4973.
16. Robb, V.A., Karbowniczek, M., Klein-Szanto, A.J., and Henske, E.P. (2007). Activation of the mTOR signaling pathway in renal clear cell carcinoma. *J Urol* 177:346–352.
17. Pantuck, A.J., *et al.* (2007). Prognostic relevance of the mTOR pathway in renal cell carcinoma: implications for molecular patient selection for targeted therapy. *Cancer* 109:2257–2267.

18. Lang, S.A., *et al.* (2007). Mammalian target of rapamycin is activated in human gastric cancer and serves as a target for therapy in an experimental model. *Int J Cancer* 120:1803–1810.
19. Annovazzi, L., Mellai, M., Caldera, V., Valente, G., Tessitore, L., and Schiffer, D. (2009). mTOR, S6 and AKT expression in relation to proliferation and apoptosis/autophagy in glioma. *Anticancer Res* 29:3087–3094.
20. Chen, S., *et al.* (2009). Immunohistochemical analysis of the mammalian target of rapamycin signaling pathway in extramammary Paget's disease. *Br J Dermatol* 161:357–363.
21. Follo, M.Y., *et al.* (2007). The Akt/mammalian target of rapamycin signal transduction pathway is activated in high-risk myelodysplastic syndromes and influences cell survival and proliferation. *Cancer Res* 67:4287–4294.
22. Inoki, K., Corradetti, M.N., and Guan, K.L. (2005). Dysregulation of the TSC-mTOR pathway in human disease. *Nat Genet* 37:19–24.
23. Kwiatkowski, D.J., and Manning, B.D. (2005). Tuberous sclerosis: a GAP at the crossroads of multiple signaling pathways. *Hum Mol Genet* 14(Spec No. 2):R251–R258.
24. Gau, C.L., Kato-Stankiewicz, J., Jiang, C., Miyamoto, S., Guo, L., and Tamanoi, F. (2005). Farnesyltransferase inhibitors reverse altered growth and distribution of actin filaments in Tsc-deficient cells via inhibition of both rapamycin-sensitive and -insensitive pathways. *Mol Cancer Ther* 4:918–926.
25. Uhlmann, E.J., Li, W., Scheidenhelm, D.K., Gau, C.L., Tamanoi, F., and Gutmann, D.H. (2004). Loss of tuberous sclerosis complex 1 (Tsc1) expression results in increased Rheb/S6K pathway signaling important for astrocyte cell size regulation. *Glia* 47:180–188.
26. Kenerson, H., Folpe, A.L., Takayama, T.K., and Yeung, R.S. (2007). Activation of the mTOR pathway in sporadic angiomyolipomas and other perivascular epithelioid cell neoplasms. *Hum Pathol* 38:1361–1371.
27. Robb, V.A., Astrinidis, A., and Henske, E.P. (2006). Frequent hyperphosphorylation of ribosomal protein S6 in lymphangioleiomyomatosis-associated angiomyolipomas. *Mod Pathol* 19:839–846.
28. Sanchez-Cespedes, M. (2007). A role for LKB1 gene in human cancer beyond the Peutz–Jeghers syndrome. *Oncogene* 26:7825–7832.
29. Sancak, Y., *et al.* (2007). PRAS40 is an insulin-regulated inhibitor of the mTORC1 protein kinase. *Mol Cell* 25:903–915.
30. Peterson, T.R., *et al.* (2009). DEPTOR is an mTOR inhibitor frequently overexpressed in multiple myeloma cells and required for their survival. *Cell* 137:873–886.
31. Kiaris, H., and Spandidos, D.A. (1995). Mutations of ras genes in human tumours. *Int J Oncol* 7:413–421.
32. Levine, A.J. (1997). p53, the cellular gatekeeper for growth and division. *Cell* 88:323–331.
33. Aspuria, P.J., Sato, T., and Tamanoi, F. (2007). The TSC/Rheb/TOR signaling pathway in fission yeast and mammalian cells: temperature sensitive and constitutive active mutants of TOR. *Cell Cycle* 6:1692–1695.
34. Matsuo, T., Otsubo, Y., Urano, J., Tamanoi, F., and Yamamoto, M. (2007). Loss of the TOR kinase Tor2 mimics nitrogen starvation and activates the sexual development pathway in fission yeast. *Mol Cell Biol* 27:3154–3164.
35. Urano, J., Sato, T., Matsuo, T., Otsubo, Y., Yamamoto, M., and Tamanoi, F. (2007). Point mutations in TOR confer Rheb-independent growth in fission yeast and nutrient-independent mammalian TOR signaling in mammalian cells. *Proc Natl Acad Sci USA* 104:3514–3519.
36. Sehgal, S.N., Baker, H., and Vezina, C. (1975). Rapamycin (AY-22, 989), a new antifungal antibiotic. II. Fermentation, isolation and characterization. *J Antibiot (Tokyo)* 28:727–732.

37. Shor, B., Gibbons, J.J., Abraham, R.T., and Yu, K. (2009). Targeting mTOR globally in cancer: thinking beyond rapamycin. *Cell Cycle* 8:3831–3837.
38. Thoreen, C.C., and Sabatini, D.M. (2009). Rapamycin inhibits mTORC1, but not completely. *Autophagy* 5:725–726.
39. Apsel, B., *et al.* (2009). Targeted polypharmacology: discovery of dual inhibitors of tyrosine and phosphoinositide kinases. *Nat Chem Biol* 4:691–699.
40. Feldman, M.E., *et al.* (2009). Active-site inhibitors of mTOR target of rapamycin-resistant outputs of mTORC1 and mTORC2. *PLoS Biol* 7:e38.
41. Bhagwat, S.V., *et al.* (2010). OSI-027 a potent and selective small molecule mTORC11/mTORC2 kinase inhibitor is mechanistically distinct from rapamycin. AACR Annual Meeting 2010, Abstract nr 4487.
42. Yu, K., *et al.* (2009). Biochemical, cellular, and in vivo activity of novel ATP-competitive and selective inhibitors of the mammalian target of rapamycin. *Cancer Res* 69:6232–6240.
43. K. Yu, et al., Beyond rapalog therapy: preclinical pharmacology and antitumor activity of WYE-125132, an ATP-competitive and specific inhibitor of mTORC1 and mTORC2, Cancer Res 70 621–631.
44. Garcia-Martinez, J.M., *et al.* (2009). Ku-0063794 is a specific inhibitor of the mammalian target of rapamycin (mTOR). *Biochem J* 421:29–42.
45. C.M. Chresta, et al., AZD8055 is a potent, selective, and orally bioavailable ATP-competitive mammalian target of rapamycin kinase inhibitor with in vitro and in vivo antitumor activity, Cancer Res 70 288–298.
46. Thoreen, C.C., *et al.* (2009). An ATP-competitive mammalian target of rapamycin inhibitor reveals rapamycin-resistant functions of mTORC1. *J Biol Chem* 284:8023–8032.
47. Kessler, L., *et al.* (2010). INK128, an orally active TOC1/2 kinase inhibitor, shows broad antitumor activity and enhances efficacy of cytotoxic as well as targeted agents. ACCR Annual Meeting 2010, Abstract nr 4496.
48. Gross, S.D., *et al.* (2010). *AR-mTOR-26- A potent, selective mTORC1/2 kinase inhibitor for the treatment of malignancy.* AACR Annual Meeting 2010, Abstract nr 4484.
49. M.R. Janes, *et al.* (2010). Effective and selective targeting of leukemia cells using a TORC1/2 kinase inhibitor, Nat Med 16 205–213.
50. Gokhale, P.C., *et al.* (2010). *OSI-027 a selective dual mTORC1/TORC2 kinase inhibitor displays broad spectrum anti-tumor activity in preclinical models of human cancer.* AACR Annual Meeting 2010, Abstract nr 4486.
51. Wang, S., *et al.* (2010). *INK128: an orally active TORC1/2 kinase inhibitor demonstrates potent antitumor activity in preclinical models of renal cell carcinoma.* AACR Annual Meeting 2010, Abstract nr 4488.

16

Systems Biology and TOR: Past, Present, and Future

SOYEON I. LIPPMAN • JAMES R. BROACH

Department of Molecular Biology
Princeton University, Princeton
New Jersey, USA

I. Abstract

A wide variety of genome-scale tools, such as global transcriptional analysis, systematic genetic screens, genetic interaction studies, and mass spectrometry based determination of global protein levels and modifications, have provided a comprehensive itemization of the components and interactions within signaling networks, particularly in the yeast *Saccharomyces*. Moreover, new computational tools have allowed us to integrate these genome-wide observations and organize them in an accessible and intuitive manner. Additional computational approaches are beginning to provide the means of developing predictive, *in silico* models of biological processes. Here, we describe the recent results from these genome-wide approaches as applied to the TOR signaling network in yeast. Furthermore, we attempt to integrate and reconcile genome-wide studies from multiple groups. As evident from our discussion, investigators have made substantial strides in applying these genomic tools as a means of developing a comprehensive description of the organization of the rapamycin-sensitive signaling network in yeast. However, we are clearly just on the forefront of developing predictive models of the network. Finally, we discuss future outlooks for TOR with an emphasis on the application of systems-level understanding to personalized treatment in cancer.

THE ENZYMES, Vol. XXVIII

ISSN NO: 1874-6047

DOI: 10.1016/S1874-6047(10)28016-2

II. Introduction

Since the discovery of TOR as the molecular target of the immunosuppressant rapamycin in yeast in 1991 by Hall and colleagues [1], we have gained an extensive understanding of the composition, molecular interactions, and biological functions of the TOR network. Our current knowledge of TOR rests on the solid foundation built by investigators in many laboratories who examined various facets of TOR—its molecular function, cellular localization, mode of regulation, downstream targets, etc. [2–6]. For instance, the identification that Tor proteins participate in two distinct complexes—TORC1 and TORC2—with quite distinct biological functions arose from focused biochemical studies of the Tor proteins. Subsequently, this foundation has provided a framework for assembling and interpreting the massive amount of genome-wide and high-throughput data into a coherent structure, which, when coupled with mathematical modeling, provide us with a system-level view of the TOR network.

The goal of a systems approach to the TOR network is to construct a comprehensive model representing this biological subsystem, using mathematic modeling based on experimental data. The driving motivation for such effort is to predict from model simulations an outcome in response to various inputs and perturbations within the system. However, the goal is not to replace the black box of the cell with the black box of the computer but rather to extract fundamental principles and understanding of the system that will allow us to manipulate it predictably, and perhaps therapeutically. Hence, systems biology begins at the bench, returns to the bench, and ends at the bedside. The power of systems biology is the ability to examine the basis of life as a dynamic interactive network composed of multiple and varied components—proteins, RNA, metabolites, etc—at the level of the organism and in context of its environment.

We focus our discussion primarily on studies in *Saccharomyces*. Not only was TOR first identified in *Saccharomyces* but also many of the tools used to obtain the data necessary to establish a systems view of TOR signaling have been developed and refined using this model organism. Moreover, the core structure of the TOR complexes and their downstream effectors are conserved in all eukaryotes examined, making many of the observations from the budding yeast relevant to larger eukaryotes as well. Thus, although the TOR network in *Saccharomyces* does not isomorphically map to that in mammalian cells, the feasibility of conducting genetic and high-throughput analyses coupled to an extensive and readily accessible infrastructure makes the budding yeast an ideal organism to build the foundation of TOR and systems biology. Accordingly, in this chapter, we focus on how systems biology has vastly expanded our understanding of TOR and its interconnected network in the budding yeast, with parallels to the mammalian counterparts. We discuss how systems biology can assist and advance our understanding of TOR from molecular, cellular, transcriptional, proteomic, and metabolomic levels with respect to the cell's internal and external environmental status. Finally, we extend the discussion to systems approaches in mammalian species with particular emphasis on potential applications of systems approaches to disease.

The complete understanding of an organism at system level requires three principal forms of data: (1) the molecular components of the network of pathways, (2) the localization of those components within the cell and in relation to one another, and (3) the responses of those molecular components to perturbations from the internal and external environments such as cell cycle, nutrient availability, and internal and external stresses. Finally, these requirements then have to be mathematically integrated to discover new aspects of the organism's biology, which are further experimentally examined and tested.

The availability of large-scale data from transcriptome, proteome, metabolomes, and more recently, next-generation sequencing has accelerated the progress of systems biology. In parallel, an increasing number of computational tools have been developed to understand and extract biological meaning from systems data [7]. The key aim of systems biology approach in understanding TOR network is to integrate quantitative and qualitative measurements from cell-wide data sets with observations probing the molecular mechanism of individual components contributing to the phenotypes visible at a global scale [8]. The utilization of "omics" methodologies has been widely explored due to a number of the intrinsic properties of the obtained data: it is already quantitative in nature, it results in large amount of data, global experiments can be conducted without necessarily having a hypothesis a priori, and especially in yeast, global high-throughput analysis is readily accessible.

It is important to note that just the utilization of global transcriptome, proteome, or metabolome tools does not automatically constitute such studies as being systems biology. In fact, a study can fully be deemed as being systems biology without employing any "omic" methods. The key element to the systems biology is not simply the tools utilized to answer a question, but the actual biological question itself and the integration of the obtained data [9–14]. Moreover, the value of systems biology is founded on perceiving a cell not as a separate entity, but itself a network with the environment including other cells or organisms. Hence, systems biology should also include how the expression of a genotype is shaped by the environment; in other words, the interplay between genotype and phenotype [8, 11, 12].

III. Genome-Wide Approach to Defining the TOR Network

A. Transcription Profiling

Genome-wide transcriptional profiling has proved to be valuable both in defining the biological role the TOR pathway plays in the cell and in dissecting subsystems that comprise the TOR pathway. Initial transcriptional profiling studies in yeast documented a strong correlation of the global transcriptional response of yeast cells following rapamycin addition to that from nitrogen starvation. This provided initial evidence that the rapamycin-sensitive TORC1 functioned as the primary conduit for the nitrogen sensing network [15–17]. Moreover, the various functional categories of the subsets of genes whose expression is altered by rapamycin addition suggested that TORC1 is responsible for regulating a wide cohort of processes required for cell growth that accurately reflect the nutritional status in the environment, particularly that of amino acids and usable nitrogen sources. Wek and colleagues subsequently used transcriptional profiling to examine how the general amino acid control (GAAC) and TOR pathways are coordinated to orchestrate transcriptional response to nitrogen availability [18]. They showed that transcription factors, Gcn4 and Gln3, regulate overlapping set of genes sensitive to TOR activity, namely those required for nitrogen utilization and amino acid biosynthesis. Moreover, they observed that TOR and GAAC pathways distinctly impinge on the activity of Gcn2 kinase, which inhibits translation by phosphorylating eIF2α GTPase while stimulating Gcn4 translation [19]. Whereas inhibition of TORC1 results in alleviation of inhibitory phosphorylation of Gcn2 through an undefined mechanism possibly involving Tap42 and Sit4 phosphatase, amino acid deprivation leads to Gcn2 activation by accumulation of uncharged tRNA [19]. Although by different means, both regulatory pathways result in phosphorylation of eIF2α and subsequent increase in translation of Gcn4 transcription factor and its transcriptional activity (Figure 16.1).

A second application of transcriptional profiling has been to identify specific subsystems within the TORC1 pathway. This approach defines how much of the cell's transcriptional response to rapamycin depends on particular signaling components. Using mutants lacking specific components of the pathway, Broach and colleagues examined the role of the protein phosphatase 2A catalytic subunits, Pph21/Pph22 and Sit4, and the phosphatase-associated protein, Tap42, in TORC1 signaling (Figure 16.1) [20]. This analysis demonstrated that Tap42 and both phosphatases, Sit4 and Pph21/22, are required for rapamycin-mediated induction of NDP

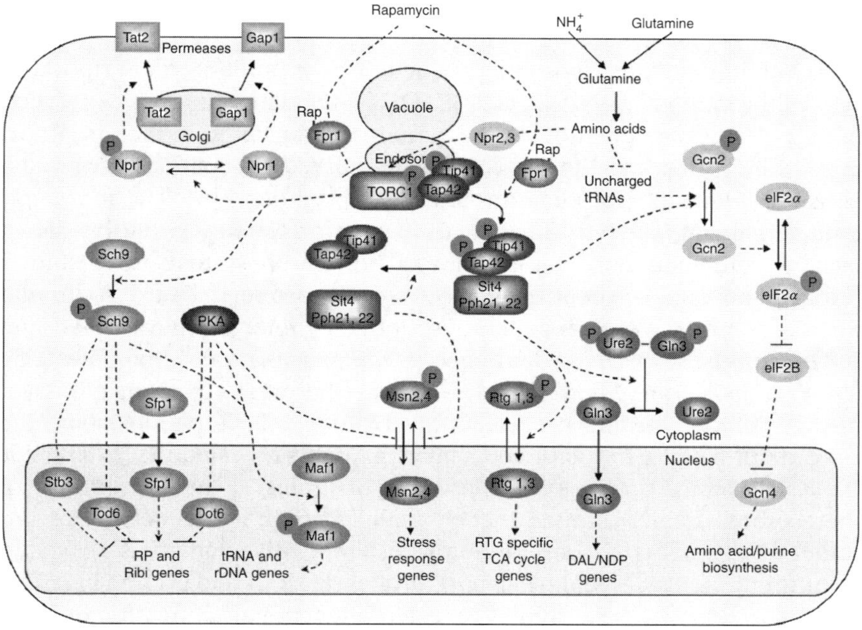

FIG. 16.1. The TORC1 Signaling Network. Shown is a simplified diagram of the TORC1 network in which the core components of the pathway, consisting of TOR complex 1 (TORC1) and associated protein phosphatase 2A and its regulators (purple components), are activated by amino acid availability and inhibited by rapamycin to positively regulate the ribosome biogenesis network (gray–green components) and negatively regulate the stress, alternate nitrogen assimilation pathways (blue components), and amino acid biosynthesis network (green components) and alter the spectrum of amino acid permeases in the membrane (orange components). See text and Ref. [6] for details of the network. (For interpretation of the references to color in this figure legend, the reader is referred to the Web version of this chapter.)

(Nitrogen Discrimination Pathway) and RTG (Retrograde) target genes. The study demonstrated that Tap42 and the PP2A subunits were also needed for rapamycin induction of stress response genes through Msn2/4 but via a mechanism distinct from that needed for NDP/RTG induction. Conversely, transcriptional profiling revealed that Tap42 and PP2A do not mediate TORC1's regulation of genes encoding ribosomal proteins (RP) or those involved in ribosome biogenesis (Ribi) [6, 20]. Rather, transcriptional profiling of strains carrying specific alleles of *SCH9*, which encodes a protein kinase homologous to mammalian S6 kinase, provided evidence that Sch9 mediates TORC1's regulation of RP and Ribi genes, but not NDP or RTG genes [21]. Additional studies based on genome-wide expression analysis suggest that Sch9 exhibits both TORC1-dependent and independent activities in governing general stress response and autophagy [22, 23].

TORC1 stimulates ribosome production by promoting transcription of RP, Ribi genes and rDNA [5, 6, 24]. TORC1 regulates RP gene expression largely through modulation of the activity of transcription factors, Rap1, Abf1, Fhl1, Ifh1, and Sfp1, a conclusion derived predominantly through classical studies of ribosomal protein gene promoters [6]. The molecular links connecting TORC1 to Ribi gene expression were more elusive. However, a combination of top-down and bottom-up systems-level studies bridged this gap by identifying DNA binding factors, Dot6, Tod6, and Stb3, as repressors of Ribi gene transcription by interacting with PAC and RRPE motifs, respectively, enriched in the promoters of Ribi genes [25–29]. Protein Binding Microarray (PBM) and Microarray Profiling of phage-display Selection (MaPS) revealed that Dot6 and Tod6 paralogs bind the PAC motif [25, 26, 29], and biochemical methods revealed that Stb3 binds the RRPE motif. Transcriptional profiling of strains lacking one or more of these proteins demonstrated that they all function, not as activators, but rather as repressors of transcription whose repressive activity is alleviated by nutrient-induced signaling, in part through TORC1 and protein kinase A (PKA). Moreover, these studies revealed a functional bifurcation of Dot6 and Tod6 paralogs, with Tod6 exclusively funneling the TORC1 signal and Dot6 primarily involved in regulation from PKA (Figure 16.1) [28].

Genome-wide screens of mutant collections described in the next section have revealed a fascinating connection between cellular aging and the TORC1/Sch9 and the Ras/protein kinase A (PKA) pathways. Both pathways appear to participate in regulating both chronological aging—the extent of persistent viability following nutrient starvation—and replicative aging—the number of generations a cell can continue dividing. These findings prompted a detailed examination of the roles played by the TOR and Ras pathways in directing the molecular mechanism of aging. By combining transcriptional profiling, promoter motif analysis, and genome-wide chromatin immunoprecipitation (ChIP) analysis, Longo and Li laboratories identified three transcription factors—Msn2/4 and Gis1—that function downstream in the Sch9, PKA, and TOR pathways to be largely responsible for the observed increased life span of *sch9*, *ras2*, and *tor1* mutants [30]. Analysis of the profiling data suggested that reduction of mitochondrial function may contribute to longevity in *sch9*, *ras2*, and *tor1* mutants after diauxic shift: whereas in wild-type cells, mitochondrial ribosomes are upregulated and cytosolic ribosomes are downregulated, the opposite is true in these mutants [31]. Li, Longo, and colleagues suggest that the reduction in mitochondrial activity in the mutants, and hence lower TCA flux and oxidative phosphorylation, results in lower generation of reactive oxygen species (ROS), and thereby contribute to life span extension [32]. In support of this finding, loss of *SOD2*, a mitochondrial superoxide dismutase, negates the extended life span of *sch9* strains [33]. Furthermore, Li, Longo, and colleagues provide evidence from microarray analysis that TORC1 and Sch9 pathways affect glycogen biosynthesis and transcriptional stress response following nutrient starvation. Notably, the loss of glycerol biosynthesis genes (such as *GPD1*, *GPD2*, or *RHR2*) reversed the chronological longevity of cells lacking *SCH9* [34].

As noted in the preceding section, transcriptional profiling has been instrumental in constructing, dissecting, and defining the TOR network. The ease in specifically inhibiting TORC1 with the macrolide drug rapamycin provided a seductive mean to examine the global transcriptional response to TORC1 inactivation. We note that while these studies were both facile and informative, subsequent observations demonstrated that rapamycin inhibited only the TORC1 complex and thus probed only one branch of the TOR network [35]. Moreover, rapamycin is not universally effective in inhibiting TORC1—*Schizosaccharomyces pombe*, and many mammalian cells are resistant to even high levels of rapamycin that inhibit TORC1 in *Saccharomyces* and T-cells, a shortcoming that has prompted the recent development of novel TOR inhibitors [36]. We anticipate that these newer compounds will prove as valuable in defining TOR signaling as rapamycin has been to date.

B. Genome-Wide Mutant Screens

Use of comprehensive yeast deletion collections has been particularly informative in identifying genes conferring altered growth fitness or sensitivity to drugs such as rapamycin [37–39]. Zheng and colleagues were the first to use this chemical-genomics approach to identify loci whose deletion altered rapamycin sensitivity (Figure 16.2A and B) [40]. By screening 2216 nonessential haploid and 50 heterozygous essential diploid mutants, they identified 73 mutants displaying rapamycin-hypersensitivity, including *ure2*, *clb5*, *hom2/3*, and 27 mutants exhibiting rapamycin resistance, including *gln3* and *rrd2* (Figure 16.2A and B). *GLN3* encodes a transcriptional activator of NDP genes and *URE2* a negative regulator of Gln3, while Rrd2 is a prolyl isomerase that activates protein phosphatase 2A (Figure 16.1). All the three have functional roles in TORC1 signaling, thus validating the screen. The majority of the genes identified in the screen clustered into eight functional processes, including protein synthesis, carbon, and nitrogen catabolite repression, and mitochondrial biogenesis.

Zheng group's initial study was substantially extended by Hughes, Boone, and colleagues, who generated chemical-genetic profiles for 12

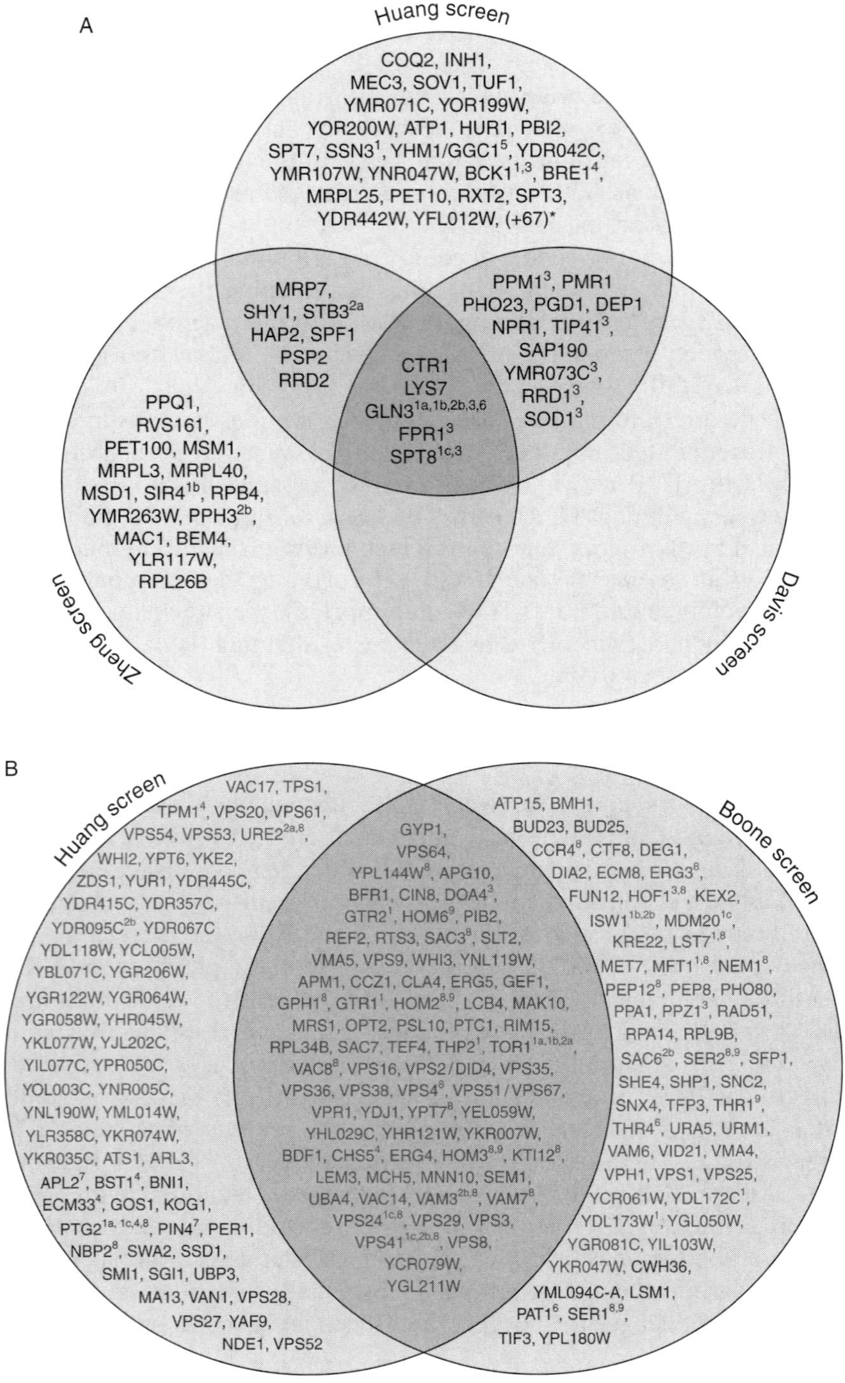
A
Huang screen
COQ2, INH1,
MEC3, SOV1, TUF1,
YMR071C, YOR199W,
YOR200W, ATP1, HUR1, PBI2,
SPT7, SSN3[1], YHM1/GGC1[5], YDR042C,
YMR107W, YNR047W, BCK1[1,3], BRE1[4],
MRPL25, PET10, RXT2, SPT3,
YDR442W, YFL012W, (+67)*
MRP7,
SHY1, STB3[2a]
HAP2, SPF1
PSP2
RRD2
PPM1[3], PMR1
PHO23, PGD1, DEP1
NPR1, TIP41[3],
SAP190
YMR073C[3],
RRD1[3],
SOD1[3]
CTR1
LYS7
GLN3[1a,1b,2b,3,6]
FPR1[3]
SPT8[1c,3]
PPQ1,
RVS161,
PET100, MSM1,
MRPL3, MRPL40,
MSD1, SIR4[1b], RPB4,
YMR263W, PPH3[2b],
MAC1, BEM4,
YLR117W,
RPL26B
Zheng screen
Davis screen
B
Huang screen
Boone screen
VAC17, TPS1,
TPM1[4], VPS20, VPS61,
VPS54, VPS53, URE2[2a,8],
WHI2, YPT6, YKE2,
ZDS1, YUR1, YDR445C,
YDR415C, YDR357C,
YDR095C[2b], YDR067C
YDL118W, YCL005W,
YBL071C, YGR206W,
YGR122W, YGR064W,
YGR058W, YHR045W,
YKL077W, YJL202C,
YIL077C, YPR050C,
YOL003C, YNR005C,
YNL190W, YML014W,
YLR358C, YKR074W,
YKR035C, ATS1, ARL3,
APL2[7], BST1[4], BNI1,
ECM33[4], GOS1, KOG1,
PTG2[1a,1c,4,8], PIN4[7], PER1,
NBP2[8], SWA2, SSD1,
SMI1, SGI1, UBP3,
MA13, VAN1, VPS28,
VPS27, YAF9,
NDE1, VPS52
GYP1,
VPS64,
YPL144W[8], APG10,
BFR1, CIN8, DOA4[3],
GTR2[1], HOM6[9], PIB2,
REF2, RTS3, SAC3[8], SLT2,
VMA5, VPS9, WHI3, YNL119W,
APM1, CCZ1, CLA4, ERG5, GEF1,
GPH1[8], GTR1[1], HOM2[8,9], LCB4, MAK10,
MRS1, OPT2, PSL10, PTC1, RIM15,
RPL34B, SAC7, TEF4, THP2[1], TOR1[1a,1b,2a],
VAC8[8], VPS16, VPS2/DID4, VPS35,
VPS36, VPS38, VPS4[8], VPS51/VPS67,
VPR1, YDJ1, YPT7[8], YEL059W,
YHL029C, YHR121W, YKR007W,
BDF1, CHS5[4], ERG4, HOM3[8,9], KTI12[8],
LEM3, MCH5, MNN10, SEM1,
UBA4, VAC14, VAM3[2b,8], VAM7[8],
VPS24[1c,8], VPS29, VPS3,
VPS41[1c,2b,8], VPS8,
YCR079W,
YGL211W
ATP15, BMH1,
BUD23, BUD25,
CCR4[8], CTF8, DEG1,
DIA2, ECM8, ERG3[8],
FUN12, HOF1[3,8], KEX2,
ISW1[1b,2b], MDM20[1c],
KRE22, LST7[1,8],
MET7, MFT1[1,8], NEM1[8],
PEP12[8], PEP8, PHO80,
PPA1, PPZ1[3], RAD51,
RPA14, RPL9B,
SAC6[2b], SER2[8,9], SFP1,
SHE4, SHP1, SNC2,
SNX4, TFP3, THR1[9],
THR4[6], URA5, URM1,
VAM6, VID21, VMA4,
VPH1, VPS1, VPS25,
YCR061W, YDL172C[1],
YDL173W[1], YGL050W,
YGR081C, YIL103W,
YKR047W, CWH36,
YML094C-A, LSM1,
PAT1[6], SER1[8,9],
TIF3, YPL180W

inhibitory compounds, including rapamycin, cycloheximide, and hydroxyurea, by screening hypersensitivities of ~4700 viable haploid mutants to these compounds (Figures 16.2b and 16.3) [41]. In addition to finding a large number of genes required for multidrug resistance, they identified drug-specific chemical–gene interactions. Two hundred and three knockout strains displayed hypersensitivity solely to rapamycin, with *tor1*, *ser1*, *gtr1*, and *gtr2* exhibiting strong sensitivity. The latter two genes have subsequently been shown to be homologs of the RagA/B and RagC/D proteins, respectively, which participate in amino acid regulation of TORC1 activity in mammalian cells [42]. In addition, these proteins are components of the EGO and GSE complexes in yeast, which are involved in recovery from rapamycin treatment and postgolgi sorting of the Gap1 permease, respectively [43].

The Huang laboratory further improved the genome-wide screening process by developing a high-density cell array method, in which up to 9600 strains are printed on a single plate using nanoliter microarraying pins (Figures 16.2b and 16.3) [44]. By employing this cell array method, Huang and colleagues screened 4850 nonessential haploid deletions and 1175 essential heterozygous diploid deletion strains to identify mutants with altered fitness to rapamycin. Their screen revealed 396 strains, of which 281 were hypersensitive, and 101 were resistant to rapamycin. Moreover, they identified that of the 14 deletion strains exhibiting growth fitness advantage

---

FIG. 16.2. Composite Analysis of Screens for Mutants with Altered Sensitivity to Rapamycin. (A) Shown is the overlap of genes whose deletion confers resistance to rapamycin from the screens reported by Chan *et al.* [40] (Zheng screen), Xie *et al.* [44] (Huang screen) and Neklesa and Davis [45] (Davis screen). For clarity, only the top 40 resistant deletion strains unique to the Xie *et al.* screen plus all deletion strains that overlap with Chan *et al.* and Neklesa and Davis, screen are shown. (B) Shown is the overlap of genes whose deletion confers increased sensitivity to rapamycin from the screens reported by Xie *et al.* [44] (Huang screen) and Parsons *et al.* [41] (Boone screen). The top 60 hypersensitive deletion strains unique to each screen as well as all the strains that overlap in the two screens are shown. All strains are listed in increasing hypersensitivity to rapamycin: least sensitive (purple) to most sensitive (blue). In addition to hypersensitivity or resistance to rapamycin, several of the strains have the following additional phenotypes: (1a) Deletion confers increased chronological life span, as reported in [53]. (1b) Deletion confers increased chronological life span, as reported in [56]. (1c) Deletion confers decreased chronological life span [56]. (2a) Deletion confers increased replicative life span [52]. (2b) Deletion confers decreased replicative life span [52]. (3) Deletion reduces rapamycin induction of P_{DAL80}-GFP [50]. (4) Deletion increases rapamycin induction of P_{DAL80}-GFP [50]. (5) Deletion is hyper-resistant to rapamycin [[74]]. (6) Rapamycin-sensitive phosphorylation [66]. (7) Deletion is hypersensitive to rapamycin, as reported in [51]. (8) Deletion is hypersensitive to rapamycin, as reported in [40]. (9) Deletion is hypersensitive to rapamycin, as reported in [45]. (For interpretation of the references to color in this figure legend, the reader is referred to the Web version of this chapter.)

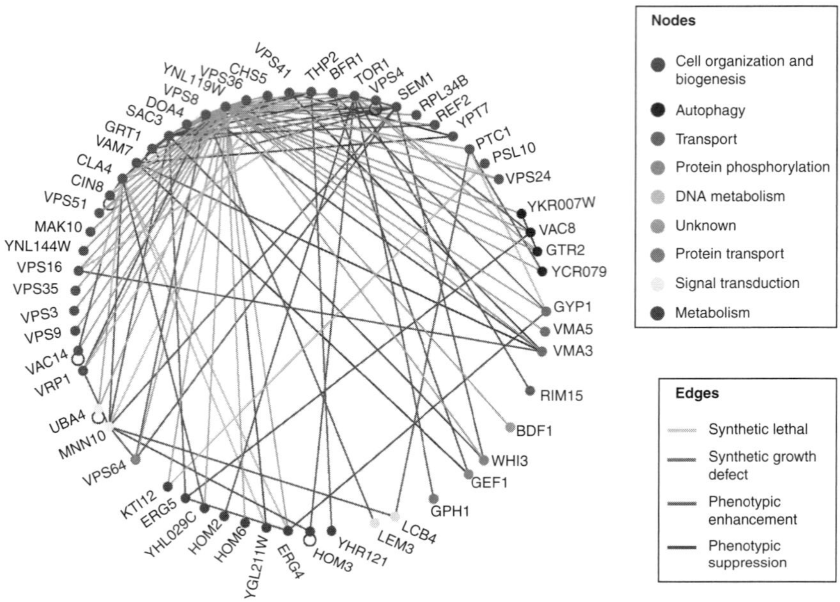

FIG. 16.3. Interaction network of overlapping set of genes in Figure 16.2B. Interaction map was assembled and generated using Osprey, a software based on Gene Ontology (GO) annotations, as maintained by bioGRID [109]. Nodes represent genes and edges represent observed interaction between nodes. Colors of nodes represent biological process, as defined by Gene Ontology, and edge colors represent the experimental system used to identify interactions. Out of 17 available experimental systems on Osprey, the indicated four were selected to query the network.). (For interpretation of the references to color in this figure legend, the reader is referred to the Web version of this chapter.)

in the presence of rapamycin (rapamycin-enhanced), nine mutants are disrupted at genes encoding mitochondrial proteins.

A more recent embodiment of this general approach relies on the genetic barcode bracketed by universal primer sites that is resident uniquely in each of the deletion strains. Accordingly, the deletion collection can be pooled, subjected to a specific environmental condition and then interrogated by microarray or sequencing of the PCR-amplified barcodes to determine those mutants that are enriched or depleted by the imposed condition. Neklesa and Davis screened a culture of pooled deletion strains for partial resistance to rapamycin. In total, they identified 16 rapamycin-resistant strains, including *fpr1*, whose functional product forms a complex with rapamycin to inactivate the TOR kinase (Figure 16.2A and B). Additionally, strains deleted in *gln3* and *rrd1*, recovered in the screens described earlier, as well as *ctr1* (copper transporter) and *sod1* (Cu,Zn-superoxide dismutase), exhibited substantial competitive fitness in the presence of low levels of rapamycin. Microarray analysis revealed that loss of Sod1 and Gln3 reduced the global transcriptional response to rapamycin. Furthermore, levels of superoxide anions were elevated in all rapamycin-resistant strains, except for *fpr1*. Although the molecular mechanism is unclear, these results suggest that the increased concentration of ROS, and thereby elevated oxidative stress, provides protection against diminished TORC1 function [45].

A variant of this method uses a barcoded yeast open reading frame (MoBY-ORF) library to readily recover the wild-type gene corresponding to a recessive mutation conferring drug resistance. The library consists of 4956 uniquely barcoded ORFs, each controlled by its native promoter and terminator, resident on a low-copy centromere-based vector. The collection is transformed into a drug-resistant mutant, the pooled transformants grown in the presence of a drug, and the single depleted gene in the pooled plasmids—that is, the plasmid that complemented the mutation and restored drug sensitivity to the mutant—determined by hybridization on a barcode specific array. As a validation of this methodology, the investigators identified *FPR1* as a gene whose inactivation confers rapamycin resistance, a well-established phenotype consistent with its role in inhibiting TORC1 function when complexed with rapamycin. Although the method did not provide new information regarding TORC1 function, it successfully identified recessive mutations, resulting in resistance to three other drugs with antifungal activities—sordarin, theopalauamide, and stichloroside—and should allow further investigation of TORC1 function in the future [46].

1. Genome-Wide Screens and New Components of TOR Network

As noted earlier, the substantial overlap in the transcriptional profile resulting from nitrogen starvation and from rapamycin treatment suggests that TORC1 functions as a conduit for transmitting information on the status of nitrogen availability. However, the mechanism connecting TORC1 to nutrient status has not been well resolved and appears distinct from that operative in mammalian cells [47–49]. Accordingly, the Davis laboratory used a genome-wide screen to identify Npr2 and Npr3 as required for TORC1 inactivation in response to amino acid starvation [50]. Their screen entailed transforming all nonessential yeast deletion strains ($\sim$5100) with a GFP reporter plasmid driven by *DAL80* promoter, which is activated by Gln3 and Gat1 in response to TORC1 inactivation. By subjecting these transformed deletion strains to rapamycin-containing or amino acid free media and recovering strains with low levels of GFP expression by flow cytometry, they found that strains lacking Npr2 or Npr3 fail to induce P_{DAL80}-GFP reporter in response to amino acid starvation. While deletion of either *GLN3* or *GAT1* blocked activation of P_{DAL80}-GFP by either rapamycin treatment or amino acid starvation, deletion of *NPR2* or *NPR3* blocked activation of P_{DAL80}-GFP by amino acid starvation but not by rapamycin treatment. This places Npr2 and Npr3 upstream of TORC1, consistent with a role in mediating TORC1 activity in response to amino acid availability. Additional studies are required to understand how the Npr2/3 complex senses amino acid availability and regulates TORC1.

In order to identify novel components of TOR regulatory network, Schreiber and colleagues screened compound libraries for novel small molecules that would antagonize rapamycin inhibition of cell growth. Their screen identified several compounds that fully restored cell growth in the presence of 100 nM rapamycin [51]. To identify the relevant targets of these compounds, the investigators used the compounds to probe a proteome chip that contained nearly the entire cohort of yeast proteins. This process identified Tep1 and the product of *YBR077c* as possible novel components of the TORC1 response pathway. Although *tep1* strains exhibit wild-type sensitivity to rapamycin, loss of one of its interacting partners, *APL2*, does result in increased sensitivity to rapamycin (Figure 16.2B). Moreover, the product of *YBR077c* has been shown to be a component of the GSE and EGO complexes in yeast mentioned earlier, both of which are associated with TORC1 function.

2. *Genome-Wide Screens and Aging*

High-throughput and genome-wide studies have revealed functional connections between TORC1 and life span. Fields and Kennedy laboratories initially used the yeast deletion collection to identify genes whose loss affected replicative life span (RLS), that is, the number of divisions a cell undergoes before ceasing to divide. They recovered ten mutants with increased RLS from a screen of 564 deletion strains. Six of the ten genes were implicated in the TOR signaling pathway. Moreover, they showed that calorie restriction did not further extend the life span in two such mutants, *tor1* or *sch9*, and that loss of Sch9 or Tor1 resulted in increased life span independently of Sir2 histone deacetylase. These results suggest that TORC1 and Sch9 pathways mediate, at least in part, increased RLS resulting from nutrient restriction [52].

A second screen also by Kennedy, Fields, and colleagues demonstrated that TORC1 activity also contributes to chronological life span (CLS), which is essentially the ability to maintain viability following entry into stationary phase [53]. Of the 90 predicted long-lived mutants out of the 4800 diploid deletion strains examined, 16 were involved in TORC1 or nutrient acquisition that diminished TORC1 activity. Moreover, inhibition of TOR activity by rapamycin or methionine sulfoximine (MSX, inhibitor of glutamine synthetase) also increased viability during stationary phase in a dose-dependent manner. One difficulty in interpreting the biological significance of this study with regard to the function of TORC1 is in understanding the nature of the stress imposed by the stationary phase. Since the mutants were all in an auxotrophic background, these experiments may not probe a normal biological process. In fact, Botstein and colleagues recently showed that cells starved specifically for nitrogen, sulfur, phosphate, or glucose maintained viability for a substantially longer time than the same cells starved for an amino acid or pyrimidine for which it is auxotrophic [54]. Moreover, inactivating mutations in *SCH9*, *TOR1*, or *PPM1* (a regulator of PP2A activity) suppressed the lethality, as well as futile consumption of glucose, arising from such auxotrophic starvation [55]. Thus, it may be the case that reduction of TORC1 pathway activity promotes a "natural" exit from the cell cycle that allows an orderly cessation of growth to preserve viability.

More recently, Smith and colleagues conducted a microarray-based screen on pooled haploid deletion mutants to identify genes contributing to CLS under nonrestricted (2% glucose concentration in medium) or calorie-restricted (0.5% glucose) conditions [56]. The short-lived mutants identified included those defective in autophagy, including *atg5*, *atg8*, and *atg12*. Interestingly, whereas the shortened life spans of autophagy mutants in yeast were rescued by caloric restriction (CR), life span extension induced by dietary restriction (DR) in *Caenorhabditis elegans* were prevented by defects in autophagy [56–58]. Moreover, the investigators observed that extended life span mediated by amino acid limitation was inhibited in mutants defective in autophagy. Additional data from this and Aris groups suggest that autophagy is required for increased CLS of cells lacking *TOR1* or cells treated by low concentrations of rapamycin [59]. Confirmed CLS-extended mutants included those lacking components of the *de novo* purine biosynthesis pathway, such as *ade2*, *ade3*, and *ade4*. Intriguingly, addition of excess adenine to the growth medium suppressed the increased CLS of *ade* mutants, and CR imposed by limited glucose concentration did not further extend CLS in these mutants.

Smith and colleagues pointed out that whereas a large fraction of identified short-lived mutants overlap between their and previous screens by the Kennedy and Fields laboratories, only a small number of long-lived mutants (*ypl056c*, *dcw1*, *ylr104w*, and *mum2*) overlap between the two screens. The poor agreement is largely attributed to differences in methodologies [56]. Whereas Kennedy and Fields groups tested individual mutants in 96-well plates, the Smith group examined a pooled collection of mutants in a single culture [53, 56]. The utilization of a mixed population of strains introduces inevitable caveats, such as growth fitness competition between the strains, and modification of growth medium composition by strains that directly challenge the viability of sensitive strains (such as sensitivity to secreted acetic acid or resultant pH changes) [56]. Additional critical difference between these screens is in growth media composition, namely the level of provided amino acids. Moreover, as mentioned earlier regarding the Kennedy and Fields screen, the approach by Smith laboratory also examined auxotrophic mutants, thereby similarly complicating the interpretation of biological significance of the identified mutants.

3. *Genome-Wide Screens and Cell Size*

The connection between nutrient status, cell cycle, and critical cell size in yeast was described by Hartwell and colleagues in the 1970s [60–63]. To explore the molecular mechanism underlining the maintenance of balance between nutrients and growth [63–65], Tyers laboratory examined 4812 haploid deletion strains during log phase growth in rich media to identify mutants with defects in size homeostasis. Approximately 500 mutants exhibited altered size distributions from that of the wild-type cells, with *sfp1* and *sch9* strains displaying one of the smallest cell-size phenotype. Global transcriptional analysis revealed that the transcriptional activator Sfp1 and the kinase Sch9 function in parallel to promote the accumulation of biosynthetic capacity in response to nutrient status by positively regulating the expression of ribosomal protein (RP) and ribosome biogenesis (Ribi) regulons. Importantly, they proposed that Sfp1 and Sch9 set the critical cell-size threshold at Start corresponding to the nutrient status by controlling the rate of ribosome production [2, 63, 64].

C. Phosphoproteome

Many of the cellular components encompassing diverse cellular functions regulated by the TORC1 signaling cascade are mediated by phosphorylation. A number of downstream effectors of TORC1 regulated by phosphorylation are known, including Sch9, Rim15, Gln3, Msn2/4, etc. In order to characterize in detail and better define the TORC1 phosphoproteome, Loewith and colleagues used label-free LC-MS phosphoproteomics to quantify relative phosphorylation patterns [66]. Comparison of phosphorylation patterns between rapamycin-treated and untreated wild-type cells identified more than one hundred proteins whose phosphorylation status was sensitive to rapamycin treatment. Notably, three transcription factors—Dot6, Tod6, and Stb3—shown by Lippman and Broach to be required for repression of ribosome biogenesis genes in response to TORC1 inactivation or nutrient starvation, became hypophosphorylated after rapamycin addition [28]. Moreover, consistent with their observation that the protein encoded by *YDL173w*, subsequently renamed *PAR32*, becomes hyperphosphorylated after rapamycin treatment, Boone and colleagues identified Par32 to mediate TORC1 regulation of Gln3, Gat1, Rtg1, and Rtg3 transcription factors based on biochemical and genetic network connectivity (see below) [67].

By using rapamycin-insensitive *tap42-11* and *SCH9*^{3E} alleles, Loewith and coworkers assigned rapamycin-sensitive phosphoproteins to either Tap42 or Sch9 effector branch. Whereas rapamycin-induced dephosphorylation of Dot6 was dependent on both Sch9 and Tap42 activity, Tod6 depended only on Sch9. This is consistent with our observations that the effects on Ribi gene regulation are mediated predominantly by Tod6. Hyperphosphorylation of Par42 after rapamycin treatment was largely inhibited in cells expressing *tap42-11*, consistent with earlier observations that rapamycin effects on NDP and RTG transcription are mediated by Tap42 and PP2A [6, 20]. In addition, the Loewith group examined rapamycin-induced dephosphorylated of Maf1 (repressor of RNA Pol III), and showed that TORC1 modulates the activity of RNA Pol III by regulating Maf1 via Sch9 [66]. Concurrently, the Willis group also demonstrated that Sch9 regulates RNA Pol III transcriptional activity [68]. These findings together with the evidence that Sch9 regulates expression of rRNA and binding of RNA Pol I to rDNA locus posit Sch9 as a major coordinator between the external environment and the generation of protein synthesis capacity through ribosome biogenesis [66].

D. Global Functional Maps

Synthetic genetic array (SGA) has provided a systematic method for identifying genetic interactions in yeast. In the SGA method, a strain containing a specific mutation of interest—such as deletion or conditional mutation—is mated to the entire haploid deletion collection, and haploids carrying both mutations are recovered from each cross. Analysis of phenotypes of the double mutant strain relative to each of the single mutant strains indicates whether the two mutations antagonize or ameliorate each other, which would indicate participation in a common process (Figure 16.4) [38].

Recently, a joint effort between Myers, Andrews, and Boone laboratories constructed an impressive functional genetic map of the yeast cell based on over five million genetic interactions obtained from crossing 1712

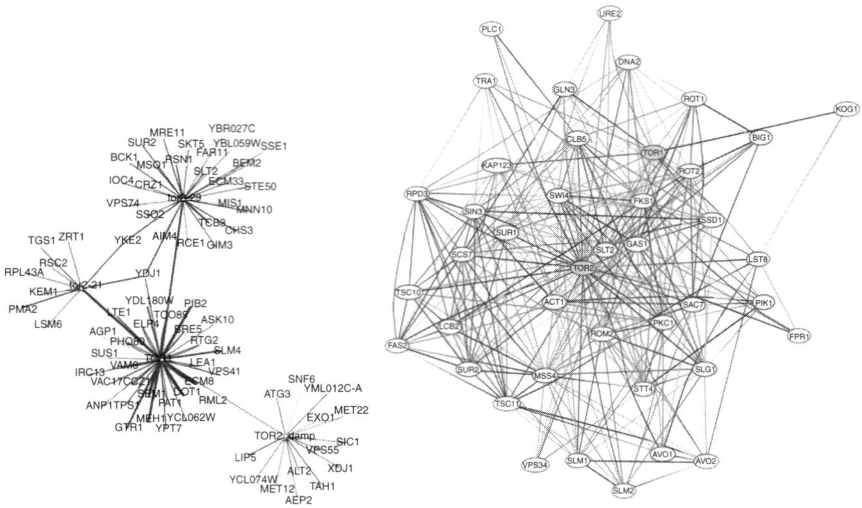

FIG. 16.4. Interaction network for Tor1 and Tor2. Left: DRYGIN map [67, 69] is based on genetic interactions derived from double-mutant SGA analysis. Nodes represent gene deletions. Positive genetic interactions are represented by green edges and negative interactions by red. The width of the edges is proportional to the magnitude of the interaction scores. SGA epsilon scores of 0.15 and – 0.15 were used as thresholds for negative and positive interactions, respectively. Network map was generated using Cytoscape [110]. Right: Interactions returned by bioPIXIE [111], which is based on Bayesian integration of genetic and physical interaction, sequence information, gene expression data, and cellular localization. Nodes indicate proteins and edge color represents the interaction confidence ranging from red (highest confidence), orange, yellow, to green (lowest confidence). (For interpretation of the references to color in this figure legend, the reader is referred to the Web version of this chapter.)

query mutant strains, composed of both deletion of nonessential genes and conditional alleles of essential genes, to 3885 array strains. Using colony size as the metric for quantifying genetic interactions, they identified genetic combinations in which the double mutant displays either a negative interaction—the fitness of the double mutant was less than the effects of single mutants—or positive interaction—the double mutant fitness is greater [67, 69]. Since genetic interactions are strongly correlated to biological functions and processes, their global network displayed organized clusters reflecting functional connections between genes. For example, three genes, *PAR32*, *ECM30*, and *UBP15*, clustered with genes whose products are associated with the Gap1 sorting pathways (the GSE cluster discussed earlier), and indeed loss of these genes resulted in defects of Gap1 localization and activity. Interestingly, information gathered from network connectivity and experimental evidence using *GAP1* and *CIT2* reporter assays in *par32* strains suggests that Par32 may mediate TORC1's regulation of Gln3, Gat1, Rtg1, and Rtg3 transcription factors. The authors emphasize that functional networks based on genetic interactions add another dimension to networks based on protein–protein interactions, since physical interaction maps provide information solely at the level of local connections regardless of the physiological context [67].

IV. Integration of Data

We as a community are just beginning to integrate data obtained from different global profiling studies. Summaries from our progress to date, including studies not directly targeted at the TOR network, are given in the following sections.

A. Transcriptome and Metabolome

We appreciate that metabolism and transcription must be interconnected: Transcription produces the proteins that catalyze production and incorporation of metabolites and, through genetic regulatory circuits, metabolites often set the level of transcription of the genes responsible for modulating metabolite levels. Nonetheless, metabolite levels and transcription levels are surprisingly uncoordinated. That is, variation in the levels of a metabolite does not precisely correlate with the variation in the levels of the transcripts encoding enzymes involved in their synthesis and decay. The Botstein group measured the levels of a large number of metabolites in yeast cells growing at different growth rates due to limitation for a variety of different nutrients and, among other analyses, compared those levels to transcriptional profiles obtained under the same conditions [70]. They found that, while a large fraction of transcripts showed growth-specific regulation independent of the limiting nutrient, the levels of most metabolites were nutrient specific. Thus, transcript levels appear to respond to growth rate input, while metabolites' levels are set primarily by mass action flow and by allosteric regulation of metabolic enzymes. This difference in the mode of regulation results in only limited correlation between transcripts and metabolites in yeast.

Although a precise correspondence between metabolites and transcripts is not readily apparent in yeast, a more subtle relationship does exist. By applying singular value decomposition (SVD) to transcript and metabolite profiles of yeast cells in response to carbon and nitrogen starvation, Troyanskaya and colleagues demonstrated a significant correlation between metabolites and transcripts involved in related biological processes [71]. Moreover, they showed that gene–metabolite relationship depends on the experimental conditions. For example, whereas metabolites involved in glycolysis and pentose-phosphate pathway display positive correlation under nitrogen starvation, they show negative relationship under carbon starvation (i.e., phosphoenolpyruvate and *ILV2*). However, metabolites from amino acid metabolic class exhibit positive correlation with corresponding genes under both nutrient starvations (i.e., methionine and *MET6*). Furthermore, they used Bayesian analysis to integrate gene–metabolite correlation under carbon and nitrogen starvation, functional relationship between the gene–metabolite interactions, and the metabolite class. This work provides evidence that global transcriptional and metabolic data can be integrated, correlated, and combined with mathematical analysis to represent a global network of gene–metabolite interaction, and describes the methodology that could be extended to the TOR network.

B. Transcriptome and Proteome

Souchelnytskyi [72] has provided an excellent review on the types of rich data sets generated by proteomics and how they can be integrated to enhance a systems-level understanding of biological networks. With specific application to the TOR network, Griffin and colleagues performed quantitative proteomic profiling on rapamycin-treated cells and correlated it to existing microarray data [73]. They observed that while most (102/127—80%) changes in protein levels correlated with those of mRNA levels, a small fraction (5/127—4%) in fact displays anticorrelation. The majority of proteins whose dynamics correlated positively with the corresponding mRNA dynamics comprise components of ribosomal biogenesis and protein translation as well as metabolic enzymes, such as those involved in the utilization of alternate nitrogen sources. Furthermore, by comparing the rapamycin-treated proteomic profile with existing transcriptional profiles in response to heat/oxidative stress, Griffin group showed that 88% of the measured proteins (112/127) also displayed changes that correlated with those of mRNA levels under these conditions [73]. The investigators noted that the considerable overlap between TORC1-regulated genes with those responsive to heat/oxidative stresses were not completely accounted for by Msn2/4 activation and proposed that the heat shock transcription factor Hsf1 may play a role in TORC1 regulation of the stress response. However, despite the proteomic correlation between TORC1 inactivation and activation of Hsf1, a mechanistic link between TORC1 signaling and Hsf1 activity has not been established.

To extend these initial findings and to identify new components regulated by TORC1, Washburn and colleagues recently performed transcriptomic and proteomic analysis of yeast cells over 6 h following rapamycin treatment [74]. The investigators demonstrated that the level of mRNA did not temporally correlate with the levels of corresponding proteins in a large subset of TORC1-regulated targets. They observed that changes in protein abundance occurred 3–5 h after changes in the corresponding mRNA abundance. This delay was particularly apparent for proteins involved in ribosome biogenesis, heat shock, autophagy, and the retrograde response. Moreover, they observed cases in which protein levels increased without an increase in the corresponding mRNA levels. A specific example was Ggc1, a mitochondrial protein involved in genome maintenance, whose stability increased in the presence of rapamycin. Through no obvious mechanistic process, deletion of *GGC1* resulted in derepression of genes involved in ribosome biogenesis and cell cycle progression, an alteration that could account for the rapamycin resistance phenotype of the mutant. In short, this comparison of transcriptional and proteomic profiles provided an insight into another mode of regulation by TOR—in coordinating the cell's biosynthetic capacity with cell cycle progression.

These integrative studies highlight insights into the TOR network that could only be achieved by doing a comparative analysis between proteomic profiles with global expression data. In addition, these studies emphasize the importance of examining the cellular processes on multiple levels of readout to accurately identify regulatory connections within a single network as well as in relation to the rest of the cell.

C. Integration of Transcriptome, Proteome, and Metabolome

The Tomita and Oliver laboratories performed groundbreaking systems-level analysis of metabolic, proteomic, and transcriptional levels of *Escherichia coli* and *Saccharomyces cerevisiae*, respectively. Tomita and colleagues obtained quantitative measurements of mRNA, protein, and metabolic levels of wild-type *E. coli* cells in glucose-limited chemostats at various growth rates and 24 single-gene knockout strains (gene-disruptions in glycolysis and pentose-phosphate pathway) in chemostats at a single growth rate [75]. Their results suggest that the metabolic levels are highly adaptive under both genetic and environmental perturbation, and that such stability is maintained by the robustness of the metabolic network, which efficiently equilibrates itself, for instance, by rerouting metabolic fluxes. Moreover, metabolic stability is provided by active regulation of gene expression, presence of isozymes, and alternate routes in catabolism of metabolites so that the cell can survive various environmental insults.

Oliver and colleagues examined the impact of growth rates on three levels of biological information—transcriptome, proteome, and metabolome—of wild-type *S. cerevisiae* cells grown in steady-state chemostat culture under four different nutrient limitations (glucose, ammonium, phosphate, and sulfate) at three different growth rates [76]. Their results reveal that the levels of 891 transcripts were determined by growth rate regardless of the specific limiting nutrient—an observation confirmed and extended by Botstein and colleagues [54]. To further examine the extent of TOR signaling in regulating these growth-rate-responsive genes, they compared expression profiles at different growth rates with that in response to TORC1 inactivation by rapamycin. They observe that 70% of growth-responsive genes overlap with rapamycin-sensitive genes. Importantly, they note that although TOR itself is a major controller of cell growth, the components of the TORC1 complex themselves are not regulated at the transcriptional level by growth rate.

D. Genetics and Transcription or Proteomics

Little correlation exists between those genes whose mutation confers enhanced resistance or sensitivity to rapamycin and those genes that exhibit altered transcription or translation in response to rapamycin treatment. For instance, only 35 genes out of the 396 whose deletion yield altered rapamycin resistance display more than threefold changes in transcriptional changes in response to rapamycin. This is consistent with systems studies in other responses suggesting that yeast cells mount a major response to a variety of stresses, but only a small number of those are directly relevant to any particular insult. Moreover, those genes involved in the processes interdicted by various stresses, including rapamycin treatment, are generally not subject to expression modulation in response to that challenge.

V. Computational Modeling and Prediction

A hallmark of quantitative computational modeling is to identify gaps in experimental data and provide insights or hypotheses for filling those gaps. Such model-based hypothesis can then be falsified or validated experimentally to correct or strengthen the generated mathematical model. To date, complete modeling of the TOR signaling network—through coupled ordinary differential equations, for instance—has not been accomplished due to lack of data on essential parameters, such as the relevant cellular concentrations of critical components or the kinetic parameters describing interactions among components. Moreover, conflicting experimental evidence complicates the determination of accurate parameters and modeling.

A. Modeling TOR and Tap42

Despite the challenges noted above, Stelling and coworkers have developed a model-based analysis of TORC1 signaling dynamics that focused on Tip41, Tap42, and PP2A, and provided key insights into the molecular mechanisms underlying their activity in response to rapamycin [77]. Tap42 was previously identified as an evolutionarily conserved substrate of TORC1 and a specificity factor for PP2A catalytic subunits that mediated rapamycin-induced phosphatase activity [78, 79]. Tip41 was identified as a binding partner for Tap42 and was proposed to be involved in a negative feedback loop inhibiting Tap42 activity [80]. Stelling and colleagues used ensemble modeling to address various unresolved or ambiguous aspects of Tap42 and Tip41 function in TORC1 signaling by representing the TOR signaling network as a dynamic system [77]. Their approach involved positing experimentally derived hypotheses as libraries of alternative mathematical models, testing the abilities of the models to accurately represent the experimental observations, and deriving targeted experiments necessary to discriminate more precisely among the alternative models. Although the actual components and much of their molecular interaction have been defined by experimental evidence, quantitative understanding of the function of Tap42–Tip41 complex, the functional implications of their phosphorylation status, and the role of Tap42 in PP2A regulation remained largely unclear.

The novel aspect of the ensemble modeling approach is that the investigators represented the known and suspected interactions of TORC1/PP2A circuit not as a single set of equations but as a collection of sets of equations. Each set consisted of the core model but described mechanistically distinct networks through the addition of one or more potential but unconfirmed interactions within the system. The optimum collection of equations could then be identified by determining which of the collections returned the least error in predicting results from previously published time course experiments. The investigators' ensemble contained 19 mathematical models based on ordinary differential equations to depict the dynamics of 24–30 molecular constituents, with rapamycin designated as input and PP2A activity measured as output. Parameters were estimated by global optimization methods to minimize the error from simulations of the models and previous experimental observations.

Overall, the ensemble approach demonstrated that, in contrast to previous suggestions, Tip41 and Tap42 function in similar fashions in TORC1 signaling. Upon phosphorylation by TORC1, both proteins form complexes with PP2A catalytic subunits. Also in contrast with earlier suggestions, such interactions do not serve to attenuate PP2A activity, since only 20% of the available PP2A could be sequestered by Tip41 and Tap42. Rather, the Tap42–Tip41 complex functions as a specificity factor that triggers rapamycin-induced and phosphatase-mediated dephosphorylation and subsequent activation of Gln3 and Rtg1 to promote transcription of the NDP and the RTG pathway, respectively. Moreover, this specificity factor has different effects on different substrates, with Tip42 having a negative effect in directing the phosphatase to the stress-activated transcription factor Msn2. Finally, the ensemble model consistently predicted that active TORC1 forms a complex with Tap42, which is abrogated by the presence of rapamycin. Both computational and experimental evidence provided by the investigators also demonstrated Tip41–PP2A complex formation in the absence of rapamycin, followed by dissolution of the complex in response to rapamycin addition.

Thus, the ensemble modeling confirmed and extended previous experimentally derived model of Tap42-associated rapamycin signaling (Figure 16.1). The model that emerged from this analysis posits that Tap42 and Tip41 bind to TORC1, become phosphorylated as a consequence of that interaction and, in their phosphorylated state, recruit the PP2A catalytic subunits, Pph21/22 and Sit4. Upon treatment with rapamycin, the Tap42/Tip41/PP2A complex is liberated from TORC1 and interacts with substrates, such as Gln3 and Rtg1, to activate rapamycin-inducible responses. Simultaneously, the cytoplasmic phosphatase 2A holoenzyme dephosphorylates Tap42 and Tip41 to disrupt their association with the PP2A catalytic subunits, thus terminating the rapamycin-induced signaling. This mechanism creates a self-limiting, transient activation of downstream pathways following rapamycin treatment, a model consistent with prior observations that the transcriptional output of the phosphatase branch of the TORC1 pathway reaches a peak and then declines following rapamycin addition.

B. Outlook for TOR

As noted in the preceding sections, a large number of genomic studies have been conducted, often with similar experiments conducted by multiple groups. Less work has been accomplished in performing meta-analyses of similar data from different groups or integrating data from different types of data, although a number of computational methods for managing and integrating global transcriptional, proteomic, metabolic data sets individually and in combination have been developed recently [81–85]. However, the full appreciation of the massive amount of global data and the derivation of biological significance from these data require mathematical approaches for modeling and predicting outcomes upon perturbation of the system. Therefore, the next step in understanding the TOR network from diverse approaches is to combine all the data and observations, and integrate them into a quantitative representation of the cell that reflects responses to environmental as well as internal perturbations. These efforts have begun but require significantly more refinement and elaboration before we realize the full potential of a systems approach to TOR signaling.

VI. Future: TOR and Cancer

The wealth of experimental data concerning TOR combined with increasing computational sophistication will soon allow us to model the TOR network in multiple dimensions: a network representing all the identified interconnecting events, the dynamic response of this network to fluctuating environmental conditions, and the variations of the network in diverse organisms and cell types. Such an interactive model could be easily adjusted to reflect a change in the pathway resulting, for instance, from mutations in TSC complex, AKT, or TOR. From this interactive model based on massive global transcript, metabolic, and protein profiles, a tailored individual "disease-profile" could be generated, with the goal of developing patient-specific therapeutics.

A. TOR and Personalized Medicine

1. *Application of Systems Biology to Cancer*

Cancer emerges from somatic mutations that alter cellular processes governing growth in response to various internal and external stimuli, such as growth stimulators and inhibitors, cell–cell contact, stromal cell interactions, and self-limitations of proliferation. However, there is no simple cause-and-effect relationship between a mutation and a cancer phenotype. Rather, a mutation results in a context-dependent dynamic phenotype determined by connectivity of multiple signaling pathways. For example, mTOR is regulated by multiple upstream cues, including amino acids via Rag GTPases, growth factors via PI3K, cellular energy status and oxygen levels via AMPK, and ROS via SESN proteins [86, 87]. How TOR responds to changes in any one of those multiple inputs to regulate cell growth depends substantially on what other inputs it receives and the activity of related growth-promoting pathways, in a manner not yet fully appreciated. Hence, the genotypic basis of a disease varies among individual patients and, accordingly, therapeutic strategies should be uniquely tailored for effectively targeting the individual patient's disease signature.

Systems biology offers an avenue to such personalized medicine. For instance, transcriptional, proteomic, and metabolic profiles of a tumor can identify the aberrant signaling interactions driving the tumor and thereby pinpoint the most effective therapeutics regimen. Such tailored therapy requires an accurate prediction of drug response of the individual patient. To begin to assemble the data necessary to make such predictions, Lander and Golub laboratories investigated the transcriptional profiles of 6817 genes in 60 human cancer cell lines (NCI-60) to 232 compounds [88]. This genomics-based approach allowed an unbiased prediction of drug response as compared to the utilization of genetic information of individual genes. More recently, the Guo and Weinstein laboratories added to the transcript-based pharmacogenomics approach by including proteomic signatures, thereby allowing transcript-protein correlation in predicting chemosensitivity [89–92]. In a similar effort, Kruglyak and colleagues examined 104 individual genotyped yeast strains to compare the predictive power of drug response between 6229 transcript levels versus 2894 genetic biomarkers [93]. Thus, large-scale databases are being created and appropriate information extracted to predict drug responses on the basis of genome-wide signatures of individuals.

Nevins and colleagues recently made important progress in predictive ability by formulating a pathway module strategy in which disease phenotypes can be derived from molecular pathway signatures [94]. The logic of their approach was that cell-signaling processes, even at the level of post-translational modification such as phosphorylation, ultimately result in changes in gene expression, which can be deconstructed to identify causal pathway-related signatures. Based on this logic, they developed a statistical factor analysis method to deconvolve the complex expression data and construct interacting pathway modules based on common gene expression patterns. Their method reduced oncogenic signaling networks to basic modules using gene expression profiles in order to predict the effectiveness of pathway-specific therapeutics. The key biological strength was the emphasis in their design that signaling pathways are not simple linear events but branched nodes with proteins influencing multiple downstream targets with multiple overlap and crosstalk among the pathways. Furthermore, application of this method to a set of adenocarcinoma and squamous cell carcinomas of the lung allowed the investigators to identify particular Ras and E2F signatures that retrospectively bifurcated the tumors into low and high survival subgroups [94]. Thus, the methodology has the potential for providing a prognostic indication of individual tumors and thus determining the level of aggressiveness of treatment for individual patients. Although the tools and experimental evidence are not yet available for predicting the efficacy of rapamycin or its derivatives on various cancer and cell types, the results of this study are promising in applying similar methodologies and predicting various outcomes of inhibiting the TOR pathway. Moreover, the findings of this study strengthen the notion that cancer is a disease of networks of interconnected signaling pathways, and therefore should be treated accordingly.

2. *Targeting TOR as a Cancer Therapy*

Additional complexity in modeling and predicting the activity and outputs of TOR network with respect to cancer biology is derived from the genetic and biochemical differences between various forms of cancer, types of inflicted cells, and the phases of the cell cycle. Different subsets of cancers have been demonstrated to show varying degrees of response to rapamycin or its analogs that inhibit mTORC1 [95, 96]. Moreover, how a cell responds to inhibition of mTOR depends on the type of the cell. It has become increasingly clear that rapamycin is not an one-for-all drug to treat cancer [97–100]. Effects of rapamycin depend on the type of tumor and type of inflicted tissue, which is partially determined by the extent of mTORC1 activity on those tissue types. Sabatini and colleagues recently demonstrated that mTORC2, although not directly inhibited by rapamycin, plays a critical role in regulating the effects of rapamycin in specific cell types [98]. For example, they show that rapamycin-induced inhibition of Akt phosphorylation depends on the cell type, and suggest that mTORC2 may partially be responsible for the high accumulation of lipids in adipose tissue where Akt response to insulin is critical. Accordingly, T lymphocytes, smooth muscle cells, and endothelial cells, in which Akt plays an integral role, are key targets of rapamycin.

Inhibition of an inappropriate point in TOR network can have unanticipated, and occasionally counterintuitive, consequences. For instance, Sabatini laboratory identified Deptor, which is overexpressed in 28% of human multiple myelomas, as a negative regulator of both mTORC1 and mTORC2 kinase activities. Inhibition of Deptor overexpression in myeloma cell lines could inhibit Akt, but while doing so, increase mTORC1 activity [101]. Likewise, inhibition of mTORC1 with rapamycin resulted in unexpected activation of Akt [102]. Inhibition of mTOR by rapamycin does not prevent all downstream readouts, and the extent of rapamycin effect varies between cell types [86, 103]. As Efeyan and Sabatini point out, incomplete inhibition of mTORC1 by rapamycin may instead result in hyperactivation of select rapamycin-resistant branches of the TOR network [86]. Furthermore, Hay and colleagues suggest that rapamycin may be an effective therapeutic in mammary and salivary tumors resultant from Akt hyperactivation, due to the downstream position of mTORC1 under Akt. However, since Akt modulates a number of mTORC1-independent regulators of cell cycle and apoptosis, the effectiveness of rapamycin would be limited to cancer cell lines reflecting specific pathway mutation or particular cell type [86, 104, 105]. A major challenge in modeling the vast TOR network using systems biology is the incorporation of the many feedback mechanisms.

The Pandolfi laboratory demonstrated that cell cycle phase also plays a critical role in determining the efficacy of mTOR inhibitor drugs [106]. They demonstrated that whereas mTOR plays an essential role in developing embryos, it is dispensable in transformed postmitotic prostate cells. In contrast, inactivation of mTOR drastically reduced the potential for tumorigenesis in cells lacking PTEN in mitotically active and proliferating cells. Another important finding from their study is that inhibition of both mTORC1 and mTORC2 resulted in greater suppression of tumor formation due to loss of PTEN than singly targeting mTORC1. These results together with findings from the Sabatini and Gera laboratories posit a greater role of mTORC2 in prostate and gliomas tumorigenesis [107, 108].

The power/foundation of systems biology comes from: (1) genome-wide quantitative examination of the cell's dynamic concentrations of metabolite, proteins, and transcripts in response to various cellular and environmental perturbations over a course of time (top-down approach); (2) targeted identification of the many components and interactions of a specific protein or pathway with focus on molecular mechanisms of the components (bottom-up approach); (3) mathematical integration and modeling of data from the two approaches; and (4) prediction of the state of cellular physiology and outputs of cellular signaling pathways. The reductionist approach has built the foundation and provided a large amount of knowledge about the many targets of TOR, their mode of regulation, and how TOR itself is regulated. With the increasing availability and capability to simultaneously conduct and integrate genome-wide studies examining multiple levels of biological information, we can now look forward to a complete systems-level understanding of the TOR network. Furthermore, with this understanding of TOR, we can computationally predict various outputs of TOR under various conditions—for instance, drug treatment of cells with misregulated signaling pathways—to effectively treat diseases such as diabetes and cancer.

References

1. Heitman, J., Movva, N.R., and Hall, M.N. (1991). Targets for cell cycle arrest by the immunosuppressant rapamycin in yeast. *Science* 253(5022):905–909.
2. Fingar, D.C., and Blenis, J. (2004). Target of rapamycin (TOR): an integrator of nutrient and growth factor signals and coordinator of cell growth and cell cycle progression. *Oncogene* 23(18):3151–3171.

3. Jacinto, E., and Lorberg, A. (2008). TOR regulation of AGC kinases in yeast and mammals. *Biochem J* 410(1):19–37.
4. Smets, B., Ghillebert, R., De Snijder, P., Binda, M., Swinnen, E., De Virgilio, C., and Winderickx, J. (2010). Life in the midst of scarcity: adaptations to nutrient availability in *Saccharomyces cerevisiae*. *Curr Genet* 56(1):1–32.
5. Wullschleger, S., Loewith, R., and Hall, M.N. (2006). TOR signaling in growth and metabolism. *Cell* 124(3):471–484.
6. Zaman, S., Lippman, S.I., Zhao, X., and Broach, J.R. (2008). How *Saccharomyces* responds to nutrients. *Annu Rev Genet* 42:27–81.
7. Gehlenborg, N., O'Donoghue, S.I., Baliga, N.S., Goesmann, A., Hibbs, M.A., Kitano, H., Kohlbacher, O., Neuweger, H., Schneider, R., Tenenbaum, D., and Gavin, A.C. (2010). Visualization of omics data for systems biology. *Nat Methods* 7(3 Suppl):S56–S68.
8. Kirschner, M.W. (2005). The meaning of systems biology. *Cell* 121(4):503–504.
9. Auffray, C., Chen, Z., and Hood, L. (2009). Systems medicine: the future of medical genomics and healthcare. *Genome Med* 1(1):2.
10. Joyce, A.R., and Palsson, B.O. (2006). The model organism as a system: integrating 'omics' data sets. *Nat Rev Mol Cell Biol* 7(3):198–210.
11. Kitano, H. (2002). Computational systems biology. *Nature* 420(6912):206–210.
12. Kitano, H. (2002). Systems biology: a brief overview. *Science* 295(5560):1662–1664.
13. Nielsen, J. (2009). Systems biology of lipid metabolism: from yeast to human. *FEBS Lett* 583(24):3905–3913.
14. Sauer, U., Heinemann, M., and Zamboni, N. (2007). Genetics. Getting closer to the whole picture. *Science* 316(5824):550–551.
15. Cardenas, M.E., Cutler, N.S., Lorenz, M.C., Di Como, C.J., and Heitman, J. (1999). The TOR signaling cascade regulates gene expression in response to nutrients. *Genes Dev* 13(24):3271–3279.
16. Hardwick, J.S., Kuruvilla, F.G., Tong, J.K., Shamji, A.F., and Schreiber, S.L. (1999). Rapamycin-modulated transcription defines the subset of nutrient-sensitive signaling pathways directly controlled by the Tor proteins. *Proc Natl Acad Sci USA* 96(26):14866–14870.
17. Shamji, A.F., Kuruvilla, F.G., and Schreiber, S.L. (2000). Partitioning the transcriptional program induced by rapamycin among the effectors of the Tor proteins. *Curr Biol* 10(24):1574–1581.
18. Staschke, K.A., Dey, S., Zaborske, J.M., Palam, L.R., McClintick, J.N., Pan, T., Edenberg, H.J., and Wek, R.C. (2010). Integration of general amino acid control and TOR regulatory pathways in nitrogen assimilation in yeast. *J Biol Chem* 285 (22):16893–16911.
19. Cherkasova, V.A., and Hinnebusch, A.G. (2003). Translational control by TOR and TAP42 through dephosphorylation of eIF2alpha kinase GCN2. *Genes Dev* 17(7):859–872.
20. Duvel, K., Santhanam, A., Garrett, S., Schneper, L., and Broach, J.R. (2003). Multiple roles of Tap42 in mediating rapamycin-induced transcriptional changes in yeast. *Mol Cell* 11(6):1467–1478.
21. Urban, J., Soulard, A., Huber, A., Lippman, S., Mukhopadhyay, D., Deloche, O., Wanke, V., Anrather, D., Ammerer, G., Riezman, H., Broach, J.R., De Virgilio, C., *et al.* (2007). Sch9 is a major target of TORC1 in *Saccharomyces cerevisiae*. *Mol Cell* 26(5):663–674.
22. Smets, B., De Snijder, P., Engelen, K., Joossens, E., Ghillebert, R., Thevissen, K., Marchal, K., and Winderickx, J. (2008). Genome-wide expression analysis reveals TORC1-dependent and -independent functions of Sch9. *FEMS Yeast Res* 8(8):1276–1288.

23. Yorimitsu, T., Zaman, S., Broach, J.R., and Klionsky, D.J. (2007). Protein kinase A and Sch9 cooperatively regulate induction of autophagy in *Saccharomyces cerevisiae*. *Mol Biol Cell* 18(10):4180–4189.
24. De Virgilio, C., and Loewith, R. (2006). Cell growth control: little eukaryotes make big contributions. *Oncogene* 25(48):6392–6415.
25. Badis, G., Chan, E.T., van Bakel, H., Pena-Castillo, L., Tillo, D., Tsui, K., Carlson, C.D., Gossett, A.J., Hasinoff, M.J., Warren, C.L., Gebbia, M., Talukder, S., *et al.* (2008). A library of yeast transcription factor motifs reveals a widespread function for Rsc3 in targeting nucleosome exclusion at promoters. *Mol Cell* 32(6):878–887.
26. Freckleton, G., Lippman, S.I., Broach, J.R., and Tavazoie, S. (2009). Microarray profiling of phage-display selections for rapid mapping of transcription factor–DNA interactions. *PLoS Genet* 5(4):e1000449.
27. Liko, D., Slattery, M.G., and Heideman, W. (2007). Stb3 binds to ribosomal RNA processing element motifs that control transcriptional responses to growth in *Saccharomyces cerevisiae*. *J Biol Chem* 282(36):26623–26628.
28. Lippman, S.I., and Broach, J.R. (2009). Protein kinase A and TORC1 activate genes for ribosomal biogenesis by inactivating repressors encoded by Dot6 and its homolog Tod6. *Proc Natl Acad Sci USA* 106(47):19928–19933.
29. Zhu, C., Byers, K.J., McCord, R.P., Shi, Z., Berger, M.F., Newburger, D.E., Saulrieta, K., Smith, Z., Shah, M.V., Radhakrishnan, M., Philippakis, A.A., Hu, Y., *et al.* (2009). High-resolution DNA-binding specificity analysis of yeast transcription factors. *Genome Res* 19 (4):556–566.
30. Cheng, C., Fabrizio, P., Ge, H., Longo, V.D., and Li, L.M. (2007). Inference of transcription modification in long-live yeast strains from their expression profiles. *BMC Genomics* 8:219.
31. Ge, H., Wei, M., Fabrizio, P., Hu, J., Cheng, C., Longo, V.D., and Li, L.M. (2010). Comparative analyses of time-course gene expression profiles of the long-lived sch9Delta mutant. *Nucleic Acids Res* 38(1):143–158.
32. Cheng, C., Fabrizio, P., Ge, H., Wei, M., Longo, V.D., and Li, L.M. (2007). Significant and systematic expression differentiation in long-lived yeast strains. *PLoS ONE* 2(10):e1095.
33. Fabrizio, P., Liou, L.L., Moy, V.N., Diaspro, A., Valentine, J.S., Gralla, E.B., and Longo, V.D. (2003). SOD2 functions downstream of Sch9 to extend longevity in yeast. *Genetics* 163(1):35–46.
34. Wei, M., Fabrizio, P., Madia, F., Hu, J., Ge, H., Li, L.M., and Longo, V.D. (2009). Tor1/Sch9-regulated carbon source substitution is as effective as calorie restriction in life span extension. *PLoS Genet* 5(5):e1000467.
35. Cybulski, N., and Hall, M.N. (2009). TOR complex 2: a signaling pathway of its own. *Trends Biochem Sci* 34(12):620–627.
36. Dowling, R.J., Topisirovic, I., Fonseca, B.D., and Sonenberg, N. (2010). Dissecting the role of mTOR: lessons from mTOR inhibitors. *Biochim Biophys Acta* 1804(3):433–439.
37. Giaever, G., Chu, A.M., Ni, L., Connelly, C., Riles, L., Veronneau, S., Dow, S., Lucau-Danila, A., Anderson, K., Andre, B., Arkin, A.P., Astromoff, A., *et al.* (2002). Functional profiling of the *Saccharomyces cerevisiae* genome. *Nature* 418(6896):387–391.
38. Suter, B., Auerbach, D., and Stagljar, I. (2006). Yeast-based functional genomics and proteomics technologies: the first 15 years and beyond. *Biotechniques* 40(5):625–644.
39. Winzeler, E.A., Shoemaker, D.D., Astromoff, A., Liang, H., Anderson, K., Andre, B., Bangham, R., Benito, R., Boeke, J.D., Bussey, H., Chu, A.M., Connelly, C., *et al.* (1999). Functional characterization of the *S. cerevisiae* genome by gene deletion and parallel analysis. *Science* 285(5429):901–906.

40. Chan, T.F., Carvalho, J., Riles, L., and Zheng, X.F. (2000). A chemical genomics approach toward understanding the global functions of the target of rapamycin protein (TOR). *Proc Natl Acad Sci USA* 97(24):13227–13232.
41. Parsons, A.B., Brost, R.L., Ding, H., Li, Z., Zhang, C., Sheikh, B., Brown, G.W., Kane, P. M., Hughes, T.R., and Boone, C. (2004). Integration of chemical-genetic and genetic interaction data links bioactive compounds to cellular target pathways. *Nat Biotechnol* 22(1):62–69.
42. Gao, M., and Kaiser, C.A. (2006). A conserved GTPase-containing complex is required for intracellular sorting of the general amino-acid permease in yeast. *Nat Cell Biol* 8(7):657–667.
43. Binda, M., Peli-Gulli, M.P., Bonfils, G., Panchaud, N., Urban, J., Sturgill, T.W., Loewith, R., and De Virgilio, C. (2009). The Vam6 GEF controls TORC1 by activating the EGO complex. *Mol Cell* 35(5):563–573.
44. Xie, M.W., Jin, F., Hwang, H., Hwang, S., Anand, V., Duncan, M.C., and Huang, J. (2005). Insights into TOR function and rapamycin response: chemical genomic profiling by using a high-density cell array method. *Proc Natl Acad Sci USA* 102(20):7215–7220.
45. Neklesa, T.K., and Davis, R.W. (2008). Superoxide anions regulate TORC1 and its ability to bind Fpr1:rapamycin complex. *Proc Natl Acad Sci USA* 105(39):15166–15171.
46. Ho, C.H., Magtanong, L., Barker, S.L., Gresham, D., Nishimura, S., Natarajan, P., Koh, J.L., Porter, J., Gray, C.A., Andersen, R.J., Giaever, G., Nislow, C., *et al.* (2009). A molecular barcoded yeast ORF library enables mode-of-action analysis of bioactive compounds. *Nat Biotechnol* 27(4):369–377.
47. Byfield, M.P., Murray, J.T., and Backer, J.M. (2005). hVps34 is a nutrient-regulated lipid kinase required for activation of p70 S6 kinase. *J Biol Chem* 280(38):33076–33082.
48. Herman, P.K., and Emr, S.D. (1990). Characterization of VPS34, a gene required for vacuolar protein sorting and vacuole segregation in *Saccharomyces cerevisiae*. *Mol Cell Biol* 10(12):6742–6754.
49. Nobukuni, T., Joaquin, M., Roccio, M., Dann, S.G., Kim, S.Y., Gulati, P., Byfield, M.P., Backer, J.M., Natt, F., Bos, J.L., Zwartkruis, F.J., and Thomas, G. (2005). Amino acids mediate mTOR/raptor signaling through activation of class 3 phosphatidylinositol 3OH-kinase. *Proc Natl Acad Sci USA* 102(40):14238–14243.
50. Neklesa, T.K., and Davis, R.W. (2009). A genome-wide screen for regulators of TORC1 in response to amino acid starvation reveals a conserved Npr2/3 complex. *PLoS Genet* 5(6):e1000515.
51. Huang, J., Zhu, H., Haggarty, S.J., Spring, D.R., Hwang, H., Jin, F., Snyder, M., and Schreiber, S.L. (2004). Finding new components of the target of rapamycin (TOR) signaling network through chemical genetics and proteome chips. *Proc Natl Acad Sci USA* 101(47):16594–16599.
52. Kaeberlein, M., Powers, R.W., 3rd, Steffen, K.K., Westman, E.A., Hu, D., Dang, N., Kerr, E.O., Kirkland, K.T., Fields, S., and Kennedy, B.K. (2005). Regulation of yeast replicative life span by TOR and Sch9 in response to nutrients. *Science* 310 (5751):1193–1196.
53. Powers, R.W., 3rd, Kaeberlein, M., Caldwell, S.D., Kennedy, B.K., and Fields, S. (2006). Extension of chronological life span in yeast by decreased TOR pathway signaling. *Genes Dev* 20(2):174–184.
54. Brauer, M.J., Huttenhower, C., Airoldi, E.M., Rosenstein, R., Matese, J.C., Gresham, D., Boer, V.M., Troyanskaya, O.G., and Botstein, D. (2008). Coordination of growth rate, cell cycle, stress response, and metabolic activity in yeast. *Mol Biol Cell* 19(1):352–367.
55. Boer, V.M., Amini, S., and Botstein, D. (2008). Influence of genotype and nutrition on survival and metabolism of starving yeast. *Proc Natl Acad Sci USA* 105(19):6930–6935.

56. Matecic, M., Smith, D.L., Pan, X., Maqani, N., Bekiranov, S., Boeke, J.D., and Smith, J.S. (2010). A microarray-based genetic screen for yeast chronological aging factors. *PLoS Genet* 6(4):e1000921.
57. Hansen, M., Chandra, A., Mitic, L.L., Onken, B., Driscoll, M., and Kenyon, C. (2008). A role for autophagy in the extension of lifespan by dietary restriction in *C. elegans*. *PLoS Genet* 4(2):e24.
58. Jia, K., and Levine, B. (2007). Autophagy is required for dietary restriction-mediated life span extension in *C. elegans*. *Autophagy* 3(6):597–599.
59. Alvers, A.L., Wood, M.S., Hu, D., Kaywell, A.C., Dunn, W.A., Jr., and Aris, J.P. (2009). Autophagy is required for extension of yeast chronological life span by rapamycin. *Autophagy* 5(6):847–849.
60. Hartwell, L.H., Culotti, J., Pringle, J.R., and Reid, B.J. (1974). Genetic control of the cell division cycle in yeast. *Science* 183(4120):46–51.
61. Hartwell, L.H., and Unger, M.W. (1977). Unequal division in *Saccharomyces cerevisiae* and its implications for the control of cell division. *J Cell Biol* 75(2 Pt 1):422–435.
62. Johnston, G.C., Pringle, J.R., and Hartwell, L.H. (1977). Coordination of growth with cell division in the yeast *Saccharomyces cerevisiae*. *Exp Cell Res* 105(1):79–98.
63. Jorgensen, P., Rupes, I., Sharom, J.R., Schneper, L., Broach, J.R., and Tyers, M. (2004). A dynamic transcriptional network communicates growth potential to ribosome synthesis and critical cell size. *Genes Dev* 18(20):2491–2505.
64. Jorgensen, P., Nishikawa, J.L., Breitkreutz, B.J., and Tyers, M. (2002). Systematic identification of pathways that couple cell growth and division in yeast. *Science* 297 (5580):395–400.
65. Zhang, J., Schneider, C., Ottmers, L., Rodriguez, R., Day, A., Markwardt, J., and Schneider, B.L. (2002). Genomic scale mutant hunt identifies cell size homeostasis genes in *S. cerevisiae*. *Curr Biol* 12(23):1992–2001.
66. Huber, A., Bodenmiller, B., Uotila, A., Stahl, M., Wanka, S., Gerrits, B., Aebersold, R., and Loewith, R. (2009). Characterization of the rapamycin-sensitive phosphoproteome reveals that Sch9 is a central coordinator of protein synthesis. *Genes Dev* 23(16):1929–1943.
67. Costanzo, M., Baryshnikova, A., Bellay, J., Kim, Y., Spear, E.D., Sevier, C.S., Ding, H., Koh, J.L., Toufighi, K., Mostafavi, S., Prinz, J., St Onge, R.P., *et al.* (2010). The genetic landscape of a cell. *Science* 327(5964):425–431.
68. Lee, J., Moir, R.D., and Willis, I.M. (2009). Regulation of RNA polymerase III transcription involves SCH9-dependent and SCH9-independent branches of the target of rapamycin (TOR) pathway. *J Biol Chem* 284(19):12604–12608.
69. Koh, J.L., Ding, H., Costanzo, M., Baryshnikova, A., Toufighi, K., Bader, G.D., Myers, C.L., Andrews, B.J., and Boone, C. (2010). DRYGIN: a database of quantitative genetic interaction networks in yeast. *Nucleic Acids Res* 38(Database issue):D502–D507.
70. Boer, V.M., Crutchfield, C.A., Bradley, P.H., Botstein, D., and Rabinowitz, J.D. (2010). Growth-limiting intracellular metabolites in yeast growing under diverse nutrient limitations. *Mol Biol Cell* 21(1):198–211.
71. Bradley, P.H., Brauer, M.J., Rabinowitz, J.D., and Troyanskaya, O.G. (2009). Coordinated concentration changes of transcripts and metabolites in *Saccharomyces cerevisiae*. *PLoS Comput Biol* 5(1):e1000270.
72. Souchelnytskyi, S. (2005). Bridging proteomics and systems biology: what are the roads to be traveled? *Proteomics* 5(16):4123–4137.
73. Bandhakavi, S., Xie, H., O'Callaghan, B., Sakurai, H., Kim, D.H., and Griffin, T.J. (2008). Hsf1 activation inhibits rapamycin resistance and TOR signaling in yeast revealed by combined proteomic and genetic analysis. *PLoS ONE* 3(2):e1598.

74. Fournier, M.L., Paulson, A., Pavelka, N., Mosley, A.L., Gaudenz, K., Bradford, W.D., Glynn, E., Li, H., Sardiu, M.E., Fleharty, B., Seidel, C., Florens, L., *et al.* (2010). Delayed correlation of mRNA and protein expression in rapamycin-treated cells and a role for Ggc1 in cellular sensitivity to rapamycin. *Mol Cell Proteomics* 9(2):271–284.
75. Ishii, N., Nakahigashi, K., Baba, T., Robert, M., Soga, T., Kanai, A., Hirasawa, T., Naba, M., Hirai, K., Hoque, A., Ho, P.Y., Kakazu, Y., *et al.* (2007). Multiple high-throughput analyses monitor the response of *E. coli* to perturbations. *Science* 316(5824):593–597.
76. Castrillo, J.I., Zeef, L.A., Hoyle, D.C., Zhang, N., Hayes, A., Gardner, D.C., Cornell, M.J., Petty, J., Hakes, L., Wardleworth, L., Rash, B., Brown, M., *et al.* (2007). Growth control of the eukaryote cell: a systems biology study in yeast. *J Biol* 6(2):4.
77. Kuepfer, L., Peter, M., Sauer, U., and Stelling, J. (2007). Ensemble modeling for analysis of cell signaling dynamics. *Nat Biotechnol* 25(9):1001–1006.
78. Di Como, C.J., and Arndt, K.T. (1996). Nutrients, via the Tor proteins, stimulate the association of Tap42 with type 2A phosphatases. *Genes Dev* 10(15):1904–1916.
79. Jiang, Y., and Broach, J.R. (1999). Tor proteins and protein phosphatase 2A reciprocally regulate Tap42 in controlling cell growth in yeast. *EMBO J* 18(10):2782–2792.
80. Jacinto, E., Guo, B., Arndt, K.T., Schmelzle, T., and Hall, M.N. (2001). TIP41 interacts with TAP42 and negatively regulates the TOR signaling pathway. *Mol Cell* 8(5):1017–1026.
81. Di Ventura, B., Lemerle, C., Michalodimitrakis, K., and Serrano, L. (2006). From in vivo to in silico biology and back. *Nature* 443(7111):527–533.
82. Frazier, Z., McDermott, J., Guerquin, M., and Samudrala, R. (2009). Computational representation of biological systems. *Methods Mol Biol* 541:535–549.
83. Huttenhower, C., Mutungu, K.T., Indik, N., Yang, W., Schroeder, M., Forman, J.J., Troyanskaya, O.G., and Coller, H.A. (2009). Detailing regulatory networks through large scale data integration. *Bioinformatics* 25(24):3267–3274.
84. Kim, H.D., Shay, T., O'Shea, E.K., and Regev, A. (2009). Transcriptional regulatory circuits: predicting numbers from alphabets. *Science* 325(5939):429–432.
85. Ruan, J. (2010). A top-performing algorithm for the DREAM3 gene expression prediction challenge. *PLoS ONE* 5(2):e8944.
86. Efeyan, A., and Sabatini, D.M. (2009). mTOR and cancer: many loops in one pathway. *Curr Opin Cell Biol.*
87. Lee, J.H., Budanov, A.V., Park, E.J., Birse, R., Kim, T.E., Perkins, G.A., Ocorr, K., Ellisman, M.H., Bodmer, R., Bier, E., and Karin, M. (2010). Sestrin as a feedback inhibitor of TOR that prevents age-related pathologies. *Science* 327(5970):1223–1228.
88. Staunton, J.E., Slonim, D.K., Coller, H.A., Tamayo, P., Angelo, M.J., Park, J., Scherf, U., Lee, J.K., Reinhold, W.O., Weinstein, J.N., Mesirov, J.P., Lander, E.S., *et al.* (2001). Chemosensitivity prediction by transcriptional profiling. *Proc Natl Acad Sci USA* 98(19):10787–10792.
89. Ma, Y., Ding, Z., Qian, Y., Shi, X., Castranova, V., Harner, E.J., and Guo, L. (2006). Predicting cancer drug response by proteomic profiling. *Clin Cancer Res* 12(15):4583–4589.
90. Ma, Y., Ding, Z., Qian, Y., Wan, Y.W., Tosun, K., Shi, X., Castranova, V., Harner, E.J., and Guo, N.L. (2009). An integrative genomic and proteomic approach to chemosensitivity prediction. *Int J Oncol* 34(1):107–115.
91. O'Shaughnessy, K.M. (2006). HapMap, pharmacogenomics, and the goal of personalized prescribing. *Br J Clin Pharmacol* 61(6):783–786.
92. Shankavaram, U.T., Reinhold, W.C., Nishizuka, S., Major, S., Morita, D., Chary, K.K., Reimers, M.A., Scherf, U., Kahn, A., Dolginow, D., Cossman, J., Kaldjian, E.P., *et al.* (2007). Transcript and protein expression profiles of the NCI-60 cancer cell panel: an integromic microarray study. *Mol Cancer Ther* 6(3):820–832.

93. Ruderfer, D.M., Roberts, D.C., Schreiber, S.L., Perlstein, E.O., and Kruglyak, L. (2009). Using expression and genotype to predict drug response in yeast. *PLoS ONE* 4(9):e6907.
94. Chang, J.T., Carvalho, C., Mori, S., Bild, A.H., Gatza, M.L., Wang, Q., Lucas, J.E., Potti, A., Febbo, P.G., West, M., and Nevins, J.R. (2009). A genomic strategy to elucidate modules of oncogenic pathway signaling networks. *Mol Cell* 34(1):104–114.
95. Easton, J.B., and Houghton, P.J. (2006). mTOR and cancer therapy. *Oncogene* 25 (48):6436–6446.
96. Faivre, S., Kroemer, G., and Raymond, E. (2006). Current development of mTOR inhibitors as anticancer agents. *Nat Rev Drug Discov* 5(8):671–688.
97. Panwalkar, A., Verstovsek, S., and Giles, F.J. (2004). Mammalian target of rapamycin inhibition as therapy for hematologic malignancies. *Cancer* 100(4):657–666.
98. Sarbassov, D.D., Ali, S.M., Sengupta, S., Sheen, J.H., Hsu, P.P., Bagley, A.F., Markhard, A.L., and Sabatini, D.M. (2006). Prolonged rapamycin treatment inhibits mTORC2 assembly and Akt/PKB. *Mol Cell* 22(2):159–168.
99. Sawyers, C.L. (2003). Will mTOR inhibitors make it as cancer drugs? *Cancer Cell* 4(5):343–348.
100. Shor, B., Gibbons, J.J., Abraham, R.T., and Yu, K. (2009). Targeting mTOR globally in cancer: thinking beyond rapamycin. *Cell Cycle* 8(23):3831–3837.
101. Peterson, T.R., Laplante, M., Thoreen, C.C., Sancak, Y., Kang, S.A., Kuehl, W.M., Gray, N.S., and Sabatini, D.M. (2009). DEPTOR is an mTOR inhibitor frequently overexpressed in multiple myeloma cells and required for their survival. *Cell* 137(5):873–886.
102. O'Reilly, K.E., Rojo, F., She, Q.B., Solit, D., Mills, G.B., Smith, D., Lane, H., Hofmann, F., Hicklin, D.J., Ludwig, D.L., Baselga, J., and Rosen, N. (2006). mTOR inhibition induces upstream receptor tyrosine kinase signaling and activates Akt. *Cancer Res* 66(3):1500–1508.
103. Chen, Y., Zheng, Y., and Foster, D.A. (2003). Phospholipase D confers rapamycin resistance in human breast cancer cells. *Oncogene* 22(25):3937–3942.
104. Manning, B.D., and Cantley, L.C. (2007). AKT/PKB signaling: navigating downstream. *Cell* 129(7):1261–1274.
105. Skeen, J.E., Bhaskar, P.T., Chen, C.C., Chen, W.S., Peng, X.D., Nogueira, V., Hahn-Windgassen, A., Kiyokawa, H., and Hay, N. (2006). Akt deficiency impairs normal cell proliferation and suppresses oncogenesis in a p53-independent and mTORC1-dependent manner. *Cancer Cell* 10(4):269–280.
106. Nardella, C., Carracedo, A., Alimonti, A., Hobbs, R.M., Clohessy, J.G., Chen, Z., Egia, A., Fornari, A., Fiorentino, M., Loda, M., Kozma, S.C., Thomas, G., *et al.* (2009). Differential requirement of mTOR in postmitotic tissues and tumorigenesis. *Sci Signal* 2(55):ra2.
107. Guertin, D.A., Stevens, D.M., Saitoh, M., Kinkel, S., Crosby, K., Sheen, J.H., Mullholland, D.J., Magnuson, M.A., Wu, H., and Sabatini, D.M. (2009). mTOR complex 2 is required for the development of prostate cancer induced by Pten loss in mice. *Cancer Cell* 15(2):148–159.
108. Masri, J., Bernath, A., Martin, J., Jo, O.D., Vartanian, R., Funk, A., and Gera, J. (2007). mTORC2 activity is elevated in gliomas and promotes growth and cell motility via overexpression of rictor. *Cancer Res* 67(24):11712–11720.
109. Breitkreutz, B.J., Stark, C., and Tyers, M. (2003). Osprey: a network visualization system. *Genome Biol* 4(3):R22.
110. Shannon, P., Markiel, A., Ozier, O., Baliga, N.S., Wang, J.T., Ramage, D., Amin, N., Schwikowski, B., and Ideker, T. (2003). Cytoscape: a software environment for integrated models of biomolecular interaction networks. *Genome Res* 13(11):2498–2504.
111. Myers, C.L., Robson, D., Wible, A., Hibbs, M.A., Chiriac, C., Theesfeld, C.L., Dolinski, K., and Troyanskaya, O.G. (2005). Discovery of biological networks from diverse functional genomic data. *Genome Biol* 6(13):R114.

Author Index

Numbers in regular font are reference numbers and indicate that an author's work is referred to although the name is not cited in the text. Numbers in italics refer to the page numbers on which the complete reference appears.

D

G

H

I

J

K

L

N

P

Q

R

U

X

Index